Soft Material-Enabled Electronics for Medicine, Healthcare, and Human-Machine Interfaces

Special Issue Editors

Jae-Woong Jeong
Woon-Hong Yeo

MDPI • Basel • Beijing • Wuhan • Barcelona • Belgrade • Manchester • Tokyo • Cluj • Tianjin

Special Issue Editors
Jae-Woong Jeong
Korea Advanced Institute of
Science and Technology
Korea

Woon-Hong Yeo
Georgia Institute of Technology
USA

Editorial Office
MDPI
St. Alban-Anlage 66
4052 Basel, Switzerland

This is a reprint of articles from the Special Issue published online in the open access journal *Materials* (ISSN 1996-1944) (available at: https://www.mdpi.com/journal/materials/special_issues/soft_electronics).

For citation purposes, cite each article independently as indicated on the article page online and as indicated below:

LastName, A.A.; LastName, B.B.; LastName, C.C. Article Title. *Journal Name* **Year**, *Article Number*, Page Range.

ISBN 978-3-03928-282-1 (Pbk)
ISBN 978-3-03928-283-8 (PDF)

Cover image courtesy of Woon-Hong Yeo.

Soft Material-Enabled Electronics for Medicine, Healthcare, and Human-Machine Interfaces

Contents

About the Special Issue Editors

Jae-Woong Jeong (Dr.) is an Assistant Professor of Electrical Engineering at the Korea Advanced Institute of Science and Technology (KAIST). He received his Ph.D. in electrical engineering from Stanford University in 2012, and worked as a postdoctoral research associate at the University of Illinois at Urbana-Champaign from 2012 to 2014. Before joining KAIST, he was an Assistant Professor of Electrical, Computer and Energy Engineering at the University of Colorado, Boulder, from 2015 to 2017. Dr. Jeong's research focus is in future generation bio-integrated electronics, and systems for brain research and advanced healthcare.

Woon-Hong Yeo (Dr.) is an Assistant Professor at the George W. Woodruff School of Mechanical Engineering, and the Wallace H. Coulter Department of Biomedical Engineering at the Georgia Institute of Technology. He received his B.Sc. in Mechanical Engineering from Inha University, Korea, in 2003 and his Ph.D. in Mechanical Engineering from the University of Washington, Seattle, in 2011. From 2011–2013, he worked as a postdoctoral research fellow at the Beckman Institute and the Frederick Seitz Materials Research Center of the University of Illinois at Urbana-Champaign. His research areas include soft electronics, human-machine interfaces, nano-biosensors, and soft robotics.

Editorial

Soft Material-Enabled Electronics for Medicine, Healthcare, and Human-Machine Interfaces

Robert Herbert [1], Jae-Woong Jeong [2] and Woon-Hong Yeo [1,3,]*

[1] George W. Woodruff School of Mechanical Engineering, Institute for Electronics and Nanotechnology, Georgia Institute of Technology, Atlanta, GA 30332, USA; rherbert7@gatech.edu

[2] School of Electrical Engineering, Korea Advanced Institute of Science and Technology (KAIST), Daejeon 34141, Korea; jjeong1@kaist.ac.kr

[3] Wallace H. Coulter Department of Biomedical Engineering, Parker H. Petit Institute for Bioengineering and Biosciences, Neural Engineering Center, Institute for Materials, Institute for Robotics and Intelligent Machines, Georgia Institute of Technology, Atlanta, GA 30332, USA

* Correspondence: whyeo@gatech.edu; Tel.: +1-404-385-5710

Received: 19 January 2020; Accepted: 20 January 2020; Published: 22 January 2020

Abstract: Soft material-enabled electronics offer distinct advantages over conventional rigid and bulky devices for numerous wearable and implantable applications. Soft materials allow for seamless integration with skin and tissues due to the enhanced mechanical flexibility and stretchability. Wearable devices with multiple sensors offer continuous, real-time monitoring of biosignals and movements, which can be applied for rehabilitation and diagnostics, among other applications. Soft implantable electronics offer similar functionalities, but with improved compatibility with human tissues. Biodegradable soft implantable electronics are also being developed for transient monitoring, such as in the weeks following surgeries. New composite materials, integration strategies, and fabrication techniques are being developed to further advance soft electronics. This paper reviews recent progresses in these areas towards the development of soft material-enabled electronics for medicine, healthcare, and human-machine interfaces.

Keywords: soft materials; wearable electronics; implantable electronics; biodegradable; medical devices; diagnostics; health monitoring; human-machine interfaces

---

1. Introduction

Soft material-enabled electronics can address a wide range of applications by enabling a comfortable, continuous, and real-time monitoring of physiological signals via conformal, ergonomic interactions with human tissues when compared to conventional electronics based on bulky and rigid materials. Tissue-friendly, intimate lamination of soft wearable and implantable electronics allow for a long-term, high-fidelity recording of biological signals. An increasing number of materials, integration designs, and fabrication technologies are being developed towards realizing soft electronics due to these advantages. These electronics can enhance healthcare, diagnosis, and therapeutics by offering improved biocompatibility, signal monitoring, and wearability without sacrificing patient comfort. Overall, this special issue collects a number of recent advances in soft material-enabled electronics for medicine, healthcare, and human-machine interfaces.

2. Contributions

This special issue has collected thirteen papers with a focus on soft material-enabled electronics for wearable and implantable applications, healthcare, rehabilitation, and more. The main contributions and focused science and technology of each of these papers are outlined in the following.

The overall scope of key materials, enabling technologies, and significant applications of soft material-enabled wearable electronics is detailed in "Soft Material-Enabled, Flexible Hybrid Electronics for Medicine, Healthcare, and Human-Machine Interfaces" [1]. Substrate and sensing material properties are discussed, along with details of the applications for each material. The integration of soft materials with thin films and electronic components to form high performance electronics is included. Discussion regarding the current limitations of wearable electronics provides a view of the future direction of soft electronics.

These wearable electronics utilize a variety of active materials for electrical and sensing functionality. These materials are required to be biocompatible for medical applications. In "Flexible and Stretchable Bio-Integrated Electronics Based on Carbon Nanotube and Graphene" [2], the authors review details of soft electronics while using two biocompatible materials, carbon nanotubes and graphene. Integration strategies of these materials with soft substrates are discussed. A number of applications are detailed, such as neural mapping and human-machine interfaces.

One such example of these wearable electronics is demonstrated in "A Soft Polydimethylsiloxane Liquid Metal Interdigitated Capacitor Sensor and Its Integration in a Flexible Hybrid System for On-Body Respiratory Sensing" [3]. The proposed sensor utilizes a soft material to form channels that are filled with conductive liquid metal. This package forms a soft, interdigitated capacitor that senses both strain and proximity. The soft, conformal nature of the sensor, due to the soft backbone, enables conformal lamination on skin to monitor the breathing signals.

In addition to laminating the soft systems on the skin, wearable electronics can be integrated with clothing. In "Soft-Material-Based Smart Insoles for a Gait Monitoring System" [4], a soft conductive textile is integrated with thin capacitive sensors. The durable, lightweight system is seamlessly integrated into an insole and it provides a comfortable alternative to existing bulky devices. Gait data is wirelessly transmitted for real-time analysis of patients.

In addition to enabling wearable electronics, soft materials allow for lightweight, portable electronics to enhance healthcare procedures. In "An Optical Biosensing Strategy Based on Selective Light Absorption and Wavelength Filtering from Chromogenic Reaction" [5], a soft material-based biosensor is developed for user-friendly glucose sensing. A soft material biosensing channel uses a chromogenic reaction and it acts as a color-filtering layer. This color filtering layer modifies the image that is captured by a smartphone camera to signal different concentrations of glucose. This portable device is a vast improvement over the conventional high power and complex system.

While soft materials allow for conformal lamination of wearable electronics onto skin, these materials allow for seamless integration of implantable electronics with tissue. In "Advances in Materials for Recent Low-Profile Implantable Bioelectronics" [6], the authors provide a comprehensive review of these implantable electronics and relevant materials. A variety of materials are discussed, which include soft polymers, metals, and biodegradable materials. Challenges and recent advances in fabrication techniques are detailed. A discussion of biodegradable materials for transient electronics is also included.

One type of implantable electronics is focused on in "Biocompatible and Implantable Optical Fibers and Waveguides for Biomedicine" [7]. Implantable optical fibers and waveguides enable the delivery of light into deep tissues. This method allows for biological sensing, stimulation, and therapies, such as optogenetics. An array of materials, including natural materials, such as silk, and fabrication techniques are reviewed in this paper.

Implantable, prosthetic devices have also been developed with soft materials. Such soft materials allow for significantly higher patient comfort. In "Use of Superelastic Nitinol and Highly-Stretchable Latex to Develop a Tongue Prosthetic Assist Device and Facilitate Swallowing for Dysphagia Patients" [8], a soft material-based prosthetic tongue is fabricated for treating dysphagia. Soft materials, integrated with nitinol wires, replicate the elevation function of the tongue to move food to the back of the mouth for proper swallowing. Currently, there are no therapies or devices to replace the swallowing abilities once lost.

Biocompatible materials are studied and developed to continue the advancement of soft implantable electronics. Hydrogels is one such type of material. In "Electroactive Hydrogels Made with Polyvinyl Alcohol/Cellulose Nanocrystals" [9], a hydrogel is enhanced by varying cellulose nanocrystal concentrations. Transparency and electroactive properties are improved, which are promising improvements for biomimetic soft robots and active drug release.

Implantable electronics can also be used for transient applications and eliminating removal surgeries by utilizing biodegradable materials. These transient electronics use soft, biodegradable electronics while matching the mechanical properties of tissues, as discussed in "Materials and Devices for Biodegradable and Soft Biomedical Electronics" [10]. Specific applications include diagnostics, therapeutic, and power supplies, such as energy harvesters. Power supplies are of high interest, as they enable noninvasive powering and communication of data from implanted electronics.

For transient electronics, highly conductive materials are of significant interest in achieving high performance, bioresorbable electronics. In "Processing Techniques for Bioresorbable Nanoparticles in Fabricating Flexible Conductive Interconnects" [11], bioresorbable conductive patterns are screen printed with an optimized paste mixture. A multi-step process is used to achieve highly conductive and flexible patterns for use in transient electronics.

Soft materials can also play a key role for medical procedures, such as robot-assisted surgeries. A soft, sponge tactile sensor is developed in "Material Characterization of Hardening Soft Sponge Featuring MR Fluid and Application of 6-DOF MR Haptic Master for Robot-Assisted Surgery" [12]. This soft sponge is filled with magneto-rheological fluids to replicate the mechanical stiffness of various cells and biological organs. The tactile sensor allows for the surgeon to distinguish different types of sensors while performing robot-assisted surgery. The demonstrations show significant improvements by providing highly accurate feedback to the surgeon.

In addition to the vast number of medical applications, soft material electronics can be extended to other applications. In "Spray Deposition of Ag Nanowire—Graphene Oxide Hybrid Electrodes for Flexible Polymer—Dispersed Liquid Crystal Displays" [13], spray coating is applied to a flexible polymer sheet and then integrated into smart windows. The optical properties of the smart windows are controlled via an electric field to enhance the building efficiency. Existing designs rely on costly and fragile materials as compared to the flexible spray coated material.

In summary, new advancements in soft material-enabled wearable and implantable electronics, covered in this special issue, have significantly improved the device performance, compatibility, and reliability, which allowed for various applications in health monitoring, disease diagnostics, and human-machine interfaces. Continuous study in basic science and technology development in this area will enable smart medicine, home healthcare, and independent living in the near future.

Acknowledgments: Woon-Hong Yeo acknowledges a research fund from the Marcus Foundation, the Georgia Research Alliance, and the Georgia Tech Foundation through their support of the Marcus Center for Therapeutic Cell Characterization and Manufacturing (MC3M) at Georgia Tech. This work was performed in part at the Institute for Electronics and Nanotechnology, a member of the National Nanotechnology Coordinated Infrastructure, which is supported by the National Science Foundation (Grant ECCS-1542174). In addition, this work was supported by the Nano-Material Technology Development Program through the National Research Foundation of Korea (NRF) funded by the Ministry of Science, ICT, and Future Planning (2016M3A7B4900044). Jae-Woong Jeong acknowledges the support from the Basic Science Research Program through the National Research Foundation of Korea (NRF) funded by the Ministry of Science and ICT (NRF-2018R1C1B6001706 and NRF-2018025230).

References

1. Herbert, R.; Kim, J.-H.; Kim, Y.; Lee, H.; Yeo, W.-H. Soft material-enabled, flexible hybrid electronics for medicine, healthcare, and human-machine interfaces. *Materials* **2018**, *11*, 187. [CrossRef] [PubMed]

2. Kim, T.; Cho, M.; Yu, K. Flexible and stretchable bio-integrated electronics based on carbon nanotube and graphene. *Materials* **2018**, *11*, 1163. [CrossRef] [PubMed]

3. Li, Y.; Nayak, S.; Luo, Y.; Liu, Y.; Mohan, S.V.; Krishna, H.; Pan, J.; Liu, Z.; Heng, C.H.; Thean, A.V.-Y. A Soft Polydimethylsiloxane Liquid Metal Interdigitated Capacitor Sensor and Its Integration in a Flexible Hybrid System for On-Body Respiratory Sensing. *Materials* **2019**, *12*, 1458. [CrossRef] [PubMed]

4. Wang, C.; Kim, Y.; Min, S. Soft-Material-Based Smart Insoles for a Gait Monitoring System. *Materials* **2018**, *11*, 2435. [CrossRef] [PubMed]

5. Chun, H.; Han, Y.; Park, Y.; Kim, K.; Lee, S.; Yoon, H. An optical biosensing strategy based on selective light absorption and wavelength filtering from chromogenic reaction. *Materials* **2018**, *11*, 388. [CrossRef] [PubMed]

6. Chen, Y.; Kim, Y.-S.; Tillman, B.; Yeo, W.-H.; Chun, Y. Advances in materials for recent low-profile implantable bioelectronics. *Materials* **2018**, *11*, 522. [CrossRef] [PubMed]

7. Nazempour, R.; Zhang, Q.; Fu, R.; Sheng, X. Biocompatible and implantable optical fibers and waveguides for biomedicine. *Materials* **2018**, *11*, 1283. [CrossRef] [PubMed]

8. Shayan, M.; Gildener-Leapman, N.; Elsisy, M.; Hastings, J.T.; Kwon, S.; Yeo, W.-H.; Kim, J.-H.; Shridhar, P.; Salazar, G.; Chun, Y. Use of Superelastic Nitinol and Highly-Stretchable Latex to Develop a Tongue Prosthetic Assist Device and Facilitate Swallowing for Dysphagia Patients. *Materials* **2019**, *12*, 3555. [CrossRef] [PubMed]

9. Jayaramudu, T.; Ko, H.-U.; Kim, H.; Kim, J.; Muthoka, R.; Kim, J. Electroactive hydrogels made with polyvinyl alcohol/cellulose nanocrystals. *Materials* **2018**, *11*, 1615. [CrossRef] [PubMed]

10. Li, R.; Wang, L.; Yin, L. Materials and devices for biodegradable and soft biomedical electronics. *Materials* **2018**, *11*, 2108. [CrossRef] [PubMed]

11. Li, J.; Luo, S.; Liu, J.; Xu, H.; Huang, X. Processing Techniques for Bioresorbable Nanoparticles in Fabricating Flexible Conductive Interconnects. *Materials* **2018**, *11*, 1102. [CrossRef] [PubMed]

12. Oh, J.-S.; Sohn, J.; Choi, S.-B. Material Characterization of Hardening Soft Sponge Featuring MR Fluid and Application of 6-DOF MR Haptic Master for Robot-Assisted Surgery. *Materials* **2018**, *11*, 1268. [CrossRef] [PubMed]

13. Choi, Y.; Kim, C.; Jo, S. Spray deposition of Ag nanowire–graphene oxide hybrid electrodes for flexible polymer–dispersed liquid crystal displays. *Materials* **2018**, *11*, 2231. [CrossRef] [PubMed]

Review

Soft Material-Enabled, Flexible Hybrid Electronics for Medicine, Healthcare, and Human-Machine Interfaces

Robert Herbert [1], Jong-Hoon Kim [2], Yun Soung Kim [1], Hye Moon Lee [3] and Woon-Hong Yeo [1,4,*]

[1]   George W. Woodruff School of Mechanical Engineering, College of Engineering,
      Georgia Institute of Technology, Atlanta, GA 30332, USA; rherbert7@gatech.edu (R.H.);
      ysk@me.gatech.edu (Y.S.K.)
[2]   School of Engineering and Computer Science, Washington State University, Vancouver, WA 98686, USA;
      jh.kim@wsu.edu
[3]   Functional Materials Division, Korea Institute of Materials Science (KIMS), 797 Changwondaero,
      Seongsan-gu, Changwon, Gyeongnam 641-831, Korea; hyelee@kims.re.kr
[4]   Center for Flexible Electronics, Institute for Electronics and Nanotechnology, Bioengineering Program, Petit
      Institute for Bioengineering and Biosciences, Neural Engineering Center, Georgia Institute of Technology,
      Atlanta, GA 30332, USA
*   Correspondence: whyeo@gatech.edu; Tel.: +1-404-385-5710; Fax: +1-404-894-1658

Received: 5 January 2018; Accepted: 23 January 2018; Published: 24 January 2018

Abstract: Flexible hybrid electronics (FHE), designed in wearable and implantable configurations, have enormous applications in advanced healthcare, rapid disease diagnostics, and persistent human-machine interfaces. Soft, contoured geometries and time-dynamic deformation of the targeted tissues require high flexibility and stretchability of the integrated bioelectronics. Recent progress in developing and engineering soft materials has provided a unique opportunity to design various types of mechanically compliant and deformable systems. Here, we summarize the required properties of soft materials and their characteristics for configuring sensing and substrate components in wearable and implantable devices and systems. Details of functionality and sensitivity of the recently developed FHE are discussed with the application areas in medicine, healthcare, and machine interactions. This review concludes with a discussion on limitations of current materials, key requirements for next generation materials, and new application areas.

Keywords: soft materials; flexible hybrid electronics; wearable electronics; stretchable electronics; medicine; healthcare; human-machine interfaces

---

1. Introduction

Flexible hybrid electronics (FHE), configured in low-modulus, compliant materials via hard-soft materials integration, offer distinct advantages over the conventional electronic systems that are made of bulky and rigid materials. Recent advancements in the advanced materials and soft mechanics have enabled a successful integration of rigid, yet miniaturized chips with flexible/stretchable circuit interconnects, while maintaining low levels of effective moduli [1–3]. Thus, such FHE in wearable or implantable configurations can achieve a wide range of device functionalities via incorporation of capabilities of signal filter/amplification, analog-to-digital conversion, microcontroller, memory, and wireless power transfer in the systems [4–6]. In particular, FHE allow in vivo, continuous, and real-time monitoring of human health via conformal and tissue-friendly lamination on biological tissues, such as soft human skins [7–9] and internal organs with time-dynamic motions [10–12].

A successful development of high performance FHE requires thorough understanding of key materials, mechanics design, and advanced manufacturing processes. Specifically, engineering of material properties such as adhesion, flexibility, stretchability, and biocompatibility is critical for conformal and intimate integration with targeted human tissues. In general, FHE involve various types of sensors, circuits, and substrate materials; the combination of multiple components decides the overall mechanical and material properties of the electronics. Recent advances in soft materials, nano-microfabrication, and low-profile electronics have enabled the development of wearable and implantable electronics. The progress in materials processing and engineering of low-modulus functional materials allowed to design mechanically compliant electronics. Furthermore, hybrid integration of nanomaterials enhanced the electrical properties of the flexible and stretchable devices. As an example, carbon nanotubes (CNTs), first discovered in 1991 and synthesized in 1993, were embedded in soft elastomers for improved conductivity [13–15]. Graphene, a more recently observed material, is an emerging choice for creating conductive composites, while offering enhanced stretchability [16]. These materials, and others to be discussed in this review, offered versatility to manipulate a device's electrical and mechanical properties, depending on target applications. In particular, the first generation of skin-like electronics, introduced in 2011 [17], opened a new era to develop unobtrusive and wearable sensors, actuators, antennas, and other components. These advances in materials and associated integration technologies have ignited the development of new FHE where high-performance electronic components are hybridized with soft materials.

Existing reviews of wearable and implantable electronics typically offer narrow scopes. For example, recent reviews [18,19] focus only on specific wearable devices such as strain or tactile sensors, while others [20,21] include expansion of wearable sensor types but still miss the coverage of important aspects of implantable FHE. Similarly, a few reviews summarize limited areas of implantable devices for neural applications [22,23], while emphasizing characteristics of particular materials, such as graphene, nanomaterials, hydrogels, and typical polymers [24–28]. Here, we provide a broader view of FHE via the comprehensive summary of the widely used functional materials to design soft wearable and implantable electronics, device integration strategies, and their applications in the areas of medicine, healthcare, and human-machine interfaces. In addition, we review the characteristics of those soft-material based electronics in terms of device type, sensing material, target signal, sensitivity and more. Overall, this review delivers a comprehensive summary of key materials, design criteria, and associated device performance to design high performance, wearable and implantable FHE.

2. Flexible Hybrid Electronics (FHE)

2.1. Definition

FHE refer to mechanically flexible and stretchable electronic devices and systems, enabled by hybrid combination of soft functional materials, compliant membranes, and integrated functional chip components. With the use of such materials, mechanical reliability with high flexibility and stretchability in the electronic devices can be achieved unlike traditional rigid electronics. As a result, FHE have found new applications in medicine, healthcare, and machine interfaces, via intimate and conformal interactions with soft and deformable human tissues. Recent advancement and current research in the development, characterization, and applications of a variety of FHE are discussed in the following sections.

2.2. Material Characteristics for Wearable and Implantable Electronics

Low-profile, wearable and implantable electronics can realize continuous, real-time monitoring of biomarkers for rapid, point-of-care disease diagnosis and therapeutics [29,30]. In terms of the device lamination on the human skin, poor contact leads to unreliable signal quality, contamination, and discomfort [7,9]. Thus, a careful selection of materials along with the mechanical analysis is critical to ensure the conformal and intimate contact of the device with the skin. Utilization of soft materials offers enhancement of the device's overall flexibility. A systematic engineering of a thickness, stiffness, and modulus of materials provides proper mechanical compliance to make the device lamination on contoured and deformable geometries with enough adhesion. Mechanical flexibility and stretchability is generally achieved by either a deterministic structure design or hybrid material's composition [31,32]. In a deterministic system, serpentine and wavy-structured interconnects of rigid materials can accommodate applied strain, while an island-bridge geometry utilizes these interconnects to isolate strain on the rigid-material embedded island. A randomly oriented composite integrates functional materials with intrinsically soft organic materials such elastomers [33]. Deterministic systems are advantageous in the design of high-performance electronics, while composite material-enabled systems offer higher mechanical compliance. In addition, electrical properties such as conductivity, linearity, sensitivity, and selectivity are important factors for achieving high-quality and reliable sensor performance. High sensitivity of wearable and implantable devices is necessary to detect subtle changes of target analytes or physiological signals. Sensitivity relies on the electrical properties, material's composition, and structural design of the sensing materials. Similar to mechanical integrity, minimizing hysteresis is also required to prevent changes in electrical properties or responses under mechanical deformation on the living body. Another important factor to consider for flexible bioelectronics is biocompatibility, hemocompatibility, and biodegradability of materials. Adverse reactions between electronic materials and human tissues should be avoided for safe and long-term use. In particular, bioresorbable capability of transient materials can make a next-generation implantable device that avoids follow-up surgeries.

2.3. Sensing Materials

In wearable and implantable sensors, flexibility and stretchability are often limited by the sensing material, as these materials are significantly more rigid than substrate materials. As a result, informed selection of materials and structures for the sensing component are necessary to improve mechanical and electrical characteristics. Table 1 contains a summary of common and emerging sensing or active structures and materials for FHE.

As one well-known sensing material, CNTs are cylindrical tubes that are categorized as single-walled (SWCNTs) and multi-walled (MWCNTs). Due to high conductivity and flexibility, CNTs are used in a variety of structures and composites to improve mechanical or electrical properties. Soft elastomers with poor conductivity can be altered to highly conductive composites with the addition of randomly oriented CNTs. The functional elastomers can act as conductive substrates to improve signal reception or other functions to build a soft actuator; an example is shown in Figure 1a [34]. CNTs can also be aligned in films to improve functionality in the desired direction [2]. When subjected to external strain, randomly oriented CNT structures first align in the direction of external stimulation, resulting in minimal initial resistance changes. An example of aligned CNT films is shown in Figure 1b. Furthermore, CNTs can be doped for specific applications, such as an ionic liquid doping of SWCNTs to make it more sensitive to light [63]. The SWCNT-based microwire in Figure 1c undergoes an electrical resistance change when strain, pressure, or torsion is applied [35]. Although this microwire is fragile, it is capable of rejoining to produce and recover electrical property changes. In addition to wires, CNTs are used to develop thin films. Such films, particularly when formed wrinkled on a soft substrate via heating of the film or a prestrained substrate, are highly stretchable and flexible. A single layer of thin film or a layer-by-layer assembly can be used [64]. Additionally, CNTs can be modified in multiple ways to achieve biocompatibility and biodegradability [44].

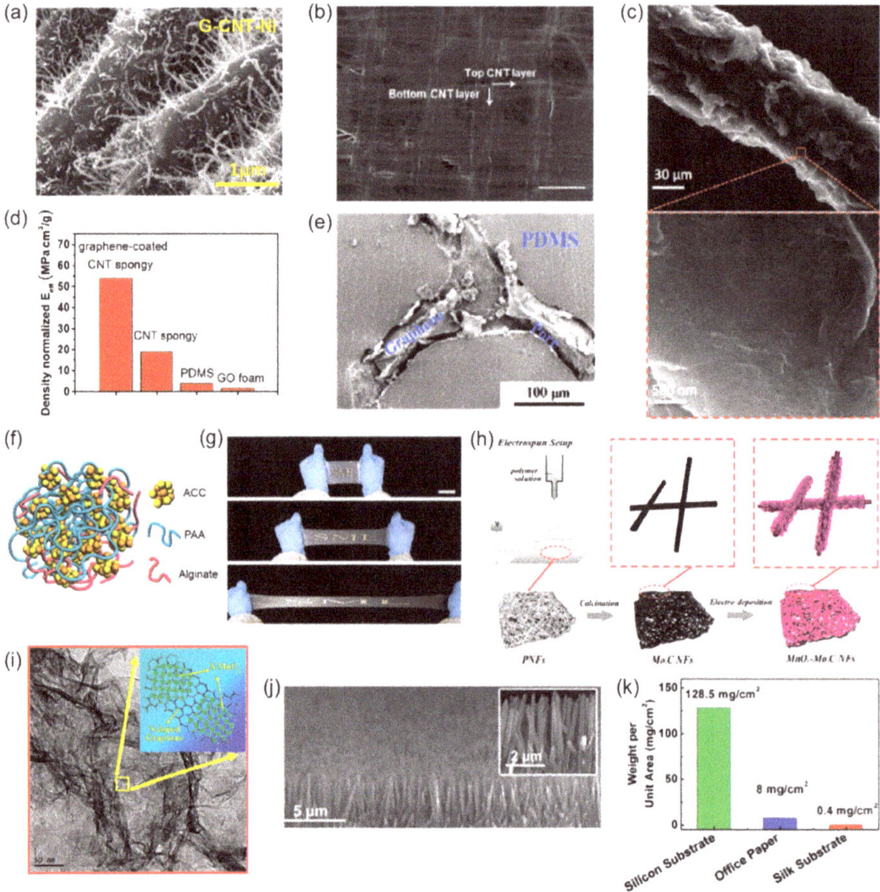

Figure 1. Soft functional materials. (**a**) Graphene-CNT-nickel hetero-nanostructure. Reprinted with permission from Reference [34], Copyright 2017, John Wiley and Sons; (**b**) Cross-stacked graphene-CNT (carbon nanotube) films. Reprinted from Reference [2], Copyright (2017), with permission from Elsevier; (**c**) SWCNT-based nanowire. Reproduced from Reference [35] with permission of The Royal Society of Chemistry; (**d**) Graphene oxide foam with low-effective elastic modulus for high sensitivity. Reprinted from Reference [36], Copyright (2017), with permission from Elsevier; (**e**) Graphene network embedded in PDMS. Reprinted with permission from Reference [37], Copyright 2016 American Chemical Society; (**f**) Schematic of mineral hydrogel. Reprinted with permission from Reference [38], Copyright 2017, John Wiley and Sons; (**g**) Elongation of 0%, 250%, and 500% of dielectric, liquid metal embedded elastomer. Reprinted with permission from Reference [39], Copyright 2016, John Wiley and Sons; (**h**) Process of developing MnO_2-Mo_2C nanofiber film. Reprinted with permission from Reference [40], Copyright 2016 American Chemical Society; (**i**) Nitrogen-doped graphene-MnO_2 nanosheet composite. Reprinted with permission from Reference [41], Copyright 2016 American Chemical Society; (**j**) Cross-sectional view of aligned NW array. Reprinted with permission from Reference [42], Copyright 2016, John Wiley and Sons; (**k**) Chart indicating silk fibroin as an ultra-lightweight substrate. Reprinted with permission from Reference [43], Copyright 2016, John Wiley and Sons.

Table 1. Representative sensing materials for wearable and implantable FHE (flexible hybrid electronics).

Sensing Material	Type	Biocompatible/Biodegradable [1]
Carbon Nanotube	Organic	Yes/Yes [2] [44]
Graphene	Inorganic	Yes/No [3] [45–47]
Hydrogel	Organic/Inorganic	Yes/Yes [4] [48–50]
Liquid Metal (EGaIn)	Inorganic	Yes/No [5] [51]
Nanosheet and Thin Film (MnO_2, Mn, Mg, Si)	Inorganic	Yes/Yes [6] [52–55]
Nanowire (Ag, ZnO, Si, Au, $BaTiO_3$, Ni)	Inorganic	Yes/No [7] [56–61]
Conducting Polymer (PEDOT:PSS)	Organic	Yes/No [8] [62]

[1] Biocompatibility is defined as the ability to biologically interact without adverse reactions, inflammatory responses, and toxicity. Biodegradability is considered being degradable, via different methods, within the human body without toxic effects [44]; [2] Surface and chemical functionalization, and designing structural defects in CNTs can be applied to create biocompatible and biodegradable CNTs [44]; [3] Biocompatibility studies of graphene composites have indicated bacterial and cell viability decrease slightly after exposure and in vivo studies, although limited, show resulting inflammation [45]. A second review suggests similar results, but states biocompatibility can be improved with surface functionalization [46]. Graphene surfaces have been shown to be biocompatible with neuronal cells and do not impact neuronal signal properties [47]; [4] Hydrogels are typically biocompatible and biodegradable, and can be tailored for applications [48–50]; [5] The common liquid metal of EGaIn is biocompatible [51]; [6] MnO_2: MnO_2 has shown low cytotoxicity, as Mn is an essential element in the body, and nanosheets degraded in the presence of glutathione [52,53]. Mg: Mg alloys are established biocompatible materials and biodegrade via corrosion [54]. Si: Silicon nanomembranes indicated no cytotoxicity during controlled dissolution in aqueous solutions [55]; [7] Ag: AgNWs exhibited no cytotoxicity with lung adenocarcinoma cells and lung normal fibroblasts [56]. ZnO: Biocompatibility was tested with a calcein and propidium iodide assay and less than 4% of cells died [57]. BaTiO3: Foam structure exhibited no short term cytotoxicity to mouse osteoblasts and did not promote a significant inflammatory response [58]. Si: Porous silicon NWs degraded in PBS [59]. Au: For different surface modifications and aspect ratios, Au NW toxicity was tested on fibroblasts and HeLa cells, and indicated potential for cell viability and low toxicity dependent on design [60]. Ni: Nickel NWs showed minimal toxicity to THP-1 cell macrophages over short timespans [61]; [8] Cell viability with PEDOT:PSS was determined for a variety of cell types [62].

Another emerging active material is graphene. Graphene can be applied in a variety of forms, including porous foams, flakes, and thin films, while showing favorable electrical and mechanical properties. In general, graphene-based wearable electronics are less stretchable and flexible than CNT-based electronics. However, graphene is more biocompatible [47,65]. Figure 1d shows a graphene oxide (GO) foam developed with a low modulus for high sensitivity [36]. A number of such foams are embedded in soft substrates to develop piezoresistive elastomers (Figure 1e) [37]. Doping with GO improves electrical conductivity and mechanically stiffen structures [63]. Graphite flakes can act as a coating layer to improve properties as well [66]. Micro-fluid, based on GO sheets, is also used as sensing materials [67]. Printable, conductive designs have been configured with graphene nanoflake inks [68]. Hydrogels can be used as biodegradable sensing materials or substrate capable of high strain deformations and self-healing. As shown in Figure 1f, a hydrogel consists of amorphous calcium carbonate (ACC) nanoparticles, polyacrylic acid (PAA), and alginate chain crosslinks [38]. Another composite sensing material consists of a polyvinyl alcohol (PVA) and borax hydrogel containing SWCNTs, graphene, and AgNW [69]. Hydrogels can also provide ionic connections between electrodes [70].

Highly flexible and stretchable properties of the sensing materials may also be achieved with liquid metals. The most commonly used liquid metal is the inorganic and biocompatible eutectic gallium indium (EGaIn), which can maintain electrical conductivity under high strain conditions. EGaIn is often selected due to its favorable properties, including low toxicity and high conductivity [71]. As a sensing material, liquid metals are enclosed in microtubules or microchannels within soft substrates. Liquid metals can be dispersed in soft substrates to create conductive or dielectric elastomers (Figure 1g). EGaIn mixed with a low-modulus elastomer (Ecoflex) produces a high dielectric constant [39]. Magnesium oxide (MnO_2) is a common material choice for nanosheets in supercapacitor applications. The process of creating an MnO_2 nanosheet is shown in Figure 1h, and the structure of a nitrogen-doped MnO_2 nanosheet is in Figure 1i. Mg films offer biodegradable electronic devices. Thin metal films, such as Au or Cu, can also be applied as electrodes using fractal, serpentine designs for improved flexibility and stretchability [72]. NWs are often adopted as sensing materials due to unique electrical and mechanical properties. Silver nanowires (AgNWs) patterning and films are demonstrated [63,73]. As shown in Figure 1j, aligned NWs can be grown on the substrate similar to CNTs. Carbon sponge has been incorporated into a PDMS substrate [74]. Conducting polymers, such as biocompatible PEDOT:PSS, allow textile-based wearable devices [75].

2.4. Substrate Materials

Unlike sensing materials, highly stretchable and flexible materials are generally used as substrates where sensing materials are embedded. Organic materials, compared to rigid inorganic ones, offer high flexibility, stretchability, and good adhesion for conformal lamination onto human tissues.

Table 2 summarizes a list of soft substrate materials and their mechanical properties, used for the development of wearable and implantable FHE. Silicone elastomers including Ecoflex, Sylgard, Dragon Skin, and Silbione are biocompatible in general and they are highly compliant with maximum elongation up to 900%. Strong adhesion of such elastomers onto target surfaces can be achieved in engineering of thin film formation. As noted in the previous section, elastomers may be combined with active sensing materials to improve electrical functionality while maintaining favorable mechanical properties. In addition to silicone elastomers, a variety of flexible polymers have been used, such as parylene, PET, PI, and poly(lactic-co-glycolic acid) (PLGA). In particular, PLGA is a common choice for biodegradable device applications [10]. These materials can also be applied as interlayers for better integration of specific materials with a substrate [76]. Additional substrates include paper, sellotape, and silk fibroin (biodegradable protein fiber). Among those, silk fibroin is used in ultra-lightweight and biodegradable applications, as indicated in Figure 1k [43].

Table 2. Substrate materials and their properties for wearable and implantable FHE.

Substrate Material	Organic/Inorganic	Young's Modulus/% Elongation at Break	Biocompatible/ Biodegradable
Silicone elastomer (Ecoflex 00-30)	Organic	0.07 MPa/900% [77]	Y/N [1] [78]
Silicone elastomer (Sylgard 184)	Organic	1.32–2.97 MPa [79]/120% [80]	Y/N [2] [81]
Silicone elastomer (Silbione LSR 4330)	Organic	1.38 MPa/750% [82]	Y/N [3] [83]
Parylene (VSI Parylene C)	Organic	2800 MPa/200% [84]	Y/N [4] [85,86]
Polyethylene terephthalate (PET)	Organic	230 MPa/120% [87]	Y/N [5] [88–90]
Polycaprolactone (PCL)	Organic	340.2 MPa/853.8% [91]	Y/Y [6] [92,93]
Polyimide (PI)	Organic	280 MPa/80% [87]	Y/N [7] [94]
Polyethylene naphthalate (PEN)	Organic	280 MPa/90% [87]	Y/N [8] [95]
Polyethersulfone (PES)	Organic	2654.5 MPa/100% [96]	Y/N [9] [97]
Polytetrafluoroethylene (PTFE)	Organic	0.06 MPa/400% [98]	Y/N [10] [99]
Poly(lactic-co-glycolic acid) (PLGA)	Organic	2000 MPa/3–10% [100]	Y/Y [11] [101]
Cyclic olefin polymer (Zeonor 1020R)	Organic	2100 MPa/90% [102]	Y/N [12] [103]
Silk fibroin	Organic	2500 MPa/2.1% (dry) 16.7 MPa/127.8% (wet) [104]	Y/Y [13] [105,106]

[1] Ecoflex was biocompatible with neuronal-like cells, indicating no cell death or cytotoxic effects [78]; [2] Biocompatibility of a Sylgard coating was tested with fibroblast cells and in vivo, and exhibited high cell viability and minimal inflammation [81]; [3] Biocompatibility indicated on material data sheet [83]; [4] Parylene C has proven biocompatibility, including intraocular, and can be improved with specific treatments and processes [85,86]; [5] One study found electrospun PET meshes caused large foreign body reaction, but indicated that improving the nanofiber dimensions and surface-to-volume ratio may diminish the reaction [88]. Treating the surface of PET prevented toxic effects on human endothelial cells and fibroblast cells [89,90]; [6] PCL indicated biocompatibility, including with mouse fibroblasts, and is biodegradable via hydrolytic degradation [92,93]; [7] PI exhibited similar biocompatibility as PCL when it was implanted in rat sciatic nerve and resulted in inflammation up to 4–8 weeks before tissue recovered [94]; [8] PEN biocompatibility was tested with culturing of osteoblast-like cells [95]; [9] Cytocompatibility of PES membranes with human liver cells was confirmed via cell viability and adhesion [97]; [10] Biocompatibility, including low cytotoxicity and cell activation, of PTFE was confirmed with endothelial cells and macrophages [99]; [11] PLGA indicated no cytotoxicity, although high concentrations of degradation products, due to increased degradation rates, resulted in toxic effects [101]; [12] COP is considered to be biocompatible, although studies are lacking [103]; [13] Silk fibroin indicates no cytotoxicity and has adjustable degradation rates by altering crystallinity and molecular weights [105,106].

2.5. Wearable FHE

Soft material-enabled, wearable electronic devices and systems include skin-mounted non-invasive sensors, soft actuators, flexible displays, low-profile energy storage systems, and miniaturized wireless communication devices. Figure 2 highlights a variety of wearable FHE that were used for conformal integration with human skins for monitoring of mechanical strain, pressure, and biopotentials and displaying information on contoured geometries. Table 3 reviews and summarizes device type, sensing materials, substrate materials, applications, target signals, and their mechanical and functional characteristics for wearable applications.

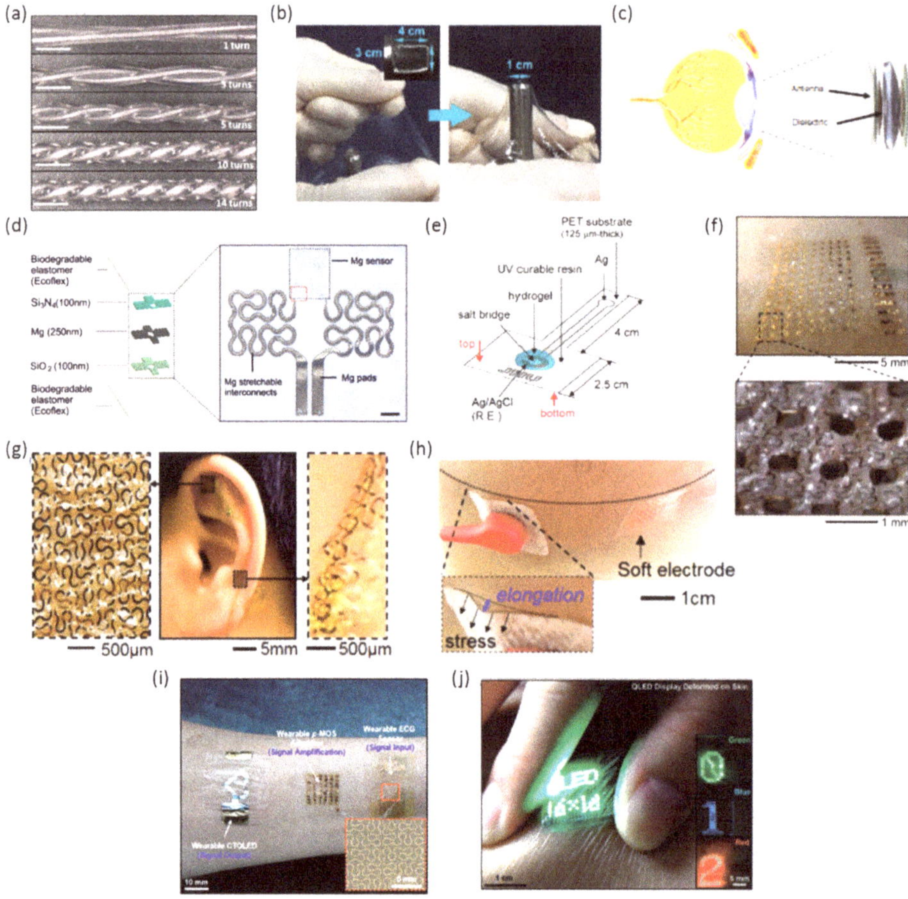

Figure 2. Wearable electronic systems. (**a**) Microtubule strain sensor and interconnect. Reprinted with permission from Reference [107]. Copyright 2017, Scientific Reports; (**b**) Self-healing strain sensor. Reprinted with permission from Reference [108]. Copyright 2017 American Chemical Society; (**c**) Wireless glucose and intraocular sensor. Reprinted with permission from Reference [29]. Copyright 2017, Nature Communications; (**d**) Biodegradable temperature sensor. Reprinted with permission from Reference [78], Copyright 2017, John Wiley and Sons; (**e**) Schematic illustration of a sweat chloride sensor. Reprinted from Reference [70], Copyright (2017), with permission from Elsevier; (**f**) EOG (electrooculography) electrode mounted on the skin. Reprinted from Reference [8], Copyright (2017), with permission from Elsevier; (**g**) EEG electrodes and interconnects on the auricle area [72]; (**h**) Comparison of rigid electrode with associated stress and soft material-enabled skin-like electrode. Reprinted with permission from Reference [109]. Copyright 2017, Scientific Reports; (**i**) Skin electrodes with amplifier. Reprinted with permission from Reference [110]. Copyright 2017 American Chemical Society; (**j**) QLED (quantum dot light-emitting diode) display on the human skin. Reprinted with permission from Reference [111], Copyright 2017, John Wiley and Sons.

Table 3. Summary of wearable FHE, material characteristics, and functionality.

Device Type	Sensing Material	Application	Substrate Material	Target Signal	Sensitivity	Flexibility	Stretchability	Reference (Year)
Strain Sensor	MWCNT	Motion, Bending	Ecoflex	Resistance	1.5 GF	-	300%	[112] (2017)
	EGaIn Liquid	Motion, Contact	Ecoflex Microtubules	Resistance	-	-	750%	[107] (2017)
	CS-PDMS	Blood Pulse, Breathing,	PDMS	Resistance, Temperature	GF 1.78	180°	228%	[74] (2016)
	Graphite Flake Sheath and Silk Fiber Core	Joint Motion, Multiaxial	Ecoflex	Resistance	14.5 GF	-	15%	[66] (2016)
	Self-healing SWCNT-Hydrogel	Bending	VHB Mounting Tape	Resistance	GF 0.24 (100% Strain), GF 1.51 (1000% Strain)	540° Twisting, 150° Bending	1000%	[69] (2017)
Pressure Sensor	GPN	Blood Pressure	PDMS	Resistance	0.09/kPa	-	40%	[37] (2016)
Light Sensor	Ionic Liquid, PU fiber, SWCNT, Au film	Electronic Skin	Ecoflex	Conductivity	2.4 mW	90°	50%	[63] (2017)
Temperature	PEIE/CNT-PDMS. Ag electrode	Healthcare Patch	PET	Resistance, Voltage	0.85%/°C,	-	-	[113] (2017)
Sweat Sensor	InGaZnO ISFET, PI, CNT/PEDOT:PSS	Healthcare and Sports	PET	Current, Resistance	51.2 mV/pH	10 mm Radius	-	[114] (2017)
	Hydrogel, Ag/AgCl Electrode	Fitness Monitoring	PET	Voltage	52.8 mV/decade	-	-	[70] (2017)
Electrode	Au	EOG, Eye Movement	PI	Voltage	13.3 µV/°	0.5 mm Radius	30%	[8] (2017)
Antenna	Ag-PDMS	Wireless Communication		Conductivity	-	-	20%	[115] (2017)
QD Display	QDs	Sensor Display, Touch Sensor	Parylene	Intensity	-	180°	-	[111] (2017)
Cooling Device	BaSrTiO Nanowires	Cooling	PDMS	-	-	5 mm Radius	25%	[42] (2016)
Supercapacitor	MnO$_2$ Nanosheet, Carbon Fiber, Graphene, PVA	Energy Storage	Cotton Textile					[116] (2016)

2.5.1. Strain Sensors

Numerous strain sensors in wearable configurations have been developed, as highlighted in Table 3. Soft substrates and functional materials offer high stretchability and sensitivity of the designed strain sensors. They typically function by undergoing a change in electrical resistance when experiencing strain change. As a result, those sensors rely on the sensing material's inherent characteristics. To design a highly sensitive resistance sensor, one study utilized SWCNT wires via wet spinning on a polydimethylsiloxane (PDMS) membrane [35]. This sensor also functioned as a pressure and torsion sensor. Although these microwires were fragile, the randomly oriented SWCNT networks fractured and rejoined under deformation, resulting in measurable electrical resistance changes with a high gauge factor (GF) of 10^5 at 15% strain. The soft PDMS substrate aided in recovering the initial electrical resistance.

Another study achieved a very high GF by using graphene in a GO woven fabric (GWF) design based on a GO-coated cotton bandage template [117]. This strain sensor, encapsulated by a natural rubber latex, indicated a maximum elongation of 57% and a GF of 416 and 3667 at lower and higher strains, respectively. One work coated silk fibers with graphite flakes to develop a fiber-shaped strain sensor [66]. With a silk fiber, the sensor showed a GF of 14.5 up to 15% strain while a spandex fiber showed a GF of 14.0 up to 30% strain. As discussed previously, graphene can be manipulated in a variety of forms; one study used it as a force sensing material and contact electrode with PMMA coating [118]. This design achieved a GF of 42.2 for up to 20% elongation with ability to sense tension, bending, and torsion. Higher GFs, between 47.74 and 98.66, and a maximum elongation of 30% were reached with a graphene foam and PDMS composite [37].

To improve stretchability and maintain a high sensitivity, silver-coated polystyrene spheres (PS@Ag) were mixed with liquid PDMS to form printable electrodes and strain sensors [119]. With a 60 wt.% of PS@Ag, the sensor achieved a maximum elongation of about 80% and GFs, from lower to higher strains, of 17.5, 6, and 78.6. Whereas the previous CNT structure resulted in limited stretchability, the LBL assembly of SWCNT and thin polymer layers on a polycaprolactone (PCL) membrane created a piezoresistive, biocompatible strain sensor with a maximum strain of 100% and adjustable GFs between 5–13 [64]. While these previous strain sensors aim for high GFs, they sacrifice stretchability. To improve maximum elongation, a carbon sponge-PDMS composite was developed as a strain sensor to monitor pulse, breathing, and walking [74]. This composite sensor endured up to 200% strain and repeatable bending of 180°, and indicated a GF of 1.78 in tension. CNTs can also be designed to realize highly stretchable strain sensors. MWCNTs mixed with Ecoflex were printed into a partially cured substrate to produce a strain sensor capable of stretching 300% with a linear electrical response and GF of 1 [112]. Similarly, an Ecoflex and MWCNT nanocomposite configured a multidirectional strain and pressure mapping sensor with measurements based on anisotropic electrical impedance tomography [120]. This work applied the device as a human-machine interface, and possesses potential to develop large, irregularly shaped sensors.

In addition to embedding sensing materials into soft substrates, stretchable strain sensors can be developed by manipulating the material's structure. By heating a CNT membrane, a wrinkled thin film was produced [121]. This wrinkled structure, integrated on an Ecoflex substrate, allowed conductivity up to 750% elongation, an approximate 60 times increase versus non-wrinkled films. GFs were 0.65 below 400% strain and increased to 48 above 400% strain, due to film fracture. Another material applied in stretchable strain sensors is liquid metal (EGaIn). One method of integrating EGaIn is to fill microtubules formed by soft polymers with the liquid metal [107]. These microtubules shown in Figure 2a can be used for contact measurements via resistance changes without electric failure upto 750% elongation. The use of other polymers, such as a PDMS nanocomposite, may improve the stretchability. Similarly, another study presents soft polymer microtubules filled with EGaIn to measure strain, torsion, and contact via capacitive changes [122].

Combining a self-healing hydrogel with SWCNTs, graphene, and AgNW realized a more stretchable piezoresistive sensor capable of enduring up to 1000% strain [69]. The hydrogel used here consisted of polyvinyl alcohol (PVA) and borax. This configuration achieved a GF of 0.24 at strains below 100% and of 1.51 at 1000% strain. Another strain sensor based on hydrogel used a PVA/polyvinylpyrrolidone (PVP) hydrogel with Fe^{3+} cross-linked cellulose nanocrystals (CNCs) to produce a strain sensor with 830% maximum elongation and self-healing ability [108]. The stretchability of this hydrogel is observed in Figure 2b. The CNCs-Fe^{3+} bonds act to dissipate energy when stretched and to improve toughness of the hydrogel. This strain sensor measured breathing and blood pulse. Wearable biaxial strain sensors have also been developed. As mentioned in the CNT discussion, aligned CNT arrays offer advantages over random orientation. With this understanding, one study developed a biaxial strain sensor with a GF of 3 via cross-stacked, aligned CNT films [2]. These CNT films were hybridized, or welded, with graphene to improve the biaxial sensing ability and increase the GF by at least 5 times existing cross-stacked CNT sensors. Similarly, aligned CNTs in polymer substrates improved functionality over randomly oriented composites [123]. This method produced a poly(methyl methacrylate) (PMMA) and PDMS composites for a multifunctional flexion- and tensile-sensitive electronic skin.

2.5.2. Pressure Sensors

Pressure sensors can use capacitive or resistive-based measurement, depending on materials and design selections. Similar to strain sensors, graphene can be applied in a variety of forms. A capacitive pressure sensor used a GO-based low elastic modulus foam as the dielectric material to achieve a high sensitivity of 0.8/kPa [36]. Another capacitive pressure sensor used a dielectric layer of polyethylene (PE) film placed between two hydrogel films formed a capacitive pressure sensor and resulted in a sensitivity of 0.17/kPa [38]. This hydrogel can self-heal when two parts of hydrogel contact and is able to fully recover if dehydrated. One study configured a single layer piezo-capacitive pressure sensor with an AgNW pattern on polyurethane-urea (PUU) [124]. This sensor showed stability under 35% elongation and 10,000 cycles at 30% elongation. The Bao group developed a highly sensitive pressure sensor using flexible transistors [125,126]. Here, a microstructured PDMS acted as a dielectric layer, where the micropatterns were varied to tune pressure sensitivity as high as 8.4/kPa. Resistive pressure sensors have also been studied, and often function as strain sensors as well. A graphene porous network embedded in PDMS resulted in a pressure and strain sensor with sensitivities of 0.09/kPa and GFs of above 2.6, respectively [37]. Additionally, this sensor determined the frequency and degree of bending. Graphene-based micro-fluids can also be sensing materials, such as that formed by dispersing GO nanosheets in distilled water [67]. The microfluidic channel placed in an Ecoflex layer and PDMS substrate acted as a resistive strain and force sensor, with a sensitivity of 3.38^{-2}/kPa. Similarly, other liquid conductive metals have been developed, such as that used in a resistive pressure sensor based on a microfluidic diaphragm [51,71] This diaphragm constructs a Wheatstone bridge circuit, enabling radial and tangential strain detection based on the microchannel resistance changes with a sensitivity of 0.0834/kPa. One study indicated a method of remarkably increasing low pressure sensitivity of a pressure sensor from 02.57/kPa to 20.6/kPa by using GO to dope an electrospun polyurethane (PU) nanofiber membrane with PEDOT and PDMS encapsulation [63]. Graphene also can improve NW functionality when integrated with soft contact lenses (Figure 2c) to measure glucose and intraocular pressure with 25% stretchability and 91% transparency [29]. By adding graphene to the structure, resistance reduced during mechanical bending. The field effect transistor (FET) to sense glucose utilized a graphene channel as well. Another study integrated graphene with a contact lens for eye protection from electromagnetic waves and dehydration [127].

2.5.3. Other Types of Sensors

In addition to strain and pressure sensors, many other types of wearable sensors exist for measuring temperature, light, sweat, and vibration. A biodegradable temperature sensor was composed of a thin Mg film active layer and Si_3N_4 and SiO_2 dielectric layers, encapsulated by Ecoflex [78]. This biodegradable sensor in Figure 2d indicated a resistive sensitivity of 0.2%/K, stretched up to 10%, and bent to a radius of 1.75 mm. Multifunctional sensors also exist for wearable applications, as one study showed for measurement of temperature, pressure, and light with ionic liquid and Au film as sensing materials [63]. By doping the ionic liquid with SWCNTs, the ionic liquid experienced an increase in light sensitivity due to the CNTs conversion of light energy to heat. By applying a InGaZnO thin film in an ion-sensitive field-effect transistor (ISFET), one study designed a sensor for simultaneous sweat pH and skin temperature monitoring to withstand bending to a 1 cm radius [114]. This ISFET on a PI substrate acted as a pH sensor while a CNT-PEDOT:PSS solution acted as the temperature sensor, both of which were incorporated onto a PET substrate. Another sweat sensor, shown in Figure 2e, measured sweat chloride and incorporated hydrogel [70]. The agarose gel acted as a salt bridge between two electrodes, creating an ionic connection between a reference solution and the environment while preventing equilibration. The Wang group applied a temporary tattoo-based sweat sensor to measure lactate [128]. Additionally, they integrated a lactate sensor with electrocardiogram (ECG) electrodes into a single patch and confirmed minimal interference [129]. In a similar way, they manufactured a tattoo-like glucose sensor and achieved a sensitivity of 23 nA/µM [130]. This group also investigated stretchable electrochemical sensors for other wearable applications [131]; PEDOT:PSS and Ag/AgCl inks, printed on an Ecoflex substrate exhibited up to 100% linear and 150% radial stretching. Additionally, a work developed a colorimetric humidity sensor via a nanowire cluster film (NWCF) [132]. Fabricating this cluster film required sputtering on a disordered anodic alumina oxide substrate, rather than a normal flat substrate, and the resulting NWCF bent to a 1.85 cm radius and indicated stability over stretching and bending cycles. The result could be visualized as NWCF exhibited an extinction spectrum that varied with water, or sweat, content on its surface. Lastly, a vibration sensor used a cellular structured graphene composite elastomer with highly soft and piezoresistive properties. This sensor detected frequencies between 300–20,000 Hz and worked as an accelerometer with a linear electrical response [133].

2.5.4. Electrodes

Another wearable device are non-invasive, skin-mounted electrodes for recording biopotentials for human health monitoring, disease diagnostics, and machine interfaces. One device used a triple-layer graphene electrode array with a graphene monolayer ground [134]. This electrode array allowed 8% and 15% elongation in separate directions, and detected approaching objects up to 7 cm away due to absorption of the object's electric field. Another study developed flexible, biocompatible microelectrodes by patterning AgNWs onto a hydrogel substrate of PEG, agarose, and PAAM, allowing flexibility to bend up to a 10 mm radius [73]. By patterning thin Au on PI substrates, electrooculography (EOG) electrodes can be developed to withstand over 30% strain and 500 micron bending radius [8]. This skin-like electrode, shown in Figure 2f, aimed to allow controlling a wireless wheelchair via EOG signals. Similarly, a study implemented a fractal design of Au for conformal contact of an electroencephalogram (EEG) electrode to the skin (auricle area; Figure 2g), which was used for a persistent brain-computer interface (EEG-based text speller) [72]. Figure 2h provides a comparison of a rigid-material based electrode and a soft material-enabled, stretchable electrode for electromyography (EMG) recording on the chin [109]. Unlike the stress-inducing rigid electrode, the skin-like sensor showed comfortable, intimate contact onto the skin. This electrode, built of Au nanomembrane on a silicone-PVA substrate, endured up to 150% strain and conformed to a 500 micron bending radius. A wearable textile-based keyboard, capable of withstanding 30% elongation, used a knitted textile substrate [75]. Areas of the substrate were coated with PEDOT:PSS to produce a capacitive electrode when touched with a human finger, allowing for application as a keyboard. CNTs can be applied to improve electrode functionality and performance. Embedding CNT networks into polymers created a wearable p-MOS inverter consisting of four CNT transistors for amplification of ECG signals from a wearable electrode and is shown in Figure 2i [110]. Another application of CNT is to make a CNT-PDMS composite for enhanced interface between the skin and ECG sensor [113]. This composite maintained flexibility of pure PDMS but incorporated ethoxylated-polyethylenimine (PEIE) to improve adhesion while the CNTs improved conductivity.

2.5.5. Electrical Components, Displays, and Actuators

Beyond sensors and energy storage devices, soft materials enable wearable actuators and flexible passive electrical components. In particular, transmission lines in electronics have been developed. A conductive graphene ink consisting of graphene nanoflakes can be printed and compressed on a paper substrate as a transmission line [68]. This transmission line did not experience a change in transmission coefficients despite bending and twisting. Flexible antennas were similarly developed. Another method of creating a flexible and stretchable transmission line is via a wrinkled CNT film, formed by transferring CNT sheets to a prestrained elastomer [135]. After releasing the strained elastomer, the CNT layer produces a wavy shape. From this method, a stretchable conductive transmission line experienced minimal strain induced changes of electrical resistance up to 600%. Alumina passivation of the CNT sheets improved performance. One group developed a highly conductive printable elastomer of mixed PDMS and Ag powder for wearable wireless applications, including fabrication of a transmission line and RF antenna [115]. EGaIn can also form flexible and passive electrical components, such as resistors, inductors, and capacitors via a patterning process based on soft lithography [136]. With this structure, a 2.5 D circuit integration enabled to power two LEDs with 35% strain. Liquid metals can be embedded in soft elastomers to alter electrical properties, such as mixing EGaIn with Ecoflex to produce the stretchable dielectric elastomer in Figure 1g [39]. Silk fibroin film enabled an ultra-lightweight and biocompatible resistive switching memory device [43]. This device conformed to an 800-micron radius while maintaining functionality with a retention time above 10,000 seconds and on/off ratio of 100,000. A stretchable transistor from the Bao group [33,137] utilized a styrene-ethylene-butadiene-styrene hydrogenated elastomer as a dielectric and substrate material, along with electrodes and semiconductors that were composed of SWCNTs. This device, unlike other SWCNT transistors, showed great reproducibility. By testing different diameters of SWCNTs, the transistor indicated a maximum mobility of 15.4 cm^2/Vs and an on/off ratio above 1000.

Wearable displays have also been developed with the usage of soft materials, such as a quantum dot light-emitting diode (QLED) display of ultrathin layers, including a layer of PEDOT:PSS encapsulated by parylene C and epoxy [111]. This display, shown in Figure 2j, bent around a radius of 68 microns and endured 1000 bending cycles. Furthermore, the study integrated the display into a touch interface, temperature sensor display, and a step counter. Another wearable display incorporated polymer light-emitting diodes (PLEDs) and organic photodetectors (OPDs) [138]. Again, ultrathin layers were encapsulated by a parylene substrate and the device conformed to a radius below 100 microns, stretch to 60%, and compress 67% while maintaining functionality. Wearable actuators have also been explored, including an artificial muscle consisting of the graphene-CNT-nickel (G-CNT-Ni) hetero-nanostructure embedded in a conductive polymer (PEDOT:PSS) [34]. Additionally, an array of aligned ferroelectric barium strontium titanate (BST) NWs acted as a wearable cooling device by using the electrocaloric effect of the NW array [42]. This array, encapsulated by PDMS, withstood a radius of curvature of 5 mm and 25% strain. NWs can also be incorporated into flexible FETs. By using sellotape as a substrate, an array of FETs based on CuPc organic NWs achieved a 3 mm bending radius [139].

2.5.6. Energy Storage

Wearable electronics also include energy storage and harvesting devices. Such FHE could power and enable wireless communication. A common material for wearable energy storage applications is MnO_2. One work developed a capacitive energy storage device by encapsulating molybdenum carbon nanofibers (Mo_2C NFs) with MnO_2 nanosheets via electrospinning [40]. Applied as a supercapacitor, the device achieved a specific capacitance of 430 F/g and 302 F/g at current densities of 0.1 A/g and 1 A/g, respectively, with 92.6% capacitance retention after 5000 charging cycles. Similarly, a supercapacitor used carbon nanofibers covered with MnO_2 nanosheets and carbon nanofibers covered with graphene to build a wearable all-solid-state supercapacitor [116]. This supercapacitor indicated a specific capacitance up to 87.1 F/g and cycling voltammetry curves did not change despite different bending states. Other than fiber-shaped supercapacitors, MnO_2-based flexible supercapacitors also can be designed in the shape of a sheet. Here, an MnO_2 nanosheet grown on a nitrogen-doped graphene nanosheet with PVA-LiCl gel electrolyte created a supercapacitor capable of 305 F/g specific capacitance [41]. Another film-shaped supercapacitor involved the development of an amino-functionalized MWCNTs and MnO_2 thin film composite as electrodes separated by a cellulose layer [140]. This design retained 90% of its capacitance after 2000 bending cycles at 90°. Another nanosheet was printed via nickel hydroxide ($Ni(OH)_2$) ink coated on a carbon fiber yarn substrate in order to create a wearable energy storage device [141]. Graphite foam composites have also been applied to wearable supercapacitors which power sensors and other components. Multilevel porous graphite foams (MPGs) and MPG/Mn_3O_4 composites (MPGMs) were developed as a lightweight and flexible supercapacitor [142]. This supercapacitor indicated a capacitance of 53 F/g and exhibited a 90% and 80% capacitance retention after 10,000 charging cycles and 1000 bending cycles, respectively. In addition to capacitors, a recent study from the Wang group developed wearable energy harvesters that implemented a biofuel system using lactate in sweat [143]. This system, comprised of Au serpentine interconnects, joined CNT bioanodes and cathodes and maintained a power density of 1.2 mW/cm^2 at 0.2 V with 50% strain. Soft elastomers also allow wearable arrays of rigid electrical components. An array of chip-scale batteries with serpentine PI/Cu/PI layered interconnects, when placed on an Ecoflex substrate with a Silbione interlayer, can withstand over 30% biaxial stretching [144]. In this work, a wireless temperature sensor was powered by the flexible system. A printable Ag-Zn battery using a temporary tattoo paper accommodated strain up to 11.1% [145]. Additionally, a rechargeable $Zn-Ag_2-O$ battery on polyurethane employed a hyper-elastic binder to achieve 100% strain while powering an attached LED [146].

2.6. Implantable FHE

Similar to wearable electronics, implantable FHE enabled by soft functional materials offer a number of applications in health monitoring, diagnostics, and therapeutics. Figure 3 shows a collection of representative examples of flexible-membrane based implantable electronics. Most of recent advancements in implantable systems are on biodegradable, transient materials; such resorbable electronics physically disappear at prescribed times and at controlled and guided rates. Table 4 collects a summarized list of implantable FHE, material characteristics, and functionality.

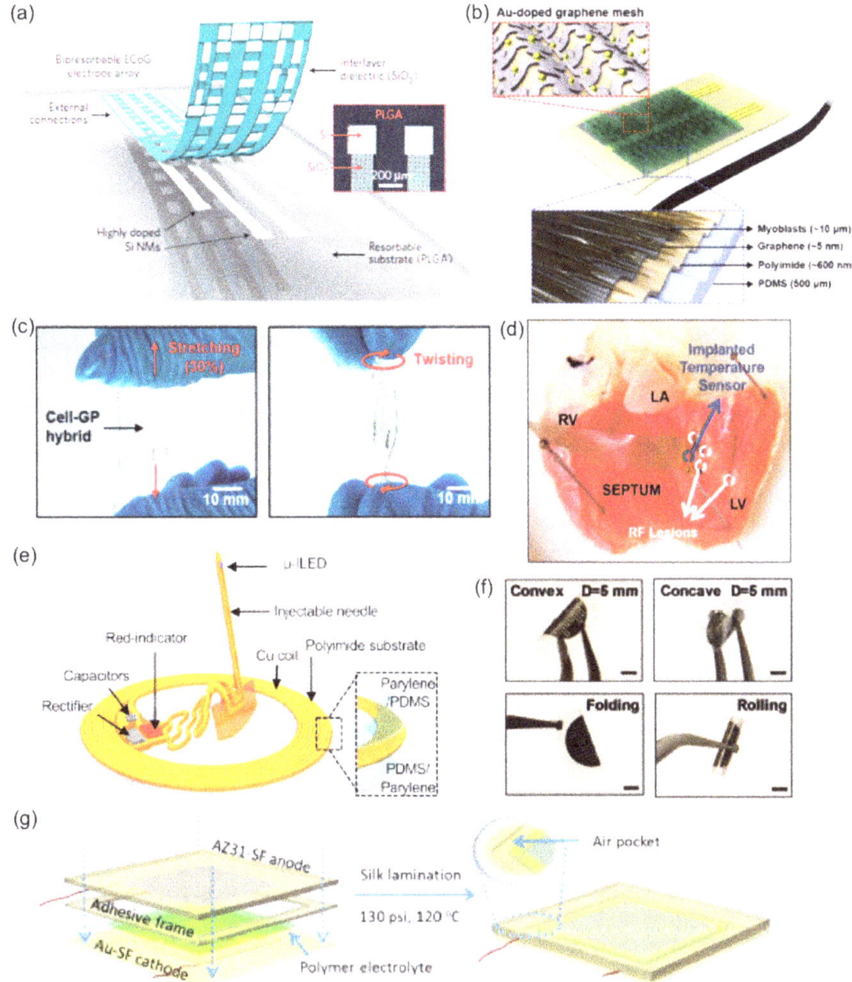

Figure 3. Implantable electronic systems. (**a**) Bioresorbable ECoG electrodes. Reprinted with permission from Macmillan Publishers Ltd.: Nature Materials [10]; (**b**) EMG electrode with Au-doped graphene mesh. Reprinted with permission from Reference [147], Copyright 2016, John Wiley and Sons; (**c**) Stretchability and flexibility of the serpentine-structured electrode in (**b**). Reprinted with permission from Reference [147], Copyright 2016, John Wiley and Sons; (**d**) Implantable cardiac sensor for monitoring temperature, thermal conductivity, and heat capacity. Reprinted from Reference [148], Copyright (2017), with permission from Elsevier; (**e**) Optogenetic device for wireless light delivery. Reprinted with permission from Reference [149], Copyright 2017, John Wiley and Sons; (**f**) Mechanically flexible, biodegradable microsupercapacitor. Reprinted with permission from Reference [49]. Copyright 2017, John Wiley and Sons; (**g**) Multi-layer illustration of a biodegradable battery with silk membrane. Reprinted with permission from Reference [150]. Copyright 2017 American Chemical Society.

Table 4. Summary of implantable FHE, material characteristics, and functionality.

Device Type	Sensing Material	Application	Substrate Material	Target Signal	Sensitivity	Flexibility	Stretchability	Reference (Year)
Electrode	Myoblasts, Au-Graphene	EMG, Stimulation, Therapy	PI, PDMS	Voltage	-	-	40%	[147] (2016)
	Si nanomembranes	Electrophysiological Mapping	PI	Voltage, Current	-	5 mm radius	-	[151] (2017)
	Doped Si Nanomembranes	Monitor Brain, Muscle, Organ Activity	PLGA	Voltage	-	1 mm radius	-	[10] (2016)
	LE-AgNW	ECG, Biventricular Pacing	SBS Rubber	Voltage, Contractility	-	-	-	[152] (2016)
Cardiac Temperature Sensor	Au	Lesion Characterization	PET	-	0.26%/°C	21 N/m Bending Stiffness	-	[148] (2016)
Optogenetic Light Delivery	Cu	Optogenetics	PI, Parylene, PDMS	Output Power	-	6 mm radius	-	[149] (2017)
Biodegradable Microsupercapacitor	W, Fe, Mo, NaCl-Hydrogel	Power Storage	PLGA	Capacitance	-	5 mm diameter	-	[49] (2017)
Biodegradable Battery	Mg	Power Supply	Silk Fibroin	-	0.06 mAh/cm^2 (Specific Capacity)	-	98%	[150] (2017)
Energy Harvester	PMN-PZT-Mn	ECG, Wireless Data Transmission	PET, PU	Voltage	-	9.95^{-5} N/m Bending Stiffness	-	[153] (2017)

2.6.1. Implantable Electrodes and Sensors

Bioresorbable ECoG electrodes have been developed with various materials. As shown in Figure 3a, phosphorus-doped Si nanomembranes (NMs) are used as an active material with a thin dielectric layer of SiO_2 and a PLGA substrate [10]. This electrode conformed to a curvilinear surface for high-fidelity signal recording, while showing long-term stability. Porous graphene and Au wires on a PI substrate are also employed for ECoG electrodes [154]. This graphene is doped with nitric acid to improve its impedance. Another study presents an ECoG electrode array on a Cyclic Olefin Polymer (COP) substrate with gold electrodes [103]. Figure 3b shows a layout of a flexible EMG sensor [147]. The multi-layers of PI, Au-doped graphene, and aligned C2C12 myoblast sheet on the PDMS substrate offer a favorable in vivo interface with the skin. The mechanical properties of the electrode reaching 40% elongation and over 90° bending are demonstrated in Figure 3c. Another EMG sensor uses capacitively coupled silicon nanomembrane transistors with an ultrathin silicon dioxide layer as a dielectric layer [151]. Through polydopamine (PDA) reduction of GO, a hydrogel can be designed to be a self-healable, conductive, and biocompatible hydrogel with an extension ratio greater than 35 [155]. This hydrogel showed a variety of applications, including a finger motion sensor, self-adhesive electrode, intramuscular electrode, and cell stimulator.

An epicardial mesh for ECG recording and biventricular pacing was made of a functional composite of biocompatible styrene-butadiene-styrene rubber and ligand exchanged AgNWs [152]. With a mesh design, the device withstood the required radial strains of 90% to stimulate the ventricles and compared well for ECG recording. Graphene FETs to record potentials also have been developed on a PI membrane via a high throughput transfer technique [156]. By using an array of ZnO NWs coated with a gold film and encapsulated by PEDOT, impedance and signal-to-noise ratio significantly decreased due to increased surface area, which improved sensor performance [57]. To enhance the mechanical flexibility, the PI and Au-graphene hybrid were used as a substrate and an interconnection line, respectively. A study created injectable cardiac sensors for real-time monitoring of temperature and thermal transport properties with a flexible PET substrate, thin Au wires, and an epoxy encapsulation [148]. This flexible cardiac sensor is implanted in a heart as shown in Figure 3d. Another implantable sensor composed of a p(AM-co-PEGDA) core and a Ca alginate cladding to detect physical and optical changes in response to glucose. [157]. This fiber bent up to 80° with a 30% light intensity loss and indicated a readout rate of 1.33 mmol/L-min, which was 17 times greater than the required rate. One study designed an Mg-based biodegradable sacrificial layer for bonding with a flexible electrode to provide a necessary rigidity for implantation [158]. Once implanted, the physiological fluid dissolved the sacrificial layer, enabling conformal integration with the tissue.

2.6.2. Actuators

Implantable actuators can be developed with soft materials. For example, a robotic sleeve to assist the heart used a low-modulus elastomer (Ecoflex) as an actuated material and hydrogel as an interlayer [76]. Additionally, a light delivering device for optogenetic purposes is shown in Figure 3e [149]. The implantable device consisting of a Cu coil and other miniaturized electronics is encapsulated by layers of parylene and PDMS, which is bendable up to 9 mm in radius and stretchable up to 51% elongation.

2.6.3. Energy Storage and Circuit Components

Similar to wearable FHE, soft materials enable compliant and implantable energy storage devices. A biodegradable and flexible microsupercapacitor is demonstrated by implementing metal thin films, agarose hydrogel, and PLGA substrate (Figure 3f) [49]. The metal films that are fabricated with gold, tungsten, iron, and molybdenum shows an areal capacitance of 0.61 mF/cm^2. Another study developed a supercapacitor to use in physiological fluids [159]. Here, aligned hydrophilic CNT sheets were applied to create CNT fibers as supercapacitor electrodes, while using the surrounding physiological fluid as an electrolyte. To develop the implantable, biodegradable battery in Figure 3g, silk fibroin films are utilized, including Mg-silk fibroin anode and Au-silk fibroin cathode [150]. The electrolyte consists of a silk fibroin-choline nitrate polymer and the device includes an encapsulated silk layer. Additionally, flexible and implantable energy harvesters have been developed; to implant an energy harvester into a heart, a PMN-PZT-Mn piezoelectric crystal is attached to a PET substrate via polyurethane [153]. This device works as a wireless ECG sensors that is bendable up to 2 cm in radius. Additionally, current research includes transient circuit components, such as the development of fully biodegradable logic circuits that utilize the biodegradability of silicon nanomembranes [160]. Using similar methods, this group demonstrated a multifunctional transient sensor [161]. A transient antenna with a tunable degradation utilized a PVA-TiO$_2$ film substrate [162]. Other works involve studies of a PVA-based substrate where the dissolution rate was controlled by adding gelatin or sucrose [163].

3. Integration Strategies of Electronic Circuits for FHE

Continued engineering efforts in the study of soft functional materials and their system implementation allow ample opportunities to manufacture various FHE that are both mechanically and organically compatible with biological tissues. While these soft materials enable FHE to achieve multiple levels of mechanical compliance, they alone cannot realize fully functional and practical FHE. Core electronic components such as amplifiers, analog-to-digital converters, filters, microprocessors, memories, and multiplexers need to be embedded into a soft FHE for data acquisition, transmission, processing and active control. In this section, we review a few strategies to integrate essential electronic components onto flexible membrane circuits for mechanically compliant hybrid electronics.

3.1. Organic Electronics

Electronic devices consisting entirely of polymers have drawn significant interests due to their intrinsic mechanical flexibility. Since most organic materials can be processed in a solution form, a large-scale fabrication can be done by printing processes. In fully printed organic systems, the devices are formed by an additive manufacturing, thus no further subtractive steps are needed for structural patterning. For example, Tokito's group has demonstrated the fabrication of fully-printed organic thin-film transistors (OTFT) and circuitry on ultra-thin parylene-C films (Figure 4a). They have showed a large-area printing (Figure 4b) on a single, continuous film, which allows the entire system to be worn over the target skin areas [164]. Complementary logic circuits (e.g., CMOS) employing both p-type and n-type OTFTs have been fabricated by a similar printing method (Figure 4c,d) [165]. Someya's group also demonstrated the integration of OTFTs as switching transistors for an ultrathin, active-matrix tactile sensor, validating the potential of organic electronics for wearable devices (Figure 4e) [166]. Despite the manufacturing flexibility and scalability of organic electronics, a complete realization of modern electronic requirements solely by the use of organic materials has been challenging. First of all, acquiring electronic functions with fast switching speeds and high on-off ratio might be challenging due to the far less carrier mobilities of organic semiconductors than that of inorganic, single crystalline semiconductors. This limits the roles of organic electronics to low-speed tasks. Moreover, both structural and chemical stability of the organic materials needs to be further improved to achieve the device reliability, especially for body-implantable applications.

3.2. Inorganic Electronics

While typical semiconducting wafers are rigid and brittle, exfoliating them into thin layers with a thickness less than 25 µm yields mechanical bendability and flexibility [167]. Therefore, an implementation of thin, single crystalline semiconductor layers into flexible substrates would provide superb electronic properties, mostly owing to the carrier mobilities that are orders of magnitude higher than those of organic counterparts, allowing the circuits to be tasked with more demanding functions. Moreover, since modern photolithography technologies would be used for patterning, extremely dense array of transistors can be packed in a small area. Rogers's group developed such materials processing method, enabling both exfoliation and manipulation of silicon nanomembranes based the use of a silicon-on-insulator (SOI) wafer [168]. As shown in Figure 4f, once high-thermal processes (oxidation and diffusion doping) are carried out on SOI wafers to define source and drain areas, the doped silicon layer can be released by removing the insulating oxide layer. The released layer is then transferred onto a flexible substrate to finalize device configuration through low-temperature processes including dielectric and metal depositions. Alternatively, as demonstrated by Hussain's group, lower-cost silicon wafers with <100> crystal orientation can be processed with directional reactive ion etching, followed by isotropic etching with XeF2 to effectively release the top silicon layer (Figure 4g) [169]. Lastly, a process utilizing a controlled fracture across a wafer with use of a stressor layer can be used to peel away a layer of wafer containing completed CMOS circuits. The Bedell group showed that controlled spalling could successfully exfoliate 10 μm-thick top layer from a 300-mm silicon wafer by using a 6 μm-thick Ni stressor layer (Figure 4h) [170]. The spalling process utilizes the intrinsic materials properties, ensuring a complete removal of residual stresses, which can adversely affect circuit properties, requires extensive processing time and fabrication complexity.

3.3. Thinned Chips

Rather than exfoliating a thin layer of a wafer, diced individual silicon chips can be chemically and mechanically ground down to achieve required mechanical properties. The thinned chips can then be bonded to flexible substrates either with the active side facing down to the pre-patterned fan-out traces using conductive adhesives or with the active side facing upward followed by spin-on film deposition and establishing electrical connections via microfabrication. Jan Vanfleteren's group demonstrated the integration of a 30 μm-thick, application-specific integrated circuit with a flexible substrate, enabling multi-channel brain recording and stimulation in small rats (Figure 4i) [171]. Similar to spalled electronic layers, the thinned chips can be sensitive to both internal and external stresses, and maturation of the technology needs to take place by further developing device validation processes.

3.4. Chip-Scale Packaging

Recent advancement in the electronic packaging technology has resulted in the production of chip packages with minimally added materials while establishing features, such as solder bumps, directly on the circuit side to enable chip integration. For example, ball grid array and chip scale package components can be considered as bare die chips with preformed solder balls either in non-array or array configurations, respectively. While this construction adds few tens of micrometers to the overall thickness, the solder-based approach provides the most robust electrical and mechanical connection between the chips and flexible substrates. Because these types of packages are commercially available, no additional processing steps are required for film integration. Moreover, multiple components, such as thick-film passive components and other solderable packages based on surface mount technology, can be assembled simultaneously by using an automatic pick-and-place tool. The manufacturing method introduced by Kim et al. [172] shows the potential of integrating commercial off-the-shelf chip components with a stretchable platform that is compatible with conventional soldering processes. The key features enabling this particular solderable and stretchable platform are (1) excellent solderability and compatibility with conventional surface mount technology, (2) ability of the assembled device to be directly integrated with a soft adhesive layer, and (3) scalability of the manufacturing method as described in Figure 4j.

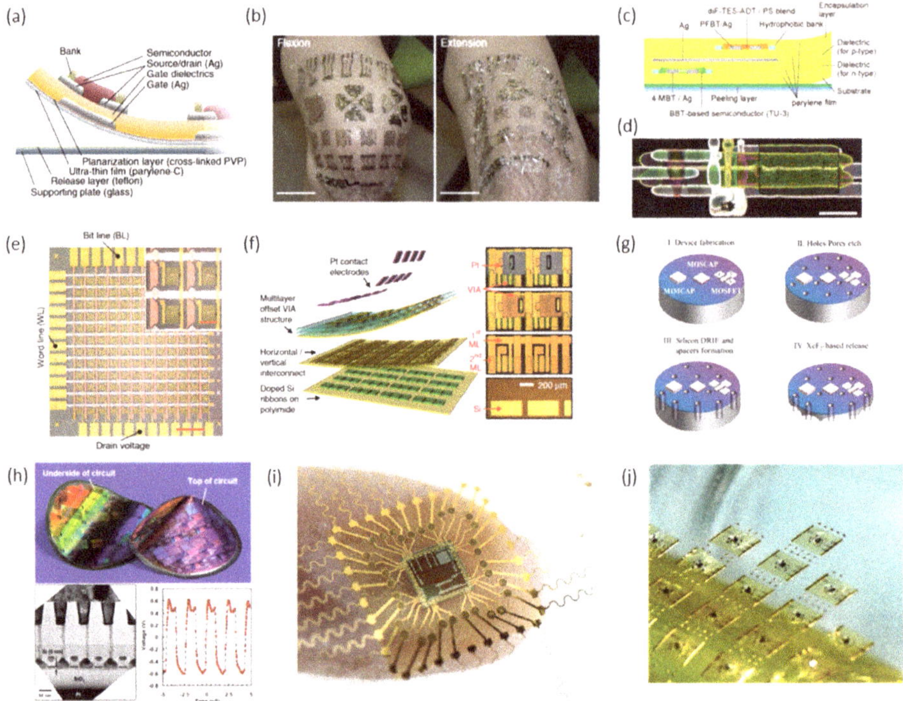

Figure 4. Integration strategies of electronic circuits for FHE. (**a**) Cross-sectional diagram of a fully-printed OTFT (organic thin-film transistors) device and (**b**) application of a thin organic film to a human knee. Scale bar, 4 cm. Reprinted with permission from Macmillan Publishers Ltd.: Nature Communications [164]; (**c**) Cross-sectional diagram and (**d**) photograph of an ultra-thin, fully-printed CMOS logic circuit. Scale bar, 500 μm. Reprint is in accordance with the Creative Commons Attribution 4.0 International License [165]; (**e**) Optical micrograph of a 12 × 12 tactile sensor array utilizing OTFTs as the switching transistors. Scale bar, 1 cm. The inset shows a magnified view of four pixels. Scale bar, 2 mm. Reprinted with permission from Macmillan Publishers Ltd.: Nature [166]; (**f**) Exploded view rendering of the flexible, high-density brain mapping device (left) and respective optical micrographs (right). Reprinted with permission from Macmillan Publishers Ltd.: Nature Communications [168]; (**g**) Process flow of XeF$_2$-based Si exfoliation. Reprinted with permission from Reference [173]. Copyright (2014) American Chemical Society; (**h**) Flexible CMOS circuits formed by controlled spalling (top). Cross-section TEM image of the flexible circuit, stressor and handle layers (bottom left). Resulting voltage waveform of a 100 stage ring oscillator (bottom right). Reprint is in accordance with the Creative Commons Attribution 3.0 International License [170]; (**i**) Integration of a thinned die in a flexible substrate. Reprinted from Reference [171], Copyright (2015), with permission from Elsevier; (**j**) Scaled production of soft-adhesive electronics with surface mount chip components. Reprinted with permission from Reference [172], Copyright 2017, John Wiley and Sons.

4. Health Monitoring and Disease Diagnostics

Non-invasive, wearable FHE enable a portable, real-time, in vivo disease diagnosis and health monitoring due to intimate and conformal integration with the target sources such as the human skin, eyes, and mouth. Unlike conventional methods, wearable FHE can be directly mounted on various human body parts to continuously and closely monitor health conditions and disease related biomarkers in timely manner, without interrupting or limiting the user's motions. This continuous physiological monitoring and intervention in a minimally invasive way would have direct benefit at early disease diagnosis and real-time monitoring of therapy, treatment, and health conditions. Figure 5 summarizes recent reports of FHE-enabled health monitoring and disease diagnosis applications.

Figure 5a shows a soft, highly compliant electrode for long-term wearable, high fidelity monitoring of difficulty in swallowing through EMG on the chin [109]. A swallow disorder is a common symptom of dysphagia. A skin-like electrode is fabricated by the combination of a wafer-scale microfabrication and material transfer printing. The resulting device mounted on the target location of the skin demonstrates clinical feasibility of the ergonomic electronics in rehabilitation for patients with dysphagia. Figure 5b captures a wearable sensor patch that brings together soft and hard electronics into a single platform using direct printing technique with gold and nickel oxide nanoparticle inks [174]. ECG electrodes and thermistor directly printed on a flexible polyimide substrate offer soft, low-modulus mechanics at the system level. The wearable sensor patch is mounted on a person's lower left rib cage to measure ECG and skin temperature. Figure 5c presents graphene-based bacteria detection on tooth enamel. Graphene is printed onto water-soluble silk that allows intimate biotransfer onto a tooth [175]. A resonant coil is incorporated to eliminate the need for onboard power and external connections. The device demonstrates the remote monitoring of respiration and bacteria detection in saliva with detection limit down to a single bacterium. Figure 5d integrates transparent, stretchable, and multifunctional sensors onto wearable soft contact lenses for the wireless detection of glucose and intraocular pressure with high-sensitivity for ocular diagnostics [29]. A multifunctional contact lens sensor is designed to monitor glucose within tears, as well as intraocular pressure using the resistance and capacitance of the electronic device. The hybrid of graphene and metal nanowires offers sufficient transparency (>91%) and stretchability (~25%) that ensure reliable, comfort, and unobstructed vision when users have the soft contact lens on. Its reliable operation is demonstrated both in vivo and in vitro studies using a live rabbit and bovine eyeball.

A flexible, electronic device shown in Figure 5e offers non-invasive mapping of pressure-induced tissue damage [176]. The flexible electrode array is placed on a wound and the electrical impedance is collected for each pair of neighboring electrodes. A map of impedance magnitude, phase angle and damage threshold (indicated in red in Figure 5e) are constructed based on the location of each measurement pair. The results show the feasibility of an automated, non-invasive 'smart-bandage' for early detection of pressure ulcers. Figure 5f shows a fully integrated FHE composed of sensor array for simultaneous and selective multiplexed analysis in sweat including glucose, lactate, electrolytes, and skin temperature [177]. The platform enables a wide range of personalized diagnostic and physiological monitoring applications. Wearable FHE sensors have the potential to play a major role in the wireless, continuous, and noninvasive monitoring of physiological conditions, as well as the detection of biomarkers associated with diseases. In particular, multiplexing sensing elements show substantial promise for next-generation medical devices to provide a tangible impact on health and wellness.

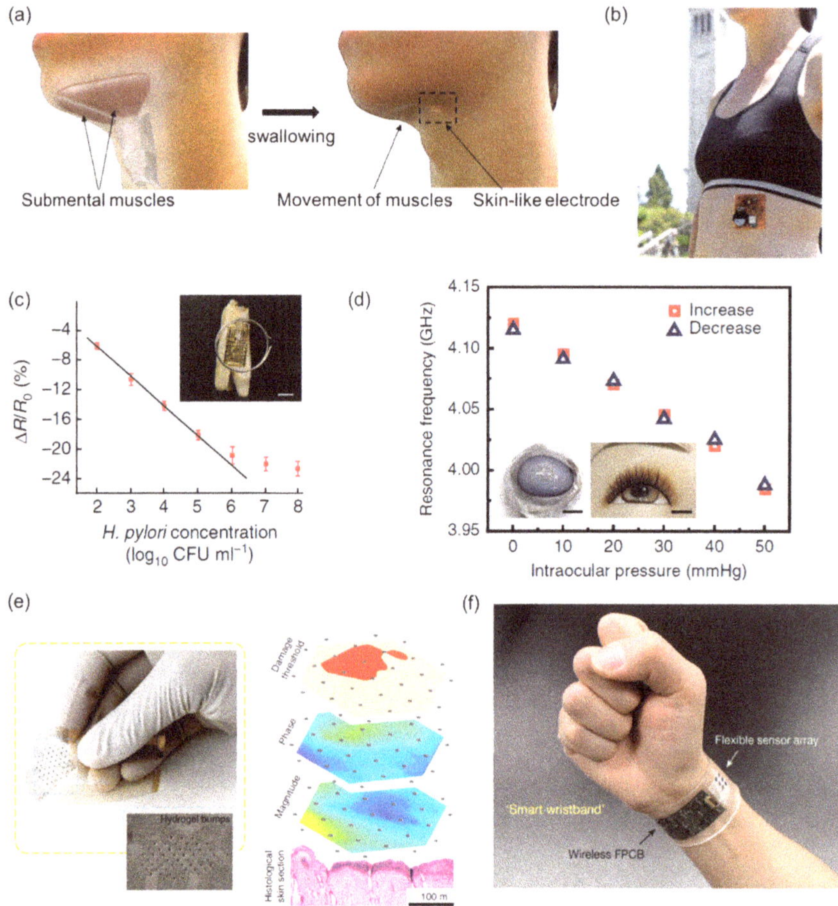

Figure 5. Health monitoring and disease diagnostic systems. (**a**) Illustration of targeted submental muscles on the chin and photos capturing the movement of the muscles upon swallowing activity. Reprinted with permission from [109], Copyright 2017, Scientific Reports; (**b**) Photographs of the wearable sensor patch mounted on a person's lower left rib cage (left) and the component side of the patch (right). Reproduced from [174], Copyright 2016, John Wiley and Sons; (**c**) Percentage change in graphene resistance versus concentration of H. pylori cells with optical image of the graphene wireless sensor biotransferred onto the surface of a tooth (inset optical image). Reproduced from [175], Copyright 2012, Macmillan Publishers Ltd.: Nature Communications; (**d**) An inkjet printed array, showing the hexagonal configuration of 55 equally spaced gold electrodes; inset shows printed hydrogel bumps on the fabricated array (left). Schematic representation of the device operation for early detection of pressure ulcers (right). Reproduced from [176], Copyright 2015, Macmillan Publishers Ltd.: Nature Communications; (**e**) Frequency response of the sensor during a pressure cycle for ocular diagnostics. Inset shows photographs of the sensor transferred onto the contact lens worn by a bovine eyeball (left) and a mannequin eye (right). Scale bar, 1 cm. Reproduced from [29], Copyright 2017, Macmillan Publishers Ltd.: Nature Communications; (**f**) Photograph of a wearable flexible integrated sensing array on a subject's writs, integrating the multiplexed sweat sensor array and the wireless flexible printed circuit board (left). Simultaneous system-level measurements (right). Reproduced from [177], Copyright 2016, Macmillan Publishers Ltd.: Nature.

5. Human-Machine Interfaces (HMI)

Recent remarkable progress in the development of wearable FHE has offered non-invasive, highly sensitive interactions between human and machines. Mechanically compliant FHE enable conformal contact to the human skin in a non-invasive way, while recording important physiological data such as EMG, EOG, and EEG. The biopotentials, measured from the skin-mounted electrodes, can be used to control a humanoid robot, drone, prosthetic hand, display interface, electronic wheelchair, and more. Figure 6 captures representative examples of recent HMI applications, controlled by FHE-enabled biopotentials. The main advantage of FHE in such examples is their portable, comfortable, and ergonomic arrangements via a low-profile, miniaturized electronic circuit, conformal electrodes, and data recording and management system.

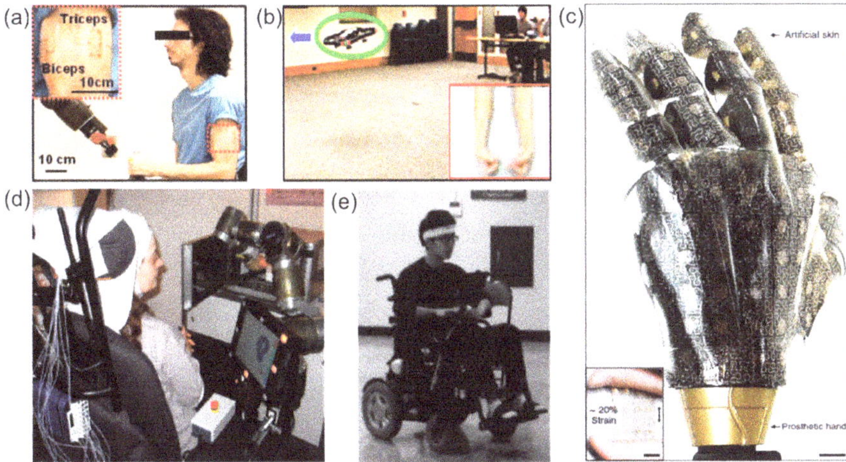

Figure 6. Flexible hybrid electronics for applications in human-machine interfaces. (**a**) EMG-enabled control of a humanoid robot. Reprinted with permission from Reference [178], Copyright 2016, John Wiley and Sons; (**b**) Bimanual gestures and their EMG signals, interfacing with a quadcopter. Reprinted with permission from Reference [7], Copyright 2013, John Wiley and Sons; (**c**) Sensor-laden bionic hand, instrumented with silicon nanoribbon. Reprinted with permission from Macmillan Publishers Ltd.: Nature Communications [179]; (**d**) Wearable headset and EEG (electroencephalogram) recording for a brain-interfaced system. Reprinted with permission from Reference [180]. Copyright 2015, MDPI; (**e**) Recording of EOG via a wearable forehead system for a wheelchair control. Reprinted with permission from Reference [181]. Copyright 2017, MDPI.

A soft, skin-like sensor system in Figure 6a [178], directly mounted on the skin via van der Waals interactions, offers unobtrusive, comfortable HMI for sensorimotor prosthetic control of a humanoid robot. The multifunctional device has open-mesh structured, microfabricated electrode and sensor, integrated on an elastomeric membrane, which is mechanically flexible and stretchable to accommodate the strain from the skin deformation. The gold nanomembrane-based system has advantages of simultaneous sensing and electrical stimulation, which allows sensorimotor control of a robot arm. With the stimulation feedback, a subject who wears the device can grip a water bottle in a controlled manner to prevent collapse. A similar device [7] that makes an intimate contact to the skin (Figure 6b) offers a high-fidelity recording of surface EMG on forearms. Bimanual gestures, recorded by two sets of electrodes on forearms generate different signal commands to control a drone (e.g., quadcopter). There were four commands, generated by a signal classification algorithm that used EMG data on the skin. With an optimized electrode design and placement on the target muscle, this work demonstrated 91.1% classification accuracy with four classes.

Figure 6c presents a "smart" prosthetic hand [179] with artificial skin and embedded soft sensors, enabled by stretchable silicon nanoribbon on a silicone elastomer (PDMS). A high performance, single crystalline silicon designed a highly sensitive device with strain, pressure, and temperature sensors. The electronic skin on the prosthetic hand with the soft sensor package demonstrated capabilities of hand shaking, keyboard typing, ball grasping, and feeling surface temperature in daily lives. Another wearable set of FHE embedded in a headset [180] with EEG electrodes demonstrated the feasibility of hybrid brain-controlled computer (Figure 6d). In this study, the wearable head set recorded steady-state visually evoked potentials to control a computer interface, which can be directly usable for severely disabled people who cannot use their arms and hands. The event-related synchronization of EEG can be used in many applications such as text spelling interface, computer control, wheelchair control, and more. Among those, hands-free control of an electronic wheelchair has gained a great interest in rehabilitation and aging societies. Patients with Parkinson's disease or amyotrophic lateral sclerosis experience paralyzed voluntary muscles and significantly reduced motor strength. In addition to the EEG-base method, eye movement can be utilized to control a wheelchair. A wearable FHE shown in Figure 6e [181] enables a non-invasive, comfortable arrangements to measure eye movements via EOG recording. In this work, the wearable system included forehead EOG electrodes on a headband along with Bluetooth-based wireless telemetry. A subject wearing the electrode system successfully drove a power wheelchair through an 8-shaped driving course without collision with obstacles. Collectively, soft, lightweight materials play a key role to design wearable, comfortable FHE. The mechanical compliance of the soft sensors makes conformal and intimate contact to the skin for high-fidelity recording of non-invasive biopotentials for various HMI applications.

6. Conclusions and Outlook

Recent progress in soft functional materials has enabled advances in wearable and implantable electronics. In this review, soft materials and designs for sensing and substrate components, along with the facilitated electronics and applications, have been summarized. Silicone elastomers and other soft, organic substrate materials, such as silk fibroin, have resulted in increasingly flexible and stretchable electronics, improving device functionality and increasing the breadth of potential applications. Additionally, a number of these materials have indicated necessary biocompatibility and biodegradability, allowing for applications in transient electronic needs. Likewise, sensing materials, such as CNTs, graphene, hydrogels, and nanostructures, have improved the mechanical and electrical characteristics of FHE. Sensing materials generally limit the mechanical compliance, but integration with soft substrates and recent advances have improved mechanical properties while maintaining favorable electrical properties. These advancements in soft material-enabled wearable and implantable electronics have improved functionality and allowed applications in different areas, including healthcare, disease diagnostics, and human-machine interfaces.

To continue progress in the advancement of wearable and implantable FHE, a variety of challenges and opportunities is still being addressed. Further development of soft material-based sensing components is required to offer smaller form factors of functional devices. New addition of self-healing materials will expand the current mechanical and application limits of FHE. For hard-soft device configurations that combine both compliant elastomers and rigid electronic components, advanced manufacturing methods are needed to provide robust mechanical and electrical hybridization of two types of materials. Improved technologies of multi-scale integration of hybrid materials will reduce the current gaps in performance variation between materials. While currently limited, further development of miniaturized and more efficient wireless powering and communication technologies will allow continuous and long-term usage of wearable and implantable systems. Additionally, continued investigation of biodegradable materials will broaden the applicability of transient electronics for implantable systems. Direct coupling of soft devices with cells and tissues will improve in vivo interfaces. Moving forward, continued development and integration of soft, functional sensing and substrate materials will heighten the functionality and applicability of FHE.

Acknowledgments: Woon-Hong Yeo acknowledges a seed grant from the Institute for Electronics and Nanotechnology (IEN) at Georgia Institute of Technology, a research grant from the Fundamental Research Program (PNK5061) of Korea Institute of Materials Science (KIMS) and startup funding from the Woodruff School of Mechanical Engineering at Georgia Institute of Technology. Jong-Hoon Kim acknowledges a New Faculty Seed Grant (131078) from Washington State University.

Author Contributions: Robert Herbert and Woon-Hong Yeo conceived and designed the materials in the paper; Robert Herbert and Woon-Hong Yeo conducted reviews of the FHE overview; Jong-Hoon Kim, Yun Soung Kim, Hye Moon Lee, and Woon-Hong Yeo reviewed and summarized integration strategies and FHE applications; all authors wrote the paper together.

Conflicts of Interest: The authors declare no conflict of interest.

References

1. Xu, S.; Zhang, Y.; Jia, L.; Mathewson, K.E.; Jang, K.-I.; Kim, J.; Fu, H.; Huang, X.; Chava, P.; Wang, R. Soft microfluidic assemblies of sensors, circuits, and radios for the skin. *Science* **2014**, *344*, 70–74. [CrossRef] [PubMed]

2. Shi, J.; Hu, J.; Dai, Z.; Zhao, W.; Liu, P.; Zhao, L.; Guo, Y.; Yang, T.; Zou, L.; Jiang, K. Graphene welded carbon nanotube crossbars for biaxial strain sensors. *Carbon* **2017**, *123*, 786–793. [CrossRef]

3. Stokes, A.A.; Shepherd, R.F.; Morin, S.A.; Ilievski, F.; Whitesides, G.M. A hybrid combining hard and soft robots. *Soft Robot.* **2014**, *1*, 70–74. [CrossRef]

4. Valentine, A.D.; Busbee, T.A.; Boley, J.W.; Raney, J.R.; Chortos, A.; Kotikian, A.; Berrigan, J.D.; Durstock, M.F.; Lewis, J.A. Hybrid 3D printing of soft electronics. *Adv. Mater.* **2017**, *29*. [CrossRef] [PubMed]

5. Chen, Y.; Howe, C.; Lee, Y.; Cheon, S.; Yeo, W.-H.; Chun, Y. Microstructured thin film nitinol for a neurovascular flow-diverter. *Sci. Rep.* **2016**, *6*, 23698. [CrossRef] [PubMed]

6. Roshan, Y.M.; Park, E.J. Design approach for a wireless power transfer system for wristband wearable devices. *IET Power Electron.* **2017**. [CrossRef]

7. Jeong, J.W.; Yeo, W.H.; Akhtar, A.; Norton, J.J.; Kwack, Y.J.; Li, S.; Jung, S.Y.; Su, Y.; Lee, W.; Xia, J. Materials and optimized designs for human-machine interfaces via epidermal electronics. *Adv. Mater.* **2013**, *25*, 6839–6846. [CrossRef] [PubMed]

8. Mishra, S.; Norton, J.J.; Lee, Y.; Lee, D.S.; Agee, N.; Chen, Y.; Chun, Y.; Yeo, W.-H. Soft, conformal bioelectronics for a wireless human-wheelchair interface. *Biosens. Bioelectron.* **2017**, *91*, 796–803. [CrossRef] [PubMed]

9. Yeo, W.H.; Kim, Y.S.; Lee, J.; Ameen, A.; Shi, L.K.; Li, M.; Wang, S.D.; Ma, R.; Jin, S.H.; Kang, Z.; et al. Multifunctional epidermal electronics printed directly onto the skin. *Adv. Mater.* **2013**, *25*, 2773–2778. [CrossRef] [PubMed]

10. Yu, K.J.; Kuzum, D.; Hwang, S.-W.; Kim, B.H.; Juul, H.; Kim, N.H.; Won, S.M.; Chiang, K.; Trumpis, M.; Richardson, A.G. Bioresorbable silicon electronics for transient spatiotemporal mapping of electrical activity from the cerebral cortex. *Nat. Mater.* **2016**, *15*, 782–791. [CrossRef] [PubMed]

11. Gutbrod, S.R.; Sulkin, M.S.; Rogers, J.A.; Efimov, I.R. Patient-specific flexible and stretchable devices for cardiac diagnostics and therapy. *Prog. Biophys. Mol. Biol.* **2014**, *115*, 244–251. [CrossRef] [PubMed]

12. Xu, L.; Gutbrod, S.R.; Bonifas, A.P.; Su, Y.; Sulkin, M.S.; Lu, N.; Chung, H.-J.; Jang, K.-I.; Liu, Z.; Ying, M. 3D multifunctional integumentary membranes for spatiotemporal cardiac measurements and stimulation across the entire epicardium. *Nat. Commun.* **2014**, *5*. [CrossRef] [PubMed]

13. Iijima, S. Helical microtubules of graphitic carbon. *Nature* **1991**, *354*, 56. [CrossRef]

14. Iijima, S.; Ichihashi, T. Single-shell carbon nanotubes of 1-nm diameter. *Nature* **1993**, *363*, 603–605. [CrossRef]

15. Bethune, D.; Klang, C.; De Vries, M.; Gorman, G.; Savoy, R.; Vazquez, J.; Beyers, R. Cobalt-catalysed growth of carbon nanotubes with single-atomic-layer walls. *Nature* **1993**, *363*, 605–607. [CrossRef]

16. Novoselov, K.S.; Geim, A.K.; Morozov, S.V.; Jiang, D.; Zhang, Y.; Dubonos, S.V.; Grigorieva, I.V.; Firsov, A.A. Electric field effect in atomically thin carbon films. *Science* **2004**, *306*, 666–669. [CrossRef] [PubMed]

17. Kim, D.-H.; Lu, N.; Ma, R.; Kim, Y.-S.; Kim, R.-H.; Wang, S.; Wu, J.; Won, S.M.; Tao, H.; Islam, A. Epidermal electronics. *Science* **2011**, *333*, 838–843. [CrossRef] [PubMed]

18. Amjadi, M.; Kyung, K.U.; Park, I.; Sitti, M. Stretchable, skin-mountable, and wearable strain sensors and their potential applications: A review. *Adv. Funct. Mater.* **2016**, *26*, 1678–1698. [CrossRef]

19. Yang, T.; Xie, D.; Li, Z.; Zhu, H. Recent advances in wearable tactile sensors: Materials, sensing mechanisms, and device performance. *Mater. Sci. Eng. R Rep.* **2017**, *115*, 1–37. [CrossRef]

20. An, B.W.; Shin, J.H.; Kim, S.-Y.; Kim, J.; Ji, S.; Park, J.; Lee, Y.; Jang, J.; Park, Y.-G.; Cho, E. Smart sensor systems for wearable electronic devices. *Polymers* **2017**, *9*, 303. [CrossRef]

21. Jin, H.; Abu-Raya, Y.S.; Haick, H. Advanced materials for health monitoring with skin-based wearable devices. *Adv. Healthc. Mater.* **2017**. [CrossRef] [PubMed]

22. Lacour, S.P.; Courtine, G.; Guck, J. Materials and technologies for soft implantable neuroprostheses. *Nat. Rev. Mater.* **2016**, *1*. [CrossRef]

23. Lee, J.H.; Kim, H.; Kim, J.H.; Lee, S.-H. Soft implantable microelectrodes for future medicine: Prosthetics, neural signal recording and neuromodulation. *Lab Chip* **2016**, *16*, 959–976. [CrossRef] [PubMed]

24. Reina, G.; González-Domínguez, J.M.; Criado, A.; Vázquez, E.; Bianco, A.; Prato, M. Promises, facts and challenges for graphene in biomedical applications. *Chem. Soc. Rev.* **2017**, *46*, 4400–4416. [CrossRef] [PubMed]

25. Choi, S.; Lee, H.; Ghaffari, R.; Hyeon, T.; Kim, D.H. Recent advances in flexible and stretchable bio-electronic devices integrated with nanomaterials. *Adv. Mater.* **2016**, *28*, 4203–4218. [CrossRef] [PubMed]

26. Choi, C.; Choi, M.K.; Hyeon, T.; Kim, D.H. Nanomaterial-based soft electronics for healthcare applications. *ChemNanoMat* **2016**, *2*, 1006–1017. [CrossRef]

27. Lin, S.; Yuk, H.; Zhang, T.; Parada, G.A.; Koo, H.; Yu, C.; Zhao, X. Stretchable hydrogel electronics and devices. *Adv. Mater.* **2016**, *28*, 4497–4505. [CrossRef] [PubMed]

28. Someya, T.; Bao, Z.; Malliaras, G.G. The rise of plastic bioelectronics. *Nature* **2016**, *540*, 379–385. [CrossRef] [PubMed]

29. Kim, J.; Kim, M.; Lee, M.-S.; Kim, K.; Ji, S.; Kim, Y.-T.; Park, J.; Na, K.; Bae, K.-H.; Kim, H.K. Wearable smart sensor systems integrated on soft contact lenses for wireless ocular diagnostics. *Nat. Commun.* **2017**, *8*. [CrossRef] [PubMed]

30. Coffey, J.W.; Corrie, S.R.; Kendall, M.A. Early circulating biomarker detection using a wearable microprojection array skin patch. *Biomaterials* **2013**, *34*, 9572–9583. [CrossRef] [PubMed]

31. Lipomi, D.J. Stretchable figures of merit in deformable electronics. *Adv. Mater.* **2016**, *28*, 4180–4183. [CrossRef] [PubMed]

32. Mohan, A.; Kim, N.; Gu, Y.; Bandodkar, A.J.; You, J.M.; Kumar, R.; Kurniawan, J.F.; Xu, S.; Wang, J. Merging of thin-and thick-film fabrication technologies: Toward soft stretchable "island–bridge" devices. *Adv. Mater. Technol.* **2017**, *2*. [CrossRef]

33. Lipomi, D.J.; Bao, Z. Stretchable and ultraflexible organic electronics. *MRS Bull.* **2017**, *42*, 93–97. [CrossRef]

34. Kim, J.; Bae, S.H.; Kotal, M.; Stalbaum, T.; Kim, K.J.; Oh, I.K. Soft but powerful artificial muscles based on 3D Graphene-CNT-Ni heteronanostructures. *Small* **2017**, *13*. [CrossRef] [PubMed]

35. Zhou, J.; Xu, X.; Yu, H.; Lubineau, G. Deformable and wearable carbon nanotube microwire-based sensors for ultrasensitive monitoring of strain, pressure and torsion. *Nanoscale* **2017**, *9*, 604–612. [CrossRef] [PubMed]

36. Wan, S.; Bi, H.; Zhou, Y.; Xie, X.; Su, S.; Yin, K.; Sun, L. Graphene oxide as high-performance dielectric materials for capacitive pressure sensors. *Carbon* **2017**, *114*, 209–216. [CrossRef]

37. Pang, Y.; Tian, H.; Tao, L.; Li, Y.; Wang, X.; Deng, N.; Yang, Y.; Ren, T.-L. Flexible, highly sensitive, and wearable pressure and strain sensors with graphene porous network structure. *ACS Appl. Mater. Interfaces* **2016**, *8*, 26458–26462. [CrossRef] [PubMed]

38. Lei, Z.; Wang, Q.; Sun, S.; Zhu, W.; Wu, P. A bioinspired mineral hydrogel as a self-healable, mechanically adaptable ionic skin for highly sensitive pressure sensing. *Adv. Mater.* **2017**, *29*. [CrossRef] [PubMed]

39. Bartlett, M.D.; Fassler, A.; Kazem, N.; Markvicka, E.J.; Mandal, P.; Majidi, C. Stretchable, high-*k* dielectric elastomers through liquid-metal inclusions. *Adv. Mater.* **2016**, *28*, 3726–3731. [CrossRef] [PubMed]

40. Shi, M.; Zhao, L.; Song, X.; Liu, J.; Zhang, P.; Gao, L. Highly conductive Mo_2C nanofibers encapsulated in ultrathin MnO_2 nanosheets as a self-supported electrode for high-performance capacitive energy storage. *ACS Appl. Mater. Interfaces* **2016**, *8*, 32460–32467. [CrossRef] [PubMed]

41. Liu, Y.; Miao, X.; Fang, J.; Zhang, X.; Chen, S.; Li, W.; Feng, W.; Chen, Y.; Wang, W.; Zhang, Y. Layered-MnO_2 nanosheet grown on nitrogen-doped graphene template as a composite cathode for flexible solid-state asymmetric supercapacitor. *ACS Appl. Mater. Interfaces* **2016**, *8*, 5251–5260. [CrossRef] [PubMed]

42. Zhang, G.; Zhang, X.; Huang, H.; Wang, J.; Li, Q.; Chen, L.Q.; Wang, Q. Toward wearable cooling devices: Highly flexible electrocaloric $Ba_{0.67}Sr_{0.33}TiO_3$ nanowire arrays. *Adv. Mater.* **2016**, *28*, 4811–4816. [CrossRef] [PubMed]

43. Wang, H.; Zhu, B.; Wang, H.; Ma, X.; Hao, Y.; Chen, X. Ultra-lightweight resistive switching memory devices based on silk fibroin. *Small* **2016**, *12*, 3360–3365. [CrossRef] [PubMed]

44. Bianco, A.; Kostarelos, K.; Prato, M. Making carbon nanotubes biocompatible and biodegradable. *Chem. Commun.* **2011**, *47*, 10182–10188. [CrossRef] [PubMed]

45. Pinto, A.M.; Goncalves, I.C.; Magalhães, F.D. Graphene-based materials biocompatibility: A review. *Colloids Surf. B Biointerfaces* **2013**, *111*, 188–202. [CrossRef] [PubMed]

46. Syama, S.; Mohanan, P. Safety and biocompatibility of graphene: A new generation nanomaterial for biomedical application. *Int. J. Biol. Macromol.* **2016**, *86*, 546–555. [CrossRef] [PubMed]

47. Fabbro, A.; Scaini, D.; León, V.N.; Vázquez, E.; Cellot, G.; Privitera, G.; Lombardi, L.; Torrisi, F.; Tomarchio, F.; Bonaccorso, F. Graphene-based interfaces do not alter target nerve cells. *ACS Nano* **2016**, *10*, 615–623. [CrossRef] [PubMed]

48. Ma, X.; Sun, X.; Hargrove, D.; Chen, J.; Song, D.; Dong, Q.; Lu, X.; Fan, T.-H.; Fu, Y.; Lei, Y. A biocompatible and biodegradable protein hydrogel with green and red autofluorescence: Preparation, characterization and in vivo biodegradation tracking and modeling. *Sci. Rep.* **2016**, *6*. [CrossRef] [PubMed]

49. Lee, G.; Kang, S.K.; Won, S.M.; Gutruf, P.; Jeong, Y.R.; Koo, J.; Lee, S.S.; Rogers, J.A.; Ha, J.S. Fully biodegradable microsupercapacitor for power storage in transient electronics. *Adv. Energy Mater.* **2017**. [CrossRef]

50. Caló, E.; Khutoryanskiy, V.V. Biomedical applications of hydrogels: A review of patents and commercial products. *Eur. Polym. J.* **2015**, *65*, 252–267. [CrossRef]

51. Wang, X.; Liu, J. Recent advancements in liquid metal flexible printed electronics: Properties, technologies, and applications. *Micromachines* **2016**, *7*, 206. [CrossRef]

52. Yang, K.; Zeng, M.; Fu, X.; Li, J.; Ma, N.; Tao, L. Establishing biodegradable single-layer MnO_2 nanosheets as a platform for live cell microRNA sensing. *RSC Adv.* **2015**, *5*, 104245–104249. [CrossRef]

53. Fan, W.; Bu, W.; Shen, B.; He, Q.; Cui, Z.; Liu, Y.; Zheng, X.; Zhao, K.; Shi, J. Intelligent MnO_2 nanosheets anchored with upconversion nanoprobes for concurrent pH-/H_2O_2-responsive ucl imaging and oxygen-elevated synergetic therapy. *Adv. Mater.* **2015**, *27*, 4155–4161. [CrossRef] [PubMed]

54. Song, Y.; Shan, D.; Chen, R.; Zhang, F.; Han, E.-H. Biodegradable behaviors of AZ31 magnesium alloy in simulated body fluid. *Mater. Sci. Eng. C* **2009**, *29*, 1039–1045. [CrossRef]

55. Hwang, S.-W.; Park, G.; Edwards, C.; Corbin, E.A.; Kang, S.-K.; Cheng, H.; Song, J.-K.; Kim, J.-H.; Yu, S.; Ng, J. Dissolution chemistry and biocompatibility of single-crystalline silicon nanomembranes and associated materials for transient electronics. *ACS Nano* **2014**, *8*, 5843–5851. [CrossRef] [PubMed]

56. Singh, M.; Movia, D.; Mahfoud, O.K.; Volkov, Y.; Prina-Mello, A. Silver nanowires as prospective carriers for drug delivery in cancer treatment: An in vitro biocompatibility study on lung adenocarcinoma cells and fibroblasts. *Eur. J. Nanomed.* **2013**, *5*, 195–204. [CrossRef]

57. Ryu, M.; Yang, J.H.; Ahn, Y.; Sim, M.; Lee, K.H.; Kim, K.; Lee, T.; Yoo, S.-J.; Kim, S.Y.; Moon, C. Enhancement of interface characteristics of neural probe based on graphene, zno nanowires, and conducting polymer pedot. *ACS Appl. Mater. Interfaces* **2017**, *9*, 10577–10586. [CrossRef] [PubMed]

58. Ball, J.P.; Mound, B.A.; Nino, J.C.; Allen, J.B. Biocompatible evaluation of barium titanate foamed ceramic structures for orthopedic applications. *J. Biomed. Mater. Res. Part A* **2014**, *102*, 2089–2095. [CrossRef] [PubMed]

59. Chiappini, C.; Liu, X.; Fakhoury, J.R.; Ferrari, M. Biodegradable porous silicon barcode nanowires with defined geometry. *Adv. Funct. Mater.* **2010**, *20*, 2231–2239. [CrossRef] [PubMed]

60. Kuo, C.W.; Lai, J.J.; Wei, K.H.; Chen, P. Studies of surface-modified gold nanowires inside living cells. *Adv. Funct. Mater.* **2007**, *17*, 3707–3714. [CrossRef]

61. Byrne, F.; Prina-Mello, A.; Whelan, A.; Mohamed, B.M.; Davies, A.; Gun'ko, Y.K.; Coey, J.; Volkov, Y. High content analysis of the biocompatibility of nickel nanowires. *J. Magn. Magn. Mater.* **2009**, *321*, 1341–1345. [CrossRef]

62. Berggren, M.; Richter-Dahlfors, A. Organic bioelectronics. *Adv. Mater.* **2007**, *19*, 3201–3213. [CrossRef]

63. Gui, Q.; He, Y.; Gao, N.; Tao, X.; Wang, Y. A skin-inspired integrated sensor for synchronous monitoring of multiparameter signals. *Adv. Funct. Mater.* **2017**, *27*. [CrossRef]

64. Kim, H.; Eom, T.S.; Cho, W.; Woo, K.; Shon, Y.; Wie, J.J.; Shim, B.S. Soft electronics on asymmetrical porous conducting membranes by molecular layer-by-layer assembly. *Sens. Actuators B Chem.* **2018**, *254*, 916–925. [CrossRef]

65. Kireev, D.; Seyock, S.; Ernst, M.; Maybeck, V.; Wolfrum, B.; Offenhäusser, A. Versatile flexible graphene multielectrode arrays. *Biosensors* **2016**, *7*, 1. [CrossRef] [PubMed]

66. Zhang, M.; Wang, C.; Wang, Q.; Jian, M.; Zhang, Y. Sheath–core graphite/silk fiber made by dry-meyer-rod-coating for wearable strain sensors. *ACS Appl. Mater. Interfaces* **2016**, *8*, 20894–20899. [CrossRef] [PubMed]

67. Yeo, J.C.; Yu, J.; Shang, M.; Loh, K.P.; Lim, C.T. Highly flexible graphene oxide nanosuspension liquid-based microfluidic tactile sensor. *Small* **2016**, *12*, 1593–1604.

68. Huang, X.; Leng, T.; Zhu, M.; Zhang, X.; Chen, J.; Chang, K.; Aqeeli, M.; Geim, A.K.; Novoselov, K.S.; Hu, Z. Highly flexible and conductive printed graphene for wireless wearable communications applications. *Sci. Rep.* **2015**, *5*. [CrossRef] [PubMed]

69. Cai, G.; Wang, J.; Qian, K.; Chen, J.; Li, S.; Lee, P.S. Extremely stretchable strain sensors based on conductive self-healing dynamic cross-links hydrogels for human-motion detection. *Adv. Sci.* **2017**, *4*. [CrossRef] [PubMed]

70. Choi, D.-H.; Li, Y.; Cutting, G.R.; Searson, P.C. A wearable potentiometric sensor with integrated salt bridge for sweat chloride measurement. *Sens. Actuators B Chem.* **2017**, *250*, 673–678. [CrossRef]

71. Gao, Y.; Ota, H.; Schaler, E.W.; Chen, K.; Zhao, A.; Gao, W.; Fahad, H.M.; Leng, Y.; Zheng, A.; Xiong, F. Wearable microfluidic diaphragm pressure sensor for health and tactile touch monitoring. *Adv. Mater.* **2017**, *29*. [CrossRef] [PubMed]

72. Lee, Y.; Yeo, W.-H. Skin-like electronics for a persistent brain-computer interface. *J. Nat. Sci.* **2015**, *1*, e132.

73. Ko, Y.; Kim, J.; Kim, D.; Yamauchi, Y.; Kim, J.H.; You, J. A simple silver nanowire patterning method based on poly (ethylene glycol) photolithography and its application for soft electronics. *Sci. Rep.* **2017**, *7*. [CrossRef] [PubMed]

74. Li, Y.-Q.; Zhu, W.-B.; Yu, X.-G.; Huang, P.; Fu, S.-Y.; Hu, N.; Liao, K. Multifunctional wearable device based on flexible and conductive carbon sponge/polydimethylsiloxane composite. *ACS Appl. Mater. Interfaces* **2016**, *8*, 33189–33196. [CrossRef] [PubMed]

75. Takamatsu, S.; Lonjaret, T.; Ismailova, E.; Masuda, A.; Itoh, T.; Malliaras, G.G. Wearable keyboard using conducting polymer electrodes on textiles. *Adv. Mater.* **2016**, *28*, 4485–4488. [CrossRef] [PubMed]

76. Roche, E.T.; Horvath, M.A.; Wamala, I.; Alazmani, A.; Song, S.-E.; Whyte, W.; Machaidze, Z.; Payne, C.J.; Weaver, J.C.; Fishbein, G. Soft robotic sleeve supports heart function. *Sci. Transl. Med.* **2017**, *9*. [CrossRef] [PubMed]

77. Ecoflex 00-30. Available online: https://www.smooth-on.com/products/ecoflex-00-30/ (accessed on 23 January 2018).

78. Salvatore, G.A.; Sülzle, J.; Dalla Valle, F.; Cantarella, G.; Robotti, F.; Jokic, P.; Knobelspies, S.; Daus, A.; Büthe, L.; Petti, L. Biodegradable and highly deformable temperature sensors for the internet of things. *Adv. Funct. Mater.* **2017**, *27*. [CrossRef]

79. Johnston, I.; McCluskey, D.; Tan, C.; Tracey, M. Mechanical characterization of bulk sylgard 184 for microfluidics and microengineering. *J. Micromech. Microeng.* **2014**, *24*, 035017. [CrossRef]

80. Product Information 184 Silicone Elastomer. Available online: https://www.galco.com/techdoc/dowc/sylgard-184_pg.pdf (accessed on 23 January 2018).

81. Kim, S.-J.; Lee, D.-S.; Kim, I.-G.; Sohn, D.-W.; Park, J.-Y.; Choi, B.-K.; Kim, S.-W. Evaluation of the biocompatibility of a coating material for an implantable bladder volume sensor. *Kaohsiung J. Med. Sci.* **2012**, *28*, 123–129. [CrossRef] [PubMed]

82. Silbione LSR 4330. Available online: http://www.silbione.com/wp-content/uploads/2014/02/Silbione-LSR-4330.pdf (accessed on 23 January 2018).

83. Silicones, B. Silbione Liquid Silicone Rubber (LSR) Elastomers for Healthcare and Medical Device Fabrication. Available online: http://www.silbione.com/wp-content/uploads/2014/01/Biocompatibility-LSR.pdf (accessed on 23 January 2018).

84. Parylene Properties. Available online: https://vsiparylene.com/parylene-advantages/properties/#mechanical (accessed on 24 January 2018).

85. Rodger, D.C.; Weiland, J.D.; Humayun, M.S.; Tai, Y.-C. Scalable high lead-count parylene package for retinal prostheses. *Sens. Actuators B Chem.* **2006**, *117*, 107–114. [CrossRef]

86. Song, J.S.; Lee, S.; Jung, S.H.; Cha, G.C.; Mun, M.S. Improved biocompatibility of parylene-C films prepared by chemical vapor deposition and the subsequent plasma treatment. *J. Appl. Polym. Sci.* **2009**, *112*, 3677–3685. [CrossRef]

87. Properties of Various High Performance Films. Available online: http://www.teijinfilmsolutions.jp/english/product/hi_film.html (accessed on 24 January 2018).

88. Veleirinho, B.; Coelho, D.S.; Dias, P.F.; Maraschin, M.; Pinto, R.; Cargnin-Ferreira, E.; Peixoto, A.; Souza, J.A.; Ribeiro-do-Valle, R.M.; Lopes-da-Silva, J.A. Foreign body reaction associated with pet and pet/chitosan electrospun nanofibrous abdominal meshes. *PLoS ONE* **2014**, *9*, e95293. [CrossRef] [PubMed]

89. Ramires, P.; Mirenghi, L.; Romano, A.; Palumbo, F.; Nicolardi, G. Plasma-treated pet surfaces improve the biocompatibility of human endothelial cells. *J. Biomed. Mater. Res. Part A* **2000**, *51*, 535–539. [CrossRef]

90. Swar, S.; Zajícová, V.; Rysová, M.; Lovětinská-Šlamborová, I.; Voleský, L.; Stibor, I. Biocompatible surface modification of poly(ethylene terephthalate) focused on pathogenic bacteria: Promising prospects in biomedical applications. *J. Appl. Polym. Sci.* **2017**, *134*. [CrossRef]

91. Kai, W.; Hirota, Y.; Hua, L.; Inoue, Y. Thermal and mechanical properties of a poly(Є-caprolactone)/graphite oxide composite. *J. Appl. Polym. Sci.* **2008**, *107*, 1395–1400. [CrossRef]

92. Serrano, M.; Pagani, R.; Vallet-Regı, M.; Pena, J.; Ramila, A.; Izquierdo, I.; Portoles, M. In vitro biocompatibility assessment of poly(ε-caprolactone) films using l929 mouse fibroblasts. *Biomaterials* **2004**, *25*, 5603–5611. [CrossRef] [PubMed]

93. Nair, L.S.; Laurencin, C.T. Biodegradable polymers as biomaterials. *Prog. Polym. Sci.* **2007**, *32*, 762–798. [CrossRef]

94. Del Valle, J.; de la Oliva, N.; Müller, M.; Stieglitz, T.; Navarro, X. Biocompatibility evaluation of parylene C and polyimide as substrates for peripheral nerve interfaces. In Proceedings of the 2015 7th International IEEE/EMBS Conference on Neural Engineering (NER), Montpellier, France, 22–24 April 2015; pp. 442–445.

95. Mills, C.; Escarré, J.; Engel, E.; Martinez, E.; Errachid, A.; Bertomeu, J.; Andreu, J.; Planell, J.A.; Samitier, J. Micro- and nanostructuring of poly(ethylene-2,6-naphthalate) surfaces, for biomedical applications, using polymer replication techniques. *Nanotechnology* **2005**, *16*, 369. [CrossRef]

96. Pes. Available online: https://www.plasticsintl.com/datasheets/Radel_A_PES.pdf (accessed on 23 January 2018).

97. Azadbakht, M.; Madaeni, S.S.; Sahebjamee, F. Biocompatibility of polyethersulfone membranes for cell culture systems. *Eng. Life Sci.* **2011**, *11*, 629–635. [CrossRef]

98. Teflon (ptfe). Available online: http://www.dielectriccorp.com/downloads/thermoplastics/teflon.pdf (accessed on 23 January 2018).

99. Risbud, M.V.; Hambir, S.; Jog, J.; Bhonde, R. Biocompatibility assessment of polytetrafluoroethylene/wollastonite composites using endothelial cells and macrophages. *J. Biomater. Sci. Polym. Ed.* **2001**, *12*, 1177–1189. [CrossRef] [PubMed]

100. Gentile, P.; Chiono, V.; Carmagnola, I.; Hatton, P.V. An overview of poly(lactic-*co*-glycolic) acid (PLGA)-based biomaterials for bone tissue engineering. *Int. J. Mol. Sci.* **2014**, *15*, 3640–3659. [CrossRef] [PubMed]

101. Ignatius, A.; Claes, L.E. In vitro biocompatibility of bioresorbable polymers: Poly(L,DL-lactide) and poly(L-Lactide-co-glycolide). *Biomaterials* **1996**, *17*, 831–839. [CrossRef]

102. Cyclo Olefin Polymer. Available online: http://www.zeon.co.jp/content/200181692.pdf (accessed on 23 January 2018).

103. Tubia, I.; Mujika, M.; Artieda, J.; Valencia, M.; Arana, S. Soft polymer sensor for recording surface cortical activity in freely moving rodents. *Sens. Actuators A Phys.* **2016**, *251*, 241–247. [CrossRef]

104. Lu, S.; Wang, X.; Lu, Q.; Zhang, X.; Kluge, J.A.; Uppal, N.; Omenetto, F.; Kaplan, D.L. Insoluble and flexible silk films containing glycerol. *Biomacromolecules* **2009**, *11*, 143–150. [CrossRef] [PubMed]

105. Tao, H.; Hwang, S.-W.; Marelli, B.; An, B.; Moreau, J.E.; Yang, M.; Brenckle, M.A.; Kim, S.; Kaplan, D.L.; Rogers, J.A. Silk-based resorbable electronic devices for remotely controlled therapy and in vivo infection abatement. *Proc. Natl. Acad. Sci. USA* **2014**, *111*, 17385–17389. [CrossRef] [PubMed]

106. Cao, Y.; Wang, B. Biodegradation of silk biomaterials. *Int. J. Mol. Sci.* **2009**, *10*, 1514–1524. [CrossRef] [PubMed]

107. Do, T.N.; Visell, Y. Stretchable, twisted conductive microtubules for wearable computing, robotics, electronics, and healthcare. *Sci. Rep.* **2017**, *7*. [CrossRef] [PubMed]

108. Liu, Y.-J.; Cao, W.-T.; Ma, M.-G.; Wan, P. Ultrasensitive wearable soft strain sensors of conductive, self-healing, and elastic hydrogels with synergistic "soft and hard" hybrid networks. *ACS Appl. Mater. Interfaces* **2017**, *9*, 25559–25570. [CrossRef] [PubMed]

109. Lee, Y.; Nicholls, B.; Lee, D.S.; Chen, Y.; Chun, Y.; Ang, C.S.; Yeo, W.-H. Soft electronics enabled ergonomic human-computer interaction for swallowing training. *Sci. Rep.* **2017**, *7*. [CrossRef] [PubMed]

110. Koo, J.H.; Jeong, S.; Shim, H.J.; Son, D.; Kim, J.; Kim, D.C.; Choi, S.; Hong, J.-I.; Kim, D.-H. Wearable electrocardiogram monitor using carbon nanotube electronics and color-tunable organic light-emitting diodes. *ACS Nano* **2017**, *11*, 10032–10041. [CrossRef] [PubMed]

111. Kim, J.; Shim, H.J.; Yang, J.; Choi, M.K.; Kim, D.C.; Kim, J.; Hyeon, T.; Kim, D.H. Ultrathin quantum dot display integrated with wearable electronics. *Adv. Mater.* **2017**, *29*. [CrossRef] [PubMed]

112. Giffney, T.; Bejanin, E.; Kurian, A.S.; Travas-Sejdic, J.; Aw, K. Highly stretchable printed strain sensors using multi-walled carbon nanotube/silicone rubber composites. *Sens. Actuators A Phys.* **2017**, *259*, 44–49. [CrossRef]

113. Yamamoto, Y.; Yamamoto, D.; Takada, M.; Naito, H.; Arie, T.; Akita, S.; Takei, K. Efficient skin temperature sensor and stable gel-less sticky ECG sensor for a wearable flexible healthcare patch. *Adv. Healthc. Mater.* **2017**, *6*. [CrossRef] [PubMed]

114. Nakata, S.; Arie, T.; Akita, S.; Takei, K. Wearable, flexible, and multifunctional healthcare device with an isfet chemical sensor for simultaneous sweat pH and skin temperature monitoring. *ACS Sens.* **2017**, *2*, 443–448. [CrossRef] [PubMed]

115. Chen, Z.; Xi, J.; Huang, W.; Yuen, M.M. Stretchable conductive elastomer for wireless wearable communication applications. *Sci. Rep.* **2017**, *7*. [CrossRef] [PubMed]

116. Yu, N.; Yin, H.; Zhang, W.; Liu, Y.; Tang, Z.; Zhu, M.Q. High-performance fiber-shaped all-solid-state asymmetric supercapacitors based on ultrathin MnO_2 nanosheet/carbon fiber cathodes for wearable electronics. *Adv. Energy Mater.* **2016**, *6*. [CrossRef]

117. Yin, B.; Wen, Y.; Hong, T.; Xie, Z.; Yuan, G.; Ji, Q.; Jia, H. Highly stretchable, ultrasensitive, and wearable strain sensors based on facilely prepared reduced graphene oxide woven fabrics in an ethanol flame. *ACS Appl. Mater. Interfaces* **2017**, *9*, 32054–32064. [CrossRef] [PubMed]

118. Chun, S.; Choi, Y.; Park, W. All-graphene strain sensor on soft substrate. *Carbon* **2017**, *116*, 753–759. [CrossRef]

119. Hu, Y.; Zhao, T.; Zhu, P.; Zhang, Y.; Liang, X.; Sun, R.; Wong, C.-P. A low-cost, printable, and stretchable strain sensor based on highly conductive elastic composites with tunable sensitivity for human motion monitoring. *Nano Res.* **2017**. [CrossRef]

120. Lee, H.; Kwon, D.; Cho, H.; Park, I.; Kim, J. Soft nanocomposite based multi-point, multi-directional strain mapping sensor using anisotropic electrical impedance tomography. *Sci. Rep.* **2017**, *7*. [CrossRef] [PubMed]

121. Park, S.J.; Kim, J.; Chu, M.; Khine, M. Highly flexible wrinkled carbon nanotube thin film strain sensor to monitor human movement. *Adv. Mater. Technol.* **2016**, *1*. [CrossRef]

122. Cooper, C.B.; Arutselvan, K.; Liu, Y.; Armstrong, D.; Lin, Y.; Khan, M.R.; Genzer, J.; Dickey, M.D. Stretchable capacitive sensors of torsion, strain, and touch using double helix liquid metal fibers. *Adv. Funct. Mater.* **2017**, *27*. [CrossRef]

123. Zhu, H.; Wang, X.; Liang, J.; Lv, H.; Tong, H.; Ma, L.; Hu, Y.; Zhu, G.; Zhang, T.; Tie, Z. Versatile electronic skins for motion detection of joints enabled by aligned few-walled carbon nanotubes in flexible polymer composites. *Adv. Funct. Mater.* **2017**, *27*. [CrossRef]

124. You, B.; Han, C.J.; Kim, Y.; Ju, B.-K.; Kim, J.-W. A wearable piezocapacitive pressure sensor with a single layer of silver nanowire-based elastomeric composite electrodes. *J. Mater. Chem. A* **2016**, *4*, 10435–10443. [CrossRef]

125. Mannsfeld, S.C.; Tee, B.C.; Stoltenberg, R.M.; Chen, C.V.H.; Barman, S.; Muir, B.V.; Sokolov, A.N.; Reese, C.; Bao, Z. Highly sensitive flexible pressure sensors with microstructured rubber dielectric layers. *Nat. Mater.* **2010**, *9*, 859–864. [CrossRef] [PubMed]

126. Schwartz, G.; Tee, B.C.-K.; Mei, J.; Appleton, A.L.; Kim, D.H.; Wang, H.; Bao, Z. Flexible polymer transistors with high pressure sensitivity for application in electronic skin and health monitoring. *Nat. Commun.* **2013**, *4*, 1859. [CrossRef] [PubMed]

127. Lee, S.; Jo, I.; Kang, S.; Jang, B.; Moon, J.; Park, J.B.; Lee, S.; Rho, S.; Kim, Y.; Hong, B.H. Smart contact lenses with graphene coating for electromagnetic interference shielding and dehydration protection. *ACS Nano* **2017**, *11*, 5318–5324. [CrossRef] [PubMed]

128. Jia, W.; Bandodkar, A.J.; Valdés-Ramírez, G.; Windmiller, J.R.; Yang, Z.; Ramírez, J.; Chan, G.; Wang, J. Electrochemical tattoo biosensors for real-time noninvasive lactate monitoring in human perspiration. *Anal. Chem.* **2013**, *85*, 6553–6560. [CrossRef] [PubMed]

129. Imani, S.; Bandodkar, A.J.; Mohan, A.V.; Kumar, R.; Yu, S.; Wang, J.; Mercier, P.P. A wearable chemical–electrophysiological hybrid biosensing system for real-time health and fitness monitoring. *Nat. Commun.* **2016**, *7*, 11650. [CrossRef] [PubMed]

130. Bandodkar, A.J.; Jia, W.; Yardımcı, C.; Wang, X.; Ramirez, J.; Wang, J. Tattoo-based noninvasive glucose monitoring: A proof-of-concept study. *Anal. Chem.* **2014**, *87*, 394–398. [CrossRef] [PubMed]

131. Bandodkar, A.J.; Nuñez-Flores, R.; Jia, W.; Wang, J. All-printed stretchable electrochemical devices. *Adv. Mater.* **2015**, *27*, 3060–3065. [CrossRef] [PubMed]

132. Wei, Z.; Zhou, Z.K.; Li, Q.; Xue, J.; Di Falco, A.; Yang, Z.; Zhou, J.; Wang, X. Flexible nanowire cluster as a wearable colorimetric humidity sensor. *Small* **2017**. [CrossRef] [PubMed]

133. Coskun, M.B.; Qiu, L.; Arefin, M.S.; Neild, A.; Yuce, M.; Li, D.; Alan, T. Detecting subtle vibrations using graphene-based cellular elastomers. *ACS Appl. Mater. Interfaces* **2017**, *9*, 11345–11349. [CrossRef] [PubMed]

134. Kang, M.; Kim, J.; Jang, B.; Chae, Y.; Kim, J.-H.; Ahn, J.-H. Graphene-based three-dimensional capacitive touch sensor for wearable electronics. *ACS Nano* **2017**, *11*, 7950–7957. [CrossRef] [PubMed]

135. Lee, Y.; Joo, M.-K.; Le, V.T.; Ovalle-Robles, R.; Lepró, X.; Lima, M.D.; Suh, D.G.; Yu, H.Y.; Lee, Y.H.; Suh, D. Ultrastretchable analog/digital signal transmission line with carbon nanotube sheets. *ACS Appl. Mater. Interfaces* **2017**, *9*, 26286–26292. [CrossRef] [PubMed]

136. Kim, M.G.; Alrowais, H.; Pavlidis, S.; Brand, O. Size-scalable and high-density liquid-metal-based soft electronic passive components and circuits using soft lithography. *Adv. Funct. Mater.* **2017**, *27*. [CrossRef]

137. Chortos, A.; Zhu, C.; Oh, J.Y.; Yan, X.; Pochorovski, I.; To, J.W.-F.; Liu, N.; Kraft, U.; Murmann, B.; Bao, Z. Investigating limiting factors in stretchable all-carbon transistors for reliable stretchable electronics. *ACS Nano* **2017**, *11*, 7925–7937. [CrossRef] [PubMed]

138. Yokota, T.; Zalar, P.; Kaltenbrunner, M.; Jinno, H.; Matsuhisa, N.; Kitanosako, H.; Tachibana, Y.; Yukita, W.; Koizumi, M.; Someya, T. Ultraflexible organic photonic skin. *Sci. Adv.* **2016**, *2*, e1501856. [CrossRef] [PubMed]

139. Shang, Q.; Deng, W.; Zhang, X.; Wang, L.; Huang, L.; Jie, J. An inherent multifunctional sellotape substrate for high-performance flexible and wearable organic single-crystal nanowire array-based transistors. *Adv. Electron. Mater.* **2016**, *2*. [CrossRef]

140. Li, X.; Wang, J.; Zhao, Y.; Ge, F.; Komarneni, S.; Cai, Z. Wearable solid-state supercapacitors operating at high working voltage with a flexible nanocomposite electrode. *ACS Appl. Mater. Interfaces* **2016**, *8*, 25905–25914. [CrossRef] [PubMed]

141. Shi, P.; Chen, R.; Hua, L.; Li, L.; Chen, R.; Gong, Y.; Yu, C.; Zhou, J.; Liu, B.; Sun, G. Highly concentrated, ultrathin nickel hydroxide nanosheet ink for wearable energy storage devices. *Adv. Mater.* **2017**. [CrossRef] [PubMed]

142. Li, W.; Xu, X.; Liu, C.; Tekell, M.C.; Ning, J.; Guo, J.; Zhang, J.; Fan, D. Ultralight and binder-free all-solid-state flexible supercapacitors for powering wearable strain sensors. *Adv. Funct. Mater.* **2017**. [CrossRef]

143. Bandodkar, A.; You, J.-M.; Kim, N.-H.; Gu, Y.; Kumar, R.; Vinu Mohan, A.M.; Kurniawan, J.F.; Imani, S.; Nakagawa, T.; Parish, B. Soft, stretchable, high power density electronic skin-based biofuel cells for scavenging energy from human sweat. *Energy Environ. Sci.* **2017**. [CrossRef]

144. Lee, J.W.; Xu, R.; Lee, S.; Jang, K.-I.; Yang, Y.; Banks, A.; Yu, K.J.; Kim, J.; Xu, S.; Ma, S. Soft, thin skin-mounted power management systems and their use in wireless thermography. *Proc. Natl. Acad. Sci. USA* **2016**, *113*, 6131–6136. [CrossRef] [PubMed]

145. Berchmans, S.; Bandodkar, A.J.; Jia, W.; Ramírez, J.; Meng, Y.S.; Wang, J. An epidermal alkaline rechargeable Ag–Zn printable tattoo battery for wearable electronics. *J. Mater. Chem. A* **2014**, *2*, 15788–15795. [CrossRef]

146. Kumar, R.; Shin, J.; Yin, L.; You, J.M.; Meng, Y.S.; Wang, J. All-printed, stretchable Zn-Ag$_2$O rechargeable battery via hyperelastic binder for self-powering wearable electronics. *Adv. Energy Mater.* **2017**, *7*. [CrossRef]

147. Kim, S.J.; Cho, K.W.; Cho, H.R.; Wang, L.; Park, S.Y.; Lee, S.E.; Hyeon, T.; Lu, N.; Choi, S.H.; Kim, D.H. Stretchable and transparent biointerface using cell-sheet–graphene hybrid for electrophysiology and therapy of skeletal muscle. *Adv. Funct. Mater.* **2016**, *26*, 3207–3217. [CrossRef]

148. Koh, A.; Gutbrod, S.R.; Meyers, J.D.; Lu, C.; Webb, R.C.; Shin, G.; Li, Y.; Kang, S.K.; Huang, Y.; Efimov, I.R. Ultrathin injectable sensors of temperature, thermal conductivity, and heat capacity for cardiac ablation monitoring. *Adv. Healthc. Mater.* **2016**, *5*, 373–381. [CrossRef] [PubMed]

149. Shin, G.; Gomez, A.M.; Al-Hasani, R.; Jeong, Y.R.; Kim, J.; Xie, Z.; Banks, A.; Lee, S.M.; Han, S.Y.; Yoo, C.J. Flexible near-field wireless optoelectronics as subdermal implants for broad applications in optogenetics. *Neuron* **2017**, *93*, 509–521. [CrossRef] [PubMed]

150. Jia, X.; Wang, C.; Ranganathan, V.; Napier, B.; Yu, C.; Chao, Y.; Forsyth, M.; Omenetto, F.G.; MacFarlane, D.R.; Wallace, G.G. A biodegradable thin-film magnesium primary battery using silk fibroin–ionic liquid polymer electrolyte. *ACS Energy Lett.* **2017**, *2*, 831–836. [CrossRef]

151. Fang, H.; Yu, K.J.; Gloschat, C.; Yang, Z.; Song, E.; Chiang, C.-H.; Zhao, J.; Won, S.M.; Xu, S.; Trumpis, M. Capacitively coupled arrays of multiplexed flexible silicon transistors for long-term cardiac electrophysiology. *Nat. Biomed. Eng.* **2017**, *1*. [CrossRef]

152. Park, J.; Choi, S.; Janardhan, A.H.; Lee, S.-Y.; Raut, S.; Soares, J.; Shin, K.; Yang, S.; Lee, C.; Kang, K.-W. Electromechanical cardioplasty using a wrapped elasto-conductive epicardial mesh. *Sci. Transl. Med.* **2016**, *8*, 344ra386. [CrossRef] [PubMed]

153. Kim, D.H.; Shin, H.J.; Lee, H.; Jeong, C.K.; Park, H.; Hwang, G.T.; Lee, H.Y.; Joe, D.J.; Han, J.H.; Lee, S.H. In vivo self-powered wireless transmission using biocompatible flexible energy harvesters. *Adv. Funct. Mater.* **2017**. [CrossRef]

154. Lu, Y.; Lyu, H.; Richardson, A.G.; Lucas, T.H.; Kuzum, D. *Flexible Neural Electrode Array Based-on Porous Graphene for Cortical Microstimulation and Sensing*; Scientific Reports 6; Macmillan Publishers Limited: Basingstoke, UK, 2016; Volume 9.

155. Han, L.; Lu, X.; Wang, M.; Gan, D.; Deng, W.; Wang, K.; Fang, L.; Liu, K.; Chan, C.W.; Tang, Y. A mussel-inspired conductive, self-adhesive, and self-healable tough hydrogel as cell stimulators and implantable bioelectronics. *Small* **2017**, *13*. [CrossRef] [PubMed]

156. Kireev, D.; Brambach, M.; Seyock, S.; Maybeck, V.; Fu, W.; Wolfrum, B.; Offenhäusser, A. Graphene transistors for interfacing with cells: Towards a deeper understanding of liquid gating and sensitivity. *Sci. Rep.* **2017**, *7*. [CrossRef] [PubMed]

157. Yetisen, A.K.; Jiang, N.; Fallahi, A.; Montelongo, Y.; Ruiz-Esparza, G.U.; Tamayol, A.; Zhang, Y.S.; Mahmood, I.; Yang, S.A.; Kim, K.S. Glucose-sensitive hydrogel optical fibers functionalized with phenylboronic acid. *Adv. Mater.* **2017**, *29*. [CrossRef] [PubMed]

158. Jiao, X.; Wang, Y.; Qing, Q. Scalable fabrication framework of implantable ultrathin and flexible probes with biodegradable sacrificial layers. *Nano Lett.* **2017**, *17*, 7315–7322. [CrossRef] [PubMed]

159. He, S.; Hu, Y.; Wan, J.; Gao, Q.; Wang, Y.; Xie, S.; Qiu, L.; Wang, C.; Zheng, G.; Wang, B. Biocompatible carbon nanotube fibers for implantable supercapacitors. *Carbon* **2017**, *122*, 162–167. [CrossRef]

160. Hwang, S.W.; Kim, D.H.; Tao, H.; Kim, T.I.; Kim, S.; Yu, K.J.; Panilaitis, B.; Jeong, J.W.; Song, J.K.; Omenetto, F.G. Materials and fabrication processes for transient and bioresorbable high-performance electronics. *Adv. Funct. Mater.* **2013**, *23*, 4087–4093. [CrossRef]

161. Kang, S.-K.; Murphy, R.K.; Hwang, S.-W.; Lee, S.M.; Harburg, D.V.; Krueger, N.A.; Shin, J.; Gamble, P.; Cheng, H.; Yu, S. Bioresorbable silicon electronic sensors for the brain. *Nature* **2016**, *530*, 71–76. [CrossRef] [PubMed]

162. Xu, F.; Zhang, H.; Jin, L.; Li, Y.; Li, J.; Gan, G.; Wei, M.; Li, M.; Liao, Y. Controllably degradable transient electronic antennas based on water-soluble PVA/TiO$_2$ films. *J. Mater. Sci.* **2018**, *53*, 2638–2647. [CrossRef]

163. Acar, H.; Çınar, S.; Thunga, M.; Kessler, M.R.; Hashemi, N.; Montazami, R. Study of physically transient insulating materials as a potential platform for transient electronics and bioelectronics. *Adv. Funct. Mater.* **2014**, *24*, 4135–4143. [CrossRef]

164. Fukuda, K.; Takeda, Y.; Yoshimura, Y.; Shiwaku, R.; Tran, L.T.; Sekine, T.; Mizukami, M.; Kumaki, D.; Tokito, S. Fully-printed high-performance organic thin-film transistors and circuitry on one-micron-thick polymer films. *Nat. Commun.* **2014**, *5*. [CrossRef] [PubMed]

165. Takeda, Y.; Hayasaka, K.; Shiwaku, R.; Yokosawa, K.; Shiba, T.; Mamada, M.; Kumaki, D.; Fukuda, K.; Tokito, S. Fabrication of ultra-thin printed organic tft cmos logic circuits optimized for low-voltage wearable sensor applications. *Sci. Rep.* **2016**, *6*. [CrossRef] [PubMed]

166. Kaltenbrunner, M.; Sekitani, T.; Reeder, J.; Yokota, T.; Kuribara, K.; Tokuhara, T.; Drack, M.; Schwodiauer, R.; Graz, I.; Bauer-Gogonea, S.; et al. An ultra-lightweight design for imperceptible plastic electronics. *Nature* **2013**, *499*, 458–463. [CrossRef] [PubMed]

167. Burghartz, J.N.; Appel, W.; Harendt, C.; Rempp, H.; Richter, H.; Zimmermann, M. Ultra-thin chip technology and applications, a new paradigm in silicon technology. *Solid-State Electron.* **2010**, *54*, 818–829. [CrossRef]

168. Viventi, J.; Kim, D.H.; Vigeland, L.; Frechette, E.S.; Blanco, J.A.; Kim, Y.S.; Avrin, A.E.; Tiruvadi, V.R.; Hwang, S.W.; Vanleer, A.C.; et al. Flexible, foldable, actively multiplexed, high-density electrode array for mapping brain activity in vivo. *Nat. Neurosci.* **2011**, *14*, 1599–1605. [CrossRef] [PubMed]

169. Rojas, J.P.; Syed, A.; Hussain, M.M. Mechanically flexible optically transparent porous mono-crystalline silicon substrate. In Proceedings of the 2012 IEEE 25th International Conference on Micro Electro Mechanical Systems (MEMS), Paris, France, 29 January–2 February 2012; pp. 281–284.

170. Bedell, S.W.; Fogel, K.; Lauro, P.; Shahrjerdi, D.; Ott, J.A.; Sadana, D. Layer transfer by controlled spalling. *J. Phys. D Appl. Phys.* **2013**, *46*, 152002. [CrossRef]

171. Van den Brand, J.; de Kok, M.; Koetse, M.; Cauwe, M.; Verplancke, R.; Bossuyt, F.; Jablonski, M.; Vanfleteren, J. Flexible and stretchable electronics for wearable health devices. *Solid-State Electron.* **2015**, *113*, 116–120. [CrossRef]

172. Kim, Y.S.; Lu, J.; Shih, B.; Gharibans, A.; Zou, Z.; Matsuno, K.; Aguilera, R.; Han, Y.; Meek, A.; Xiao, J. Scalable manufacturing of solderable and stretchable physiologic sensing systems. *Adv. Mater.* **2017**, *29*. [CrossRef] [PubMed]

173. Rojas, J.P.; Torres Sevilla, G.A.; Ghoneim, M.T.; Inayat, S.B.; Ahmed, S.M.; Hussain, A.M.; Hussain, M.M. Transformational silicon electronics. *ACS Nano* **2014**, *8*, 1468–1474. [CrossRef] [PubMed]

174. Khan, Y.; Garg, M.; Gui, Q.; Schadt, M.; Gaikwad, A.; Han, D.; Yamamoto, N.A.D.; Hart, P.; Welte, R.; Wilson, W.; et al. Flexible hybrid electronics: Direct interfacing of soft and hard electronics for wearable health monitoring. *Adv. Funct. Mater.* **2016**, *26*, 8764–8775. [CrossRef]

175. Mannoor, M.S.; Tao, H.; Clayton, J.D.; Sengupta, A.; Kaplan, D.L.; Naik, R.R.; Verma, N.; Omenetto, F.G.; McAlpine, M.C. Graphene-based wireless bacteria detection on tooth enamel. *Nat. Commun.* **2012**, *3*. [CrossRef] [PubMed]

176. Swisher, S.L.; Lin, M.C.; Liao, A.; Leeflang, E.J.; Khan, Y.; Pavinatto, F.J.; Mann, K.; Naujokas, A.; Young, D.; Roy, S.; et al. Impedance sensing device enables early detection of pressure ulcers in vivo. *Nat. Commun.* **2015**, *6*. [CrossRef] [PubMed]

177. Gao, W.; Emaminejad, S.; Nyein, H.Y.Y.; Challa, S.; Chen, K.; Peck, A.; Fahad, H.M.; Ota, H.; Shiraki, H.; Kiriya, D.; et al. Fully integrated wearable sensor arrays for multiplexed in situ perspiration analysis. *Nature* **2016**, *529*. [CrossRef] [PubMed]

178. Xu, B.; Akhtar, A.; Liu, Y.; Chen, H.; Yeo, W.H.; Park, S.I.; Boyce, B.; Kim, H.; Yu, J.; Lai, H.Y. An epidermal stimulation and sensing platform for sensorimotor prosthetic control, management of lower back exertion, and electrical muscle activation. *Adv. Mater.* **2016**, *28*, 4462–4471. [CrossRef] [PubMed]

179. Kim, J.; Lee, M.; Shim, H.J.; Ghaffari, R.; Cho, H.R.; Son, D.; Jung, Y.H.; Soh, M.; Choi, C.; Jung, S. Stretchable silicon nanoribbon electronics for skin prosthesis. *Nat. Commun.* **2014**, *5*. [CrossRef] [PubMed]

180. Malechka, T.; Tetzel, T.; Krebs, U.; Feuser, D.; Graeser, A. Sbci-headset—Wearable and modular device for hybrid brain-computer interface. *Micromachines* **2015**, *6*, 291–311. [CrossRef]

181. Heo, J.; Yoon, H.; Park, K.S. A novel wearable forehead eog measurement system for human computer interfaces. *Sensors* **2017**, *17*. [CrossRef] [PubMed]

Review

Advances in Materials for Recent Low-Profile Implantable Bioelectronics

Yanfei Chen [1], Yun-Soung Kim [2], Bryan W. Tillman [3,4], Woon-Hong Yeo [2,5,*] and Youngjae Chun [1,4,6,*]

[1] Department of Industrial Engineering, Swanson School of Engineering, University of Pittsburgh, Pittsburgh, PA 15261, USA; yanfeichen@pitt.edu

[2] George W. Woodruff School of Mechanical Engineering, College of Engineering, Georgia Institute of Technology, Atlanta, GA 30332, USA; ysk@me.gatech.edu

[3] Division of Vascular Surgery, University of Pittsburgh Medical Center, Pittsburgh, PA 15260, USA; tillmanbw@upmc.edu

[4] McGowan Institute for Regenerative Medicine, University of Pittsburgh, Pittsburgh, PA 15261, USA

[5] Center for Flexible Electronics, Institute for Electronics and Nanotechnology, Bioengineering Program, Petit Institute for Bioengineering and Biosciences, Neural Engineering Center, Georgia Institute of Technology, Atlanta, GA 30332, USA

[6] Department of Bioengineering, Swanson School of Engineering, University of Pittsburgh, Pittsburgh, PA 15261, USA

[*] Correspondence: whyeo@gatech.edu (W.-H.Y.); yjchun@pitt.edu (Y.C.); Tel.: +1-404-385-5710 (W.-H.Y.); +1-412-624-1193 (Y.C.)

Received: 27 February 2018; Accepted: 26 March 2018; Published: 29 March 2018

Abstract: The rapid development of micro/nanofabrication technologies to engineer a variety of materials has enabled new types of bioelectronics for health monitoring and disease diagnostics. In this review, we summarize widely used electronic materials in recent low-profile implantable systems, including traditional metals and semiconductors, soft polymers, biodegradable metals, and organic materials. Silicon-based compounds have represented the traditional materials in medical devices, due to the fully established fabrication processes. Examples include miniaturized sensors for monitoring intraocular pressure and blood pressure, which are designed in an ultra-thin diaphragm to react with the applied pressure. These sensors are integrated into rigid circuits and multiple modules; this brings challenges regarding the fundamental material's property mismatch with the targeted human tissues, which are intrinsically soft. Therefore, many polymeric materials have been investigated for hybrid integration with well-characterized functional materials such as silicon membranes and metal interconnects, which enable soft implantable bioelectronics. The most recent trend in implantable systems uses transient materials that naturally dissolve in body fluid after a programmed lifetime. Such biodegradable metallic materials are advantageous in the design of electronics due to their proven electrical properties. Collectively, this review delivers the development history of materials in implantable devices, while introducing new bioelectronics based on bioresorbable materials with multiple functionalities.

Keywords: implantable materials; low-profile bioelectronics; micro/nanofabrication; medical devices; biodegradable materials; miniaturization

---

1. Introduction

In recent years, a variety of low-profile electronics have been developed for body implantable medical devices [1], such as the pacemaker, cardiac defibrillator, bladder stimulator, cochlear implants, and biosensors for the monitoring of pressure, flow, strain, and chemical sensors [2–6]. Such implanted systems are designed for long-term use in the human body, from a few months to several years.

Therefore, a material's biocompatibility is one of the most important features to be studied [7]. While there are many reviews and studies of strategies for investigating the foreign body reaction of the materials used in these devices, no systematic review has been reported for recent trends regarding the electronic materials used in medical devices. Considering the rapid development of implantable devices, such a review is critical to understanding both the types and major properties of the materials in order to investigate strategies for enhancing their functionality and biocompatibility. The primary purpose of implanted medical devices is either to diagnose physiological conditions in health, or stimulate organs for necessary functionality in the body. For diagnostics, electronics have been designed to offer accurate and real-time monitoring of important parameters via wireless telemetry systems. For organ actuation, electronic devices such as the pacemaker, cardiac defibrillator, balder stimulator, or artificial organs, have provided either mechanical or electrical stimulation to provide functions. In past decades, the aforementioned medical devices have used rigid and bulky electronic components to meet the required performance. However, these systems with high modulus materials develop unwanted inflammation or complications in the body, which bring critical health issues [8].

Recent advances in micro/nanofabrication technologies have enabled the miniaturization of functional electronic components to develop low-profile bioelectronics [9]. Specifically, various implantable sensors have shown dramatically reduced form factors, such that they can be implanted in the human body without disturbing fluid flow or causing tissue and bone loss. While the miniaturization significantly changed the trend of implantable medical devices, there remain significant hurdles caused by rigid materials. For example, a device implanted in joints or moving regions generates complications due to a mismatch between the material's properties and the surrounding cells and tissues. In addition, minimally invasive surgeries cannot accommodate any rigid devices, since they are not deliverable with low-profile tools such as stents or catheters.

Here, new materials that offer mechanical compliance and flexibility can make great contributions for the next generation of implantable bioelectronics. Flexible materials such as functionalized polymers or thin-film metallic membranes have become a new frontier in body implantable bioelectronics, which has resulted in a significant increase in research outcomes. The most recent achievements include the finding of biodegradable metallic materials and their use in implantable systems with great electrical properties and functionalities. These devices can be programmed in a way that naturally dissolve in the body due to the controlled material's density and inherent solubility in fluid. Even though many aspects of the materials and their systemic integration still need further study and investigation, the promising outcomes are very encouraging in the design of implantable bioelectronics that disappear after a certain lifetime without a second surgery. Overall, this review focuses on trends in the development of electronic materials that are being used in implantable bioelectronics. The various types of materials that are being used in medical devices and their key properties are summarized, and device functionalities and current challenges are discussed.

2. Traditional Materials Used in Implantable Bioelectronics

In past decades, a wide range of electronic materials has been used to develop implantable bioelectronics. Among them, silicon, silicon compounds, and stainless steel were popular due to the fully established fabrication process via high-throughput and high-yield micro/nanofabrication techniques. These materials work as functional parts in electronics, while medical devices incorporate other materials such as polymers and ceramics. Although these traditional materials are intrinsically rigid, advanced technologies in materials processing and manufacturing made them sufficiently flexible for implantable applications.

2.1. Examples of Traditional Electronic Materials

Table 1 includes a representative list of well-known, traditional materials that have been used in implantable biomedical systems, including silicon, glass (silicon dioxide, or quartz), ceramic (e.g., silicon nitride), and metal/metallic oxides. They are categorized as rigid materials, since they exhibit high Young's moduli, high hardness, a high temperature processing limit, and low gas permeability. These materials are compatible with complementary metal–oxide–semiconductor (CMOS) processing, micro/nanofabrication, and integrated circuit design.

Table 1. A summary of the traditional materials with applications in implantable bioelectronics.

Materials	Properties	Device Component	Applications	References
Silicon	Compatible with microfabrication	Substrate	Intraocular pressure and cardiovascular monitoring	[10–15]
Silicon	Compatible with microfabrication	Structural diaphragm	Blood pressure and shunt pressure sensor	[16,17]
Silicon oxide	High-quality factor	Structural diaphragm and substrate	Surface acoustic wave blood pressure sensor	[18–20]
Silicon nitride	Thermally stable	Dielectric layer	Orthopedic sensor	[21–23]
Silicon nitride	Thermally stable	Insulation layer	Cerebrospinal fluid flow monitoring	[24–26]
Stainless steel	Compatible with stents	Substrate	Capacitive pressure sensor	[27,28]

Silicon is commonly used as a structural material or bulk substrate in implantable medical devices, owing to its ease of micromachining with high resolution. Various MEMS fabrication techniques have been developed, including photolithography, etching, and deposition to enable the patterning on silicon layers over the decades, which can be used to fabricate different types of sensors for in vivo diagnostics. Typical pressure sensors that utilize a thin diaphragm suspended over the hard substrate are good example of the use of hard materials in implantable bioelectronics. Basically, the design of a silicon-based pressure sensor involves a vacuum sealed cavity enclosed on at least one side of the thin membrane, and the membrane deflects in response to the applied physiological pressure, which can be measured as capacitance changes or piezoelectric signals [29]. This pressure sensor design is robust, can be low-profile, can be integrated with planar inductor coils for wireless monitoring, and has been successfully applied in intraocular pressure sensing with a thin parylene coating [10–15] and cardiovascular monitoring [16,30,31]. Another example of a device would be a highly boron-doped silicon membrane, which has been used for the deflectable diaphragm on the silicon wafer developed by Chatzandroulis et al. [16]. Silicon itself is hard and stiff material, but it can be used as a thin membrane in implantable pressure sensors, enabling the dynamic measurement of arterial blood pressure fluctuations. Yoon et al. [17] developed a micro-telemetry pressure sensor, which was designed for intracranial pressure monitoring after the implantation of the shunt system. The patterned p+ silicon membrane functions as the diaphragm, and along with two Cr/Au electrodes, constitutes a capacitor in response to applied pressure. The capacitive pressure sensor was integrated with copper coils on a Pyrex glass substrate for wireless pressure monitoring.

Glass, mainly silicon oxide (SiO_2) or single crystal quartz, is also a quite rigid material that is widely used in implantable wireless surface acoustic wave (SAW) sensors for blood pressure measurement [18–20]. Glass can be etched isotropically by hydrofluoric acid (HF), and anisotropically with deep reactive ion etching (DRIE) using SF_6 or C_4F_8 [32,33] to create the desired patterns. The high-quality factor in resonators whose natural frequencies are insensitive to temperature variations around a desired operation point makes crystal quartz a suitable candidate for implantable devices. Metallic interdigital transducer electrodes were deposited on the piezoelectric quartz substrate, and a thin quartz membrane was bonded onto the substrate to create the diaphragm. The pressure sensor was supported by two rigid polyethylene terephthalate (PET) walls and by two silicone walls to improve the sensitivity (Figure 1a).

Silicon nitride (SiN_x) is the chemical compound of the elements silicon and nitrogen; Si_3N_4 is the most thermally stable one. It is a ceramic material with a Young's modulus of 300–320 GPa, a Poisson's ratio of 0.27, a tensile strength of 350–400 MPa and a Vickers hardness of 13–16 GPa [21]. Si_3N_4 also has a non-toxic, biocompatible ceramic surface for functional human bone cells propagation in vitro, making it more attractive for orthopedic implants than other ceramic materials. In the implantable microelectromechanical (bioMEMS) sensor designed for orthopedic applications [22,23], the bioMEMS sensor is fabricated from a standard MEMS fabrication process, and Si_3N_4 is deposited using plasma-enhanced chemical vapor deposition (PECVD). Si_3N_4 was used as the dielectric layer sandwiched between two electrode layers due to its relatively high dielectric constant (~8) and low loss, resulting in a high Q-factor. The physical loading on the composite sensor layer results in the sensor's capacitance change via the resonator's structural deformation. In the implantable flow sensing design for cerebrospinal fluid flow monitoring [24–26], the capacitive sensor was fabricated from silicon wafers using the standard MEMS technologies, and silicon nitride was deposited using PECVD as the protection or insulation layer for the device. Low-stress silicon nitride (SiN_x) was deposited on both sides of the wafer, and then on the top of the bottom electrode (Cr/Ni bi-layer) as the hard mask for the etching. The capacitive pressure sensor can detect the combined effect of cerebrospinal fluid (CSF) pressure, hydrostatic pressure, and flow by measuring the flexible membrane deflection. It has the sensitivity to measure clinically relevant ranges of slow-moving fluids such as the CSF in the brain, which is typically around 20 mL/h in a healthy individual (Figure 1b).

Metallic substrates, such as type −316 L stainless steel, were also used in some implantable devices. Chen et al. [27,28] developed the MEMS capacitive pressure sensor on a stainless steel chip, which is the same as the stent for integration with inductive stents by microwelding (Figure 1c). The use of stainless steel with a cavity as the substrate enabled a direct connection between the pressure senor and the stent, and thus eliminated the need for additional electrical leads at the joint, and improved the device reliability. The pressure sensor was fabricated from a gold–polyimide composite in order to create a sufficiently sensitive to the intravascular pressures with the capacitive electrodes and diaphragm structure. The whole device was packaged with parylene-C layers for electrical and chemical insulation from the fluid environment.

Some implantable medical devices utilized rigid circuits with sensing modules on the chip level to measure the corresponding physiological signals. These often consist of two units: a sensor unit to monitor the physiologically relevant signals such as blood pressure, and a control unit to perform A/D conversion, signal processing, and transmitting. In an implantable optical sensor for long-term blood pressure measurement [34,35], the correlation between pulse transit time and blood pressure was employed with light-emitting diodes (LEDs) and a silicon photodetector on a polyimide substrate. All of the substrates and sensors were encapsulated with silicone to achieve sensor conformability to the underlying tissues. The silicone strips allowed the attachment of the sensor unit directly onto arteries with a diameter greater than 4 mm without appreciable constriction, which added the flexibility in the measurement configurations, as shown in Figure 1d. An implantable accelerometer system [36] with the function of detecting the reflected wave transit time for blood pressure measurement was also developed by mounting digital accelerometers on the flexible polyimide foils. The microcontroller and radio frequency (RF) units on the chip form the rigid part in the sensing system, and connect with the sensor units by a meander structure. A parylene-C layer was deposited to offer protection from surrounding fluids and tissues.

2.2. Challenges and Limitations of Traditional Materials

There are big challenges regarding rigid and hard materials in bioelectronics, including biocompatibility and device delivery. The biocompatibility issue always occurs at the interface between implants and tissue or blood. The sensor materials should cause very little or no long-term damage to the artery and a long-term inflammation response from the surrounding tissues. As most of the materials mentioned above may not be biocompatible, they often require additional encapsulations (e.g., silicone, parylene) to shield the body from potential harmful materials to minimize the inflammatory response from the tissues and vessels. It remains challenging on the conformal adhesions between the encapsulation materials and the implantable electronic devices, which increases the fabrication complexity for those implantable medical devices [29,37,38]. As sharp edges and corners are present in the "hard" electronic device surface, it might impose serious damages to the relatively soft tissues during the surgical placement and implantation. In addition, most of the implantable medical devices fabricated from traditional "hard" materials are quite bulky, which could cause disturbance in the blood flow or even significant coagulation or clot after being implanted in the blood vessels.

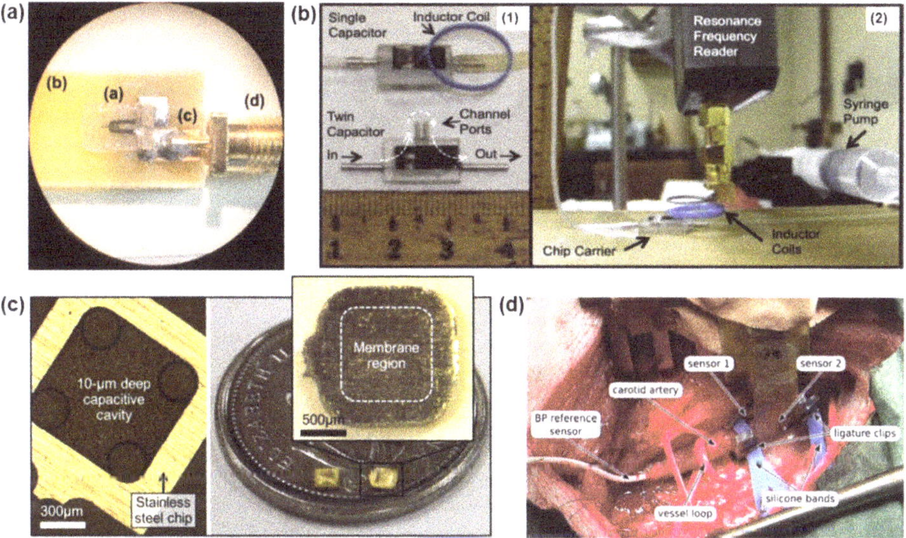

Figure 1. (**a**) Assembled blood pressure sensing device; (**a**) pressure sensor on, (**b**) FR-4 test board with, (**c**) transmission line, and (**d**) SubMiniature version A (SMA) connector. Reprinted with permission from Ref. [18], Copyright (2013), Springer Nature; (**b**) Flow sensors for cerebrospinal fluid sensing: (**1**) Test sensors with a single capacitor with inductor coil attached (**top**) and a twin-capacitor sensor without inductor coils (**bottom**); (**2**) Flow control unit (syringe pump) and spectrometer (resonance frequency reader) with a test sensor on a chip carrier. Reprinted with permission from Ref. [24], Copyright (2015), Elsevier; (**c**) Fabricated intravascular pressure sensors: (**left**) with a stainless-steel chip before membrane bonding and completed sensors; (**right**) a close-up of the diaphragm. Reprinted with permission from Ref. [27], Copyright (2014), Spring Nature; (**d**) Photograph of the optical blood pressure sensor units mounted onto the carotid artery of a domestic pig. The photo shows the operation site before measurement. Reprinted with permission from Ref. [35], Copyright (2012), Springer Nature.

Another critical issue in biomedical devices fabricated from traditional rigid materials is the mechanical property mismatch with surrounding tissues at the target site. Since human tissues and vessel walls are soft and stretchy, the high modulus materials in the device often cause excessive stress and multimodal deformation. For example, the microscale motion of an implanted device [39] caused severe damages to the tissues and exacerbated the foreign-body response of the immune system. The following consequences of the material's mismatch include unexpected infection and tissue hardening [7,29]. To avoid such abnormalities, devices are required to have an additional surface encapsulation of soft, flexible materials, which minimizes the external stress to the implanted site.

The last challenging issue is related to the delivery of bioelectronics. The new trends of minimally invasive techniques in both vascular and orthopedic applications enable the device to be delivered with a tiny puncture either in the blood vessel, tissue, or bones. These techniques reduce the hospital stay of the patients, as well as reduce the risk and pain from open invasive surgical procedures. While these techniques are a new trend and are widely used for various implantable devices, rigid and bulky devices cannot be delivered via these methods. Any implantable bioelectronics created with rigid materials, if they are bulky and stiff, cannot be delivered by minimally invasive surgical techniques, losing the benefit for the patients. Therefore, new soft and flexible materials attract a lot of attention to accommodate these trends.

3. New Materials for Soft and Flexible Electronics

With the advancement of miniaturization fabrication techniques, micro/nanopatterning capability has been significantly improved. Additionally, new soft materials have been used for various implantable bioelectronics due to their unique and excellent properties, such as large elastic range, excellent biocompatibility, and easy fabrication. Soft material-based bioelectronics can be applied for various devices that require highly flexible properties, stretchable performance, and low-profile features maintaining the electronic function [40–42]. The flexibility of electronics is defined as whether their mechanical characteristics are bendable, foldable, or stretchable. In this section, two types (organic and inorganic) of soft materials are introduced.

3.1. Organic Materials

With the development of polymeric material fabrication technologies, a set of polymeric materials including polydimethylsiloxane (PDMS), medical grade silicone, parylene, polyimide, polyvinylidene fluoride-trifluoroethylene (PVDF-TrFE), and liquid crystal polymer (LCP) have been widely used in flexible and soft electronics as substrates, sensing components, and encapsulations. Most of the polymeric materials are soft, lightweight, RF-transparent and low cost, and hence can address the current challenges associated with metallic and ceramic materials for implantable electronics. Table 2 summarizes the list of applications of several important polymeric materials in flexible electronics.

Table 2. Summary of organic materials with applications in biomedical devices. PDMS: polydimethylsiloxane; PVDF: polyvinylidene fluoride; LCP: liquid crystal polymer.

Materials	Properties	Device Component	Applications	References
PDMS	Low modulus, high dielectric strength, low chemical reactivity	Microfluidic channel	Pressure monitoring	[43,44]
		Dielectric layer	Pressure and oxygen sensor in blood	[45,46]
		Substrate layer	Physiological recording	[47–49]
Medical grade silicone	High tear strength and elasticity, transparency	Encapsulation layer	Soft contact lens sensor, intracranial and blood pressure monitoring	[50–54]
Parylene C	Chemical and biological inert, low water permeability and absorption	Structural diaphragm	Intraocular pressure monitoring	[55–58]
		Substrate layer	Neural electrode probe, hydrocephalus shunt occlusion detection	[59–64]
Polyimide	High heat resistance	Substrate layer	Intraocular and cardiovascular pressure monitoring	[15,65–69]
		Structural diaphragm	Intraocular pressure monitoring	[70,71]
PVDF	Piezoelectricity	Structural diaphragm	Intracranial and endovascular pressure monitoring	[72–75]
LCP	Low dielectric constant and low moisture absorption rate	Substrate	Intraocular pressure monitoring	[76–79]
		Encapsulation	Active intraocular pressure monitoring	[80]

PDMS, a type of silicone elastomer, offers distinct advantages over other materials in implantable pressure-sensing applications, such as (1) a unique flexibility due to a low Young's modulus (<100 MPa); (2) imperviousness to fluids; (3) high dielectric strength (~14 V·μm^{-1}); (4) low chemical reactivity; and (5) proven biocompatibility [81,82]. Therefore, PDMS is selected as the dielectric layer in capacitive pressure sensors. It was reported to be used in highly sensitive capacitive sensors designed for pressure and oxygen content measurement within the heart and blood vessels [45], and interface pressure measurement between the nerve trunk and cuff electrode for nerve tissue health monitoring (Figure 2a) [46]. PDMS polymer is also favorable for fabricating microfluidic channels in bioelectronic devices due to its optical transparency, flexibility, and suitability for soft lithography. The biocompatibility of PDMS also suggests that it might ultimately be possible to embed microfluidic devices in vivo for biomedical analysis [83]. Araci et al. [43] developed a novel passive intraocular pressure (IOP) sensor implant for glaucoma diagnosis and monitoring using standard soft lithography out of PDMS (Figure 2b). Jung et al. [44] fabricated the fluidic channel from PDMS in a resistive-type pressure sensor embedded in a microfluidic system (Figure 2c). Also, PDMS can function as a flexible substrate for soft and ultra-compliant electronic devices and encapsulation materials to mechanically and chemically decouple devices from their environment. Patterned structures of stretchable and electrically conductive materials (e.g., gold electrodes) were embedded in a thin PDMS substrate (Figure 2d–f) [47–49,84–86] to allow for high conformability over any soft/curvilinear surfaces in vivo, thereby enabling a broad range of non-invasive or minimally invasive and implantable systems to address clinical needs [87]. The PDMS-based devices are flexible enough without inducing irreversible deformations or fatigue after the devices were bended, twisted, rolled, or stretched (Figure 2e).

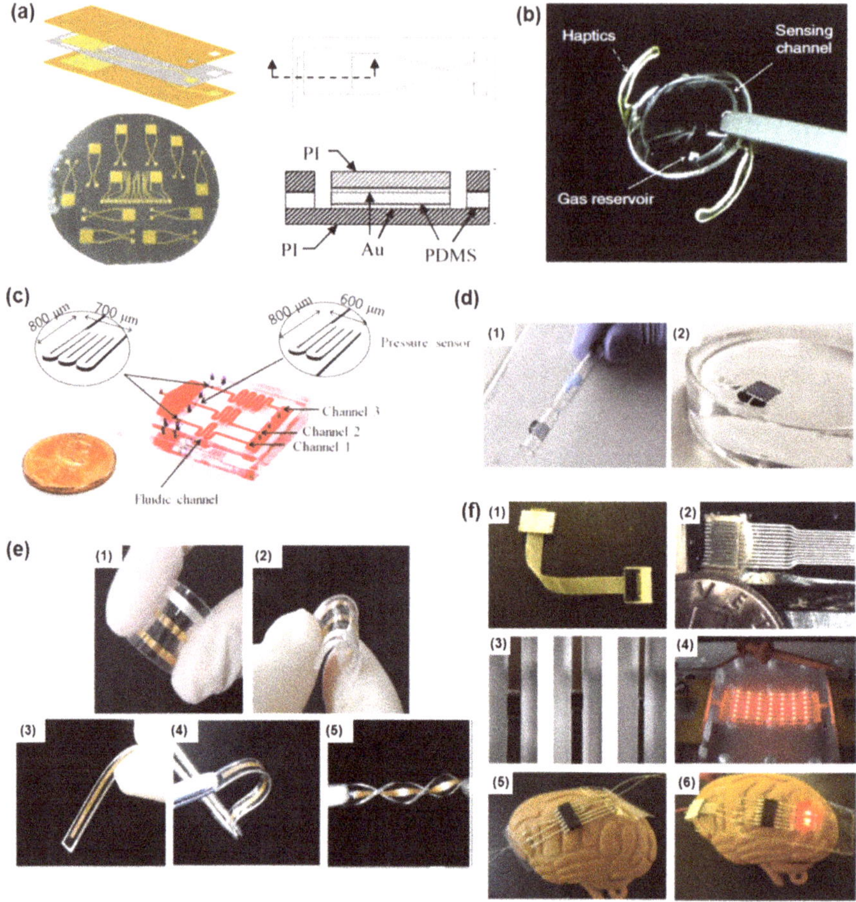

Figure 2. (**a**) Structure of the flexible capacitive pressure sensor. Reprinted with permission from Ref. [46], Copyright (2007), Elsevier; (**b**) Photograph of the microfluidic pressure sensor embedded within the intraocular lens. Reprinted with permission from Ref. [43], Copyright (2014), Springer Nature; (**c**) The intraocular pressure (IOP) sensor integrated in the microfluidic device consists of three PDMS layers: sensor, thin-film, and fluidic-channel layers. Reprinted with permission from Ref. [44], Copyright (2015), MDPI AG; (**d**) A flexible biosensor fabricated on the PDMS film showing the pattern flexibility (left, **1**) and water stability (right, **2**). Reprinted with permission from Ref. [86], Copyright (2018), Elsevier; (**e**) Pictures of a flexible PDMS-based three-electrode sensor; Top view **1** and side view **2** of the three-electrode sensor, straight **3**, bent **4** and twisted **5** working electrode. Reprinted with permission from Ref. [49], Copyright (2010), Elsevier; (**f**) Fabrication of soft electrical circuits on PDMS integrating commercially available electrical components. **1** Picture of a ribbon cable with eight conductors clamped on both ends by a commercial zero insertion force (ZIF) connector; **2** Picture of a custom-made miniature connector with 12 contacts; **3** Pictures of chip resistors bonded on conductive tracks (from left to right: 0805, 0603 and 0402 packages); **4** Picture of a large array of surface mounted device (SMD) light-emitting diodes (LEDs) bonded on a soft printed circuit board (PCB); **5** Picture of a 2-kHz clock generator produced on a 0.2-mm thick soft single-sided PCB that conforms to a plastic brain. **6** Picture of a 1-Hz clock generator with LEDs to display the output levels produced on a 0.7-mm thick double-sided soft PCB. Reprinted with permission from Ref. [84], Copyright (2014), Springer Nature.

Medical grade silicone, another type of silicone elastomer, is Food and Drug Administration (FDA)-approved for biocompatibility to be used in biomedical implants. With exceptional properties such as high tear strength and outstanding elasticity over a wide temperature range, it also exhibits a tensile strength of <10 MPa with the elongation of 300–1000% [88,89], which adds more flexibility to the silicone-based medical devices compared with PDMS. As the strong Si–O–Si (siloxane) backbone provides enhanced chemical inertness and exceptional flexibility, medical grade silicone is considered an ideal candidate for the substrate materials in implantable medical devices with ultracompliance. The important optical properties of the substrate for contact lens sensors are transparency and a sufficiently high refractive index. Therefore, soft contact lens sensors were fabricated by embedding resonance circuits in medical grade silicone layers (NuSil) for continuous IOP monitoring [52–54]. The ultracompliance of the silicone enabled IOP monitoring from the curvature of the cornea (Figure 3a,b). Aqulina et al. [50] fabricated the intracranial pressure (ICP) monitor with the flexible printed circuit board (PCB) coated with medical grade silicone rubber. Medical grade silicone was also used in a capacitive strain gauge housing strip, which can be wrapped around the artery to monitor the blood pressure changes by measuring the blood vessel deformation (Figure 3c) [51]. The Young's modulus of the device is comparable to that of the blood vessel, thus offering the flexibility to minimize the blood flow disturbance.

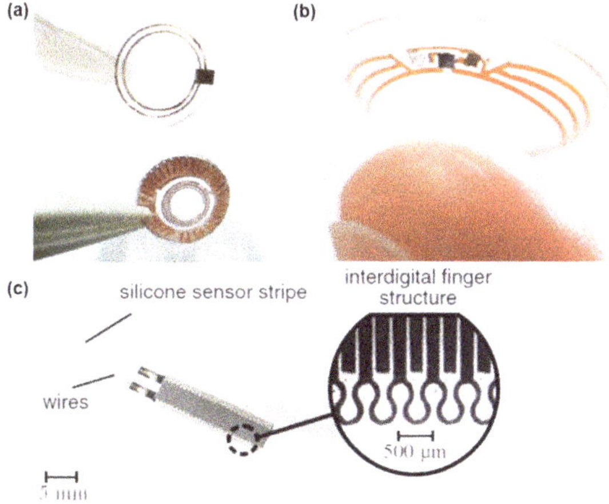

Figure 3. (**a**) A contact lens with sensing elements embedded in a silicone rubber. Reprinted with permission from Ref. [52,53], Copyright (2013, 2014), Elsevier; (**b**) The contact lens sensor under co-development by Google and Novartis. It measures glucose concentration in tears using a miniaturized electrochemical sensor embedded into a hydrogel matrix, Reprinted with permission from Ref. [54], Copyright (2014), John Wiley and Sons; (**c**) Silicone sensor strip with the dimensions of 40 mm × 5 mm × 0.5 mm. Reprinted with permission from Ref. [51], Copyright (2012), Springer Nature.

Parylene C, or poly(dichloro-*p*-xylylene), is a polymeric material that is widely used as a substrate or encapsulation material for biomedical devices due to its FDA-approved biocompatibility, chemical and biological inertness, good barrier properties with low water permeability and absorption, and its functionality as an electrical insulator. The Young's modulus is 1–4 GPa, the tensile strength is 40–110 MPa, and the elongation at break is 7.5–42% [90]. The tensile strength is sensitive to thermal treatment such as annealing, as well as deposition pressure due to the structural crystallization. Parylene-C layers shows a good adhesion to underlying materials such as Si_3N_4, platinum, and itself with an adhesion promotor, Silane A-174 (methacryloxypropyltrimethoxysilane) [90]. Parylene C was selected as the diaphragm and disk substrate for the continuous IOP monitoring of glaucoma patients. The thin parylene-C membrane is sensitive to the applied pressure, and capacitance changes can be induced by membrane deformation. The whole IOP sensor was also packaged and sealed by a thin parylene-C layer to ensure the biocompatibility in the intraocular environment for in vivo tests [55–58]. Parylene C is also a popular flexible substrate material for neural signal recording applications. In high surface-area electrode arrays for high-density simulation and recording in retinal and spinal cord prosthetics, the thin-film platinum and iridium electrodes are embedded in a flexible parylene-C substrate (Figure 4a) [62]. The three-dimensional (3D) sheath neural electrode probes for neural recordings are also constructed on the flexible parylene-C substrate (Figure 4b,c) [60,61,64,91]. The 3D sheath probe arrays were formed by thermal molding of the surface micromachined parylene-C channel, which allows for the recordings on large areas and multiple sites of interest. In a novel microbubble pressure sensor design, a pair of platinum electrodes were embedded in a parylene-C substrate for hydrocephalus treatment monitoring [63]. This unique sensing mechanism utilized the electrochemical impedance measurements of electrolytically generated microbubbles in contact with the parylene-C surface. Another example of a parylene-C based device is the patency sensor for proximal hydrocephalus shunt occlusion detection [59]. Platinum electrodes were patterned on the parylene-C substrate to measure the electrochemical impedance with respect to cerebrospinal fluid flow (Figure 4d). This inline module can be implanted into shunt to enable quantitative and accurate monitoring of shunt performance.

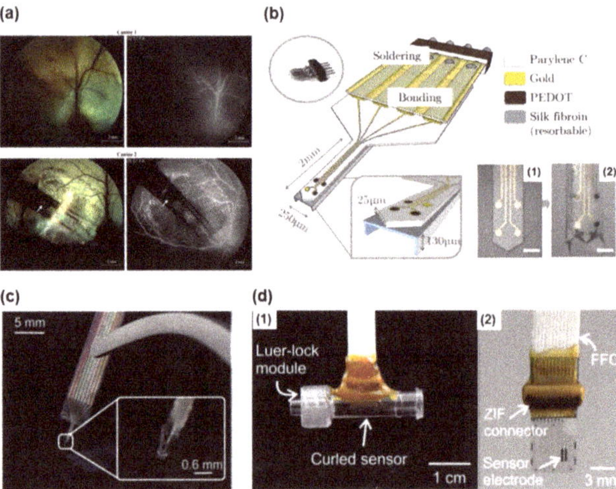

Figure 4. (**a**) Fundus photographs (**left**) showing parylene multi-electrode arrays (MEAs) tacked to the right retina of both animals, and fluorescein angiographies (FAs) (**right**) showing normal vessel perfusion under the arrays. Reprinted with permission from Ref. [62], Copyright (2008), Elsevier; (**b**) Schematic representation of a parylene C-based neural probe with three poly (3,4-ethylenedioxythiophene) (PEDOT)-nanostructured electrodes (40 μm in diameter), and one gold electrode as control. Two photographs show the probe tip **1** before and **2** after PEDOT nano-structuration and silk integration (scale bar 150 μm). Reprinted with permission from Ref. [91], Copyright (2017), Elsevier; (**c**) Fabricated flexible parylene sheath neural probe with integrated parylene cable. Reprinted with permission from Ref. [61], Copyright (2012), Royal Society of Chemistry; (**d**) **1** Packaged patency sensors in inline modules for benchtop testing; **2** Electrically packaged parylene device with final electrode design using a ZIF connector and integrated flat flexible cable (FFC). Reprinted with permission from Ref. [59], Copyright (2016), Springer Nature.

Polyimide (PI) is a polymer of imide monomers; it exhibits a Young's modulus of 1.5–3 GPa and a tensile strength of 70–100 MPa. The elongation at breakage ranges from 2% to 15%, depending on the chemical structure [92]. Polyimides show high heat resistances and high glass transition temperatures, and are stable up to a temperature of 440 °C [92]. Polyimides are widely used in electronic devices as passivation or insulation materials and substrate layers because of their excellent thermal and chemical stabilities, low dissipation factors, and low dielectric constants. Polyimide film (Kapton tape) was used as the insulation layer for copper coil patterns to form an inductor in a minimally invasive pressure sensor for continuous IOP monitoring [15]. Chen et al. [65] developed the wireless pressure monitoring and mapping system with ultrasmall sensor patterns on the flexible polyimide substrate layer (Figure 5a). Viventi et al. [69] fabricated the ultra-thin and flexible electrode arrays on a 12.5-µm thick polyimide film to record the spatial properties of brain activity in vivo (Figure 5b). The extreme flexibility of the device is achieved by reducing the array and substrate thickness to minimize the induced strain during the folding. The extreme flexibility of the device enabled the access to rarely explored cortical areas for neural activity mapping. The highly flexible polyimide substrates for electrode arrays allow for the conformal coverage to the tissues, and cause less harm to the implant site, which makes them suitable for chronic neural activity recordings (Figure 5c) [66,67]. However, the polymer may experience buckling caused by the insertion force during the device implantation [68]. Polyimides are also used in structural components in medical sensors. In a wireless IOP sensor, the copper inductor pattern was deposited on top of the flexible polyimide membrane while a high frequency NiZn ferrite was attached on the bottom [70]. The inductance was varied, with a variable distance between the ferrite material and the inductor pattern with respect to the applied pressure. Shin et al. [71] developed a dual-mode IOP sensor with two separate diaphragms (flexible polyimide and elastomer membrane) to conduct the changes in inductance and capacitance in order to enhance the device performance. A thin polyimide film was also used as the diaphragm in the capacitive pressure transducer for implantable cardiovascular applications (Figure 5d) [93,94].

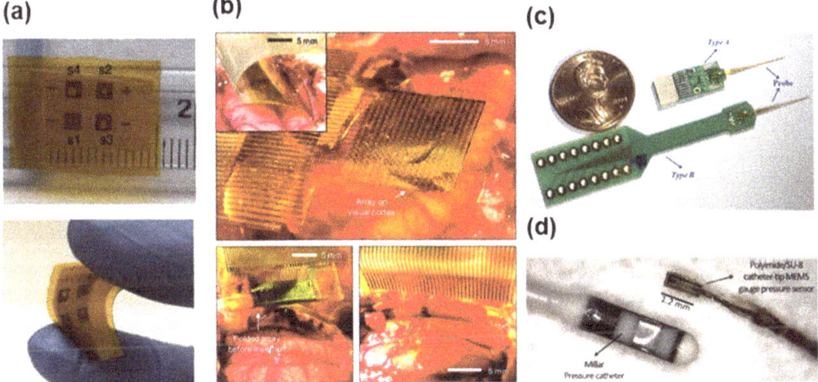

Figure 5. (**a**) Pressure mapping with a 2 × 2 flexible array of 2 × 2 mm² sensors on Kapton film showing the size (**top**) and flexibility when bent (**bottom**). Reprinted with permission from Ref. [65], Copyright (2014), Springer Nature; (**b**) Animal experiment using feline model. (**Top**) A flexible, high-density active electrode array was placed on the visual cortex. Inset, the same electrode array was inserted into the interhemispheric fissure; (**Bottom left**) Folded electrode array before insertion into the interhemispheric fissure; (**Bottom right**) Flat electrode array inserted into the interhemispheric fissure. Reprinted with permission from Ref. [69], Copyright (2011), Springer Nature; (**c**) The National Chiao Tung University (NCTU) probe was bonded onto two different types of PCB. *Type A* assembly was used for chronic recording, and *Type B* was used for acute recording in free-moving animals. Reprinted with permission from Ref. [66], Copyright (2009), Elsevier; (**d**) Assembled polyimide/SU-8 catheter-tip microelectromechanical (MEMS) gauge pressure sensor in comparison with a commercial Millar Mikro-Cath™ disposable pressure catheter. Reprinted with permission from Ref. [94], Copyright (2012), Springer Nature.

Polyvinylidene fluoride (PVDF) and its copolymers polyvinylidene fluoride trifluoroethylene (PVDF-TrFE) are attractive in a broad range of applications, including acoustic transducers and electromechanical actuators, because of their piezoelectric response (generating electrical signals while it is mechanically deformed) and thermal and chemical stability. PVDF polymer exhibits a relatively high room-temperature dielectric constant (>40) and a high electrostriction (strain >4%). Also, it is a thin, flexible, lightweight material and can sustain higher strains (40–140% elongation) compared with other ferroelectric materials [95,96]. PVDF-TrFE piezoelectric films were used to fabricate the flexible diaphragm in a dual-mode intracranial pressure sensor. The PVDF-TrFE diaphragm can operate in a capacitive and resonant mode, allowing for high linearity over small pressure changes with insensitivity to environmental temperature variations and high sensitivity with easy adaption for wireless applications [73]. A PVDF-TrFE copolymer film pressure sensor can also be integrated with a catheter for intravascular measurements. PVDF-TrFE copolymer was spin-coated into thin films (1-μm thick) to tap the near β-phase formation, and it showed no electrical pooling or mechanical stretching. A PVDF-TrFE film can be then patterned using a standard lithography process, and fabricated pressure sensors can be easily mounted on catheter surfaces for real-time measurements [74,75]. Another application of PVDF polymer is simultaneous heartbeat and respiration monitoring based on the piezoelectric response to pulsatile vibrations and periodical deformations on the chest wall [72].

Liquid crystal polymer (LCP) belongs to the family of aromatic polymers. It exhibits a Young's modulus of 2–10 GPa, a tensile strength of 270–500 MPa, and a relatively low dielectric constant (~2.9 at 1 MHz). The uniqueness of the LCP material is its much lower moisture absorption rate (<0.04%) compared with polyimide and parylene-C materials [77,97]. As the moisture-generated surface may cause reliability issues for an electrical circuit and its components, its application in LCP allows for possibly longer-term device implantation. LCP materials have been used for substrates and encapsulations in bioelectronics due to their superior characteristics in heat resistance, chemical stability, mechanical flexibility, and biocompatibility. LCP was used as a soft and flexible host substrate for a miniature capacitive pressure sensor in IOP fluctuation monitoring [58]. The patterning of LCP films is compatible with conventional silicon-based MEMS processing, and can be easily integrated with the capacitive sensing components. In brain and vagal stimulation applications, an iridium oxide electrode was deposited on LCP as a flexible substrate for long-term electrical stimulation [79]. Lee et al. [78] developed an implantable light-emitting phosphor-coated GaN light-emitting diode (LED) on an LCP substrate for prostate-specific antigen detection. Jeong et al. [76,77] fabricated a retina prosthetic implants on the LCP film substrate whose structure is conformable to the eye surface; it allows for the attachment of the whole implant to the eyeball surface. Also, the LCP encapsulation provides long-term reliability without electrical degradation. LCP is also selected as the packaging materials for an implantable active IOP monitoring system consisting of a MEMS pressure sensor, a power storage array, an application-specific integrated circuit (ASIC) for signal processing, and a monopole antenna. The in vivo studies showed the least amount of fibrous encapsulation and inflammation on LCP-packaged devices [80].

3.2. Inorganic Materials

Flexible and stretchable electronics have the capabilities to absorb high levels of strain without fracture or performance degradation. As most inorganic materials are intrinsically brittle, the strategy for achieving flexibility and stretchability is to combine the bendable designs with layouts that enable the device's out-of-plane motion. One strategy of configurations in flexibility for inorganic materials involves the construction of inorganic materials embedded in elastomer substrates, with significant applied strains absorbed by the elastomer substrates. These elastomer substrate materials have been reviewed in the previous section. Another strategy exploits coiled-spring or S-shaped (i.e., serpentine) interconnect structures to accommodate the applied strains [98]. In this section, we are going to review the inorganic materials that achieve high applied strains through serpentine interconnect layouts.

Transferrable monocrystalline silicon nanomembrane (Si NM) is a suitable candidate active material for fabricating fast flexible electronics due to its material uniformity, low interfacial stresses in bonded configurations, mechanical flexibility and durability, equivalent electrical properties to the bulk silicon, and ease of processing at a low cost [99–101]. Therefore, flexible or even stretchable devices can be designed using Si NMs. The manufacturing of Si NMs remains challenging due to their ultra-low profile and associated frangibility. Recently, transfer printing technology has been developed to retrieve the Si NM designs from the source wafers. Thus, silicon NMs can be patterned using photolithography and reactive ion etching (RIE), then released from SOI substrates completely and transferred to a versatile thin and flexible substrate with microscale precision using transfer printing techniques. With the integration of an ultraflexible substrate, the sensors fabricated from Si NMs can accommodate the extreme bending into a small radius of curvature, or even folding, without mechanical failures. An application of emerging nanomembrane technologies involves the intimation coupling of flexible or stretchable electronics with biological tissues such as heart and brain [102]. An example of bio-integrated electronic devices is a conformal bio-interfaced sensor system consisting of Si NM transistors configured to map cardiac electrical activity directly in vivo [103]. Doped silicon NMs were patterned from silicon wafer before being transfer printed onto the polyimide substrate. The whole system was insulated from the wet environment using a multilayer barrier strategy. With the applications of Si NMs, this device combined the high-performance transistors with medium-scale levels of integration to create the electrodes on a flexible substrate, which enabled the adhesion to the constantly moving epicardial surface without penetrating the tissues. Si NM transistors were also used as electrode arrays for direct brain activity mapping [69]. High-density electrode arrays allow for high spatial resolution over a larger region of brain with an improved signal-to-noise ratio.

Metallic materials such as gold and copper are rigid; however, they can be formed into wavy shaped interconnects to achieve stretchability. Deformations primarily occur at the curved edges of the serpentine interconnects to accommodate applied strains, while rigid active device regions still remain non-stretched. By optimization the thickness, shape, and curvature of the interconnects, the device can be stretched up to 70% without significantly affecting the electrical properties [98]. These stretchable electrode designs were demonstrated in the various applications such as stretchable supercapacitors, transistors, LED displays and touch-panel displays, wearable electronic devices, artificial skins, and muscles [104]. However, only a few implantable medical devices for healthcare monitoring were reported. An example of metallic interconnects in flexible bioelectronics is a miniaturized pH sensor array embedded in an elastomer substrate to achieve high surface conformal monitoring of the beating heart undergoing ischemia in a minimally invasive fashion [105]. The gold traces defined the sensing electrodes and contact pads, and a thin bilayer of Cr/Au was patterned to define the serpentine structure as interconnects to minimize the material strains. This pH sensor array can be either integrated with inflatable balloon catheters or have direct contact with an endocardial surface.

4. Biodegradable Materials for Transient Electronics

Biodegradable materials are a category of materials that can be degraded in vivo, either enzymatically or non-enzymatically, to produce biocompatible or toxicologically safe byproducts that can be eliminated by the normal metabolic pathways. Therefore, the biodegradability of transient electronics involves the breakdown of materials, which is mediated chemically and biologically into smaller fragments that can be dissolved or absorbed by the body [106].

4.1. Metallic Biodegradable Materials

Table 3 summarizes the common biodegradable metallic materials (Mg, Zn, and Fe) that are used as electrodes and interconnects in transient electronics. Conventional metallic materials in transient electronics are appealing because of their low electrical resistance, stable properties, and biocompatibility. With the exception of tungsten (W), all other metallic elements are essential for biological functions. The degradation behaviors of these metallic materials can be affected by various factors, and the electrical dissolution rates in thin film forms can be different compared with corresponding bulk materials [107]. Magnesium and magnesium alloys have been considered as potential candidate materials for short-term implants due to their high reactivity in corrosive media environments such as biofluids. As magnesium is present in large amounts in human bodies, the biodegraded magnesium materials can be absorbed. The key issue for the magnesium is the need to reduce the degradation rate in the human fluid environment for transient electronics applications, and the release of the degraded products should be within the human absorption level. It has been found that the degradation rate of magnesium can be reduced by magnesium purification, selective alloying, and anodized coating [108,109]. An example of implantable transient electronic devices that can provide thermal therapy to control surgical site infections uses magnesium as the conductors, magnesium oxide for the dielectrics, silicon nanomembrane (Si NMs) as the semiconductors, and silk for the substrate and encapsulation. After a time scale of 15 days, the device disappeared, with only remnants of silk left [110]. In another example, iron and zinc bilayers were used as the sensor conductor materials in combination with biodegradable polymers as insulation and packaging materials for a wireless pressure sensor [111]. The Zn layer allows for the formation of high-Q elements for conductors. Due to the slow degradation rate of pure Zn in a saline environment, an electrical bilayer on the biodegradable poly-L-lactide (PLLA) and polycaprolactone (PCL) layer was formed with the combination of Fe, resulting in a rapid and controllable degradation. The in vitro degradation tests revealed that the weight loss rate for the Zn/Fe bilayer in saline was 0.46 mg/(cm^2·h), which is tenfold higher than that of pure Zn. Therefore, the degradation rate of the Zn/Fe bilayer can be tailored through modulation of the Zn to Fe exposed surface ratio. Metallic foils such as Fe, Mo, Zn, and Mg can also be used as the substrates for transient *n*-channel metal oxide semiconductor field effect transistors (MOSFETs) [112]. In contrast to the polymer substrates, which can swell and crack the supported electronic structures in biofluids, thin film metallic foil substrates are more robust, compatible for direct device fabrication, and thermally stable to provide hermetic protection. The dissolution behaviors for degradable metals (Fe, Mo, Zn, and Mg) are affected by the thickness, grain structure, and surface morphology in the form of thin film foils. These biodegradable metallic foils can be used in transient electronics for temporary biomedical implants and monitors, with a performance comparable to conventional non-transient materials.

Table 3. Summary of biodegradable materials.

Materials	Degradation Rate	Device Component	Applications	References
Magnesium	High reactivity	Electrode	Thermal therapy	[110]
Zinc and Iron	0.46 mg/(cm^2·h)	Conductor	RF pressure sensor	[111]
Polylactic-co-Glycolic Acid (PLGA)	Several weeks	Substrate	Brain monitoring, wound healing, pressure monitoring	[113–115]
poly(glycerol-sebacate) (PGS)	A few months	Substrate	Cardiovascular monitoring	[116]
Silk	Several weeks	Substrate	Neural recording, drug delivery device	[86,117–119]

4.2. Polymeric Biodegradable Materials

Biodegradable polymeric materials have been attractive in fabricating bioresorbable devices over the recent decades. These categories of materials can be degraded in vivo without the need for device retrieving after implantation. Also, most of them are soft and flexible, which can be used in stretchable and flexible medical devices.

Polylactic-*co*-Glycolic Acid (PLGA) is a copolymer of polylatic acid (PLA) and polyglycolic acid (PGA), and can dissolve in a wide range of common solvents, including chlorinated solvents, tetrahydrofuran, acetone, or ethyl acetate. In water, PLGA biodegrades by hydrolysis of its ester linkages. The degradation rate of PLGA is dependent on different factors, including the initial molecular weight of the monomers, the chemical composition, and the exposure time to fluid. The degradation time varies from 1–2 weeks to 5–6 weeks depending on the lactic acid (LA): glycolic acid (GA) ratio. Also, PLGA exhibits the Young's modulus of 2 GPa with the elongation of 3–10%. It can be formed into a variety of sizes and shapes such as films, porous scaffolds, hydrogels, or microspheres. Therefore, PLGA is considered the best defined biomaterial available as a drug delivery carrier due to its design and controlled biodegradation rate [120,121]. Kang et al. [114] fabricated a silicon-based piezoresistive wireless sensing system onto a 30-μm thick PLGA substrate for intracranial pressure and temperature monitoring during traumatic brain injury (Figure 6a). The uniqueness of the sensing device is its ability to dissolve completely into biocompatible byproducts when immersed in aqueous solutions such as cerebrospinal fluid (CSF) (Figure 6b). A thin, flexible electrode array based on monocrystalline silicon (Si NM) structures with an insulation layer of SiO_2 was constructed on a substrate of PLGA as a platform for the high-speed spatiotemporal mapping of brain activities (Figure 6c,d) [115]. The flexibility of the device allows for the conformal contact and chronically stable interfaces with neural tissues. Also, the whole device and materials are estimated to dissolve in biofluids completely in two months (Figure 6e). A transient hydration sensing system consisting of phosphorous-doped silicon for electrodes, magnesium for contacts/interconnects, and silicon dioxide for the dielectric layer, was fabricated on a PLGA substrate designed for wound-healing process monitoring [122]. The whole device will disappear over several months, as PLGA is the last to dissolve among all of the materials in this device. The challenging issue is the swelling of substrate material such as PLGA, which will lead to fracture/disintegration of the supported device structure during the dissolution. Another group reported the use of poly-L-lactide (PLLA) polymer, which is a variant of PLGA, as a piezoelectric force sensor designed for intra-organ pressure monitoring by employing the piezoelectric behaviors of the material [113]. The sensor includes two piezoelectric PLLA layers, sandwiched between a molybdenum or magnesium electrode and a polylactic acid (PLA) encapsulation layer. This piezoelectric force sensor completely degrades and breaks down after about two months.

Wang et al. [123] first reported the development of a new biodegradable polymer with improved mechanical properties and biocompatibility, poly(glycerol-sebacate) (PGS). PGS exhibits a Young's modulus of 0.05–2 MPa, the tensile Young's modulus of 0.28 MPa, and elongation at breakage is larger than 260%. PGS degrades by 17% after 60 days in phosphate-buffered saline (PBS) solution, and the degradation rate can be tailored. A pressure sensor array designed for cardiovascular monitoring was developed by sandwiching a PGS elastic dielectric layer between two iron–magnesium electrode layers (Figure 7a) [116,124]. The pressure response stays stable, even at a bending radius down to 27 mm. The key element in this design is the PGS layer, which retains the device performance even after prolonged exposure to the degrading environment. The device is expected to degrade completely after a few months. Lewitus et al. [125] also synthesized a new biodegradable polymer from polyethylene glycol (PEG) and desaminotyrosyl-tyrosine (DT) to enhance the degradation as the carriers for neural probe applications.

Silk is another appealing material as a temporary and soluble supporting substrate for transient electronics. Silk can be obtained from the cocoons of the larvae of the silkworm, and it offers the advantages over other biodegradable materials, including optical transparency, mechanically flexibility in thin-film form, compatibility with aqueous processing, biocompatibility, and bioresorbability with a controlled degradation rate. A silk fibroin solution has the ability to crystallize through protein self-assembly with exposure to the air to yield a class of patterned freestanding films or a mechanically robust substrate for biodegradable devices with the ability to control thickness [119]. Kim et al. [117,126] integrated electrode arrays designed for passive neural recording on the silk fibroin substrate (Figure 7b,c). It allows for the spontaneous, conformal wrapping on the curvilinear surfaces of brain tissues once the silk was dissolved in the biofluids. Also, the ultra-thin structure minimizes the stress on the tissue while ensuring highly conformal coverage. The dissolution rate of silk can be programmed with ethanol treatment in neural mapping applications. An electrochemical biosensor designed for dopamine and ascorbic acid detection was also developed by constructing conductive Poly(3,4-ethylenedioxythiophene): poly(styrene sulfonate) (PEDOT: PSS) micropatterns as a working electrode on a silk fibroin substrate. This device is ultracompliant, and can retain its integrity without loss of performance after 150 bending cycles. In addition, the whole device is shown to degrade over four weeks under enzymatic action [118]. Silk can also be used in drug delivery devices due to its biodegradability. Tao et al. [119] developed a biodegradable, remotely controllable, and implantable therapeutic device designed for infection management at a surgical site. The magnesium heater was integrated with drug-loaded silk films on a silk substrate. The drug can be released with a wireless activated heater, and the thermally-triggered drug release profiles can be controlled. It offers an expanded perspective for a restorable therapeutic medical device such as a drug delivery device.

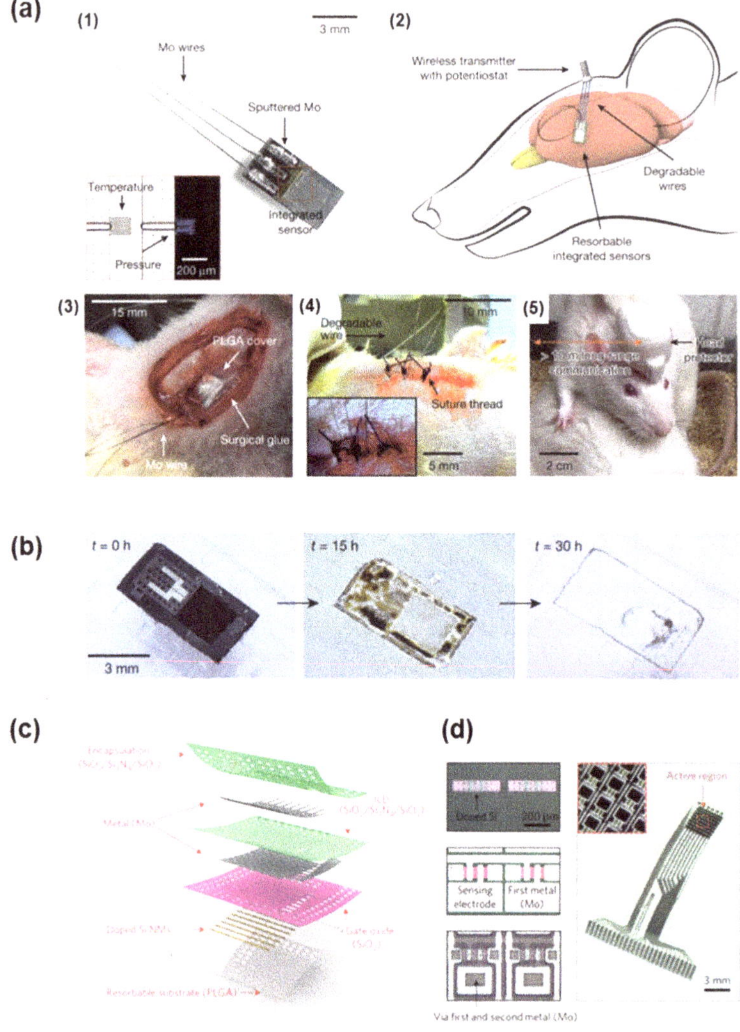

Figure 6. *Cont.*

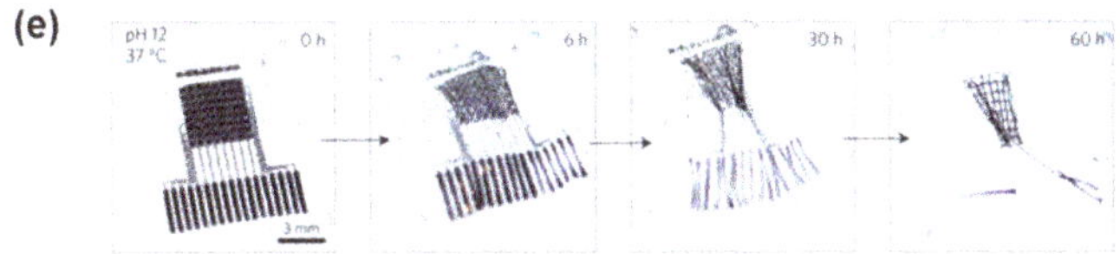

Figure 6. (**a**) Bioresorbable interfaces between intracranial sensors and external wireless data-communication modules with percutaneous wiring. (**1**) Image of bioresorbable pressure and temperature sensors integrated with dissolvable metal interconnects; (**2**) Diagram of a bioresorbable sensor system in the intracranial space of a rat; (**3**) Demonstrations of an implanted bioresorbable sensor in a rat and (**4**) sutured individual; (**5**) Healthy, freely moving rat equipped with a complete, biodegradable wireless intracranial sensor system. Reprinted with permission from Ref. [114], Copyright (2016), Springer Nature; (**b**) Images collected at several stages of accelerated dissolution of a bioresorbable pressure sensor upon insertion into an aqueous buffer solution (pH 12). Reprinted with permission from Ref. [114], Copyright (2016), Springer Nature; (**c**) Bioresorbable, actively multiplexed neural electrode array in an exploded-view rendering. Reprinted with permission from Ref. [115], Copyright (2016), Springer Nature; (**d**) Optical micrograph images of a pair of unit cells at various stages of fabrication (**left**) and a picture of a complete system (**right**). Reprinted with permission from Ref. [115], Copyright (2016), Springer Nature; (**e**) Images collected at several stages of accelerated dissolution of a system immersed into an aqueous buffer solution (pH 12) at 37 °C. Reprinted with permission from Ref. [115], Copyright (2016), Springer Nature.

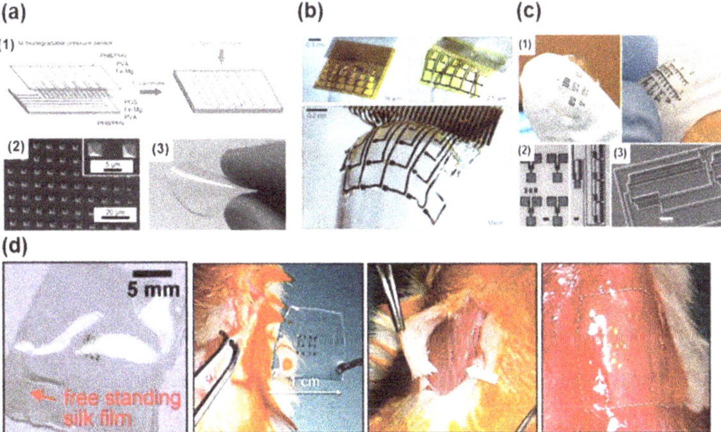

Figure 7. (**a**) Schematic design and fabrication of a fully biodegradable and flexible pressure sensor array from microstructured poly(glycerol sebacate) PGS films. (**1**) Schematic of the final fabrication step of the fully biodegradable pressure sensor and its device structure; (**2**) SEM image of the microstructured PGS film. Two-dimensional (2D) array of square pyramids are formed into PGS from a PDMS mold; (**3**) PGS biodegradable elastomer film. Reprinted with permission from Ref. [124], Copyright (2015), John Wiley and Sons; (**b**) Images of electrode arrays (76 μm sheet in left top, 2.5 μm sheet in right top and 2.5 μm mesh in bottom panel) wrapped onto a glass hemisphere. Reprinted with permission from Ref. [126], Copyright (2010), Springer Nature; (**c**) Formation of complex microstructures via photolithography. (**1**) Large area micropatterns of PEDOT:PSS can be formed on flexible and conformable silk fibroin sheets, (**2**) Optical micrographs and (**3**) SEM images of PEDOT:PSS micropatterns on glass. Scale bars = 100 μm. Reprinted with permission from Ref. [118], Copyright (2016), Elsevier; (**d**) Ultra-thin devices on a flexible silk substrate and the results of the animal toxicity test: image before (second image), shortly after (third image), and two weeks after (fourth image) implantation. Reprinted with permission from Ref. [117], Copyright (2009), AIP Publishing LLC.

5. Conclusions and Outlook

The recent development of implantable biomedical devices has been accompanied by rapid advances in both organic and inorganic functional materials. In this review, we summarized a set of three material groups, including traditional rigid materials, soft and flexible materials, and biodegradable transient materials. Silicon and silicon compounds (e.g., silicon dioxide, silicon nitride), metals, and metallic oxides have represented the first-generation materials in the development of implantable bioelectronics due to their great compatibility with well-established microfabrication processes. However, the highly stiff and rigid mechanical properties of these materials have imposed challenges in medical implantation, including damages to soft and compliant tissues and vessels. Then, soft and flexible materials were introduced, since they could improve the devices' biocompatibility and reliability. Polymeric materials such as silicone elastomer, parylene, polyimide, and PVDF-TrFE have been widely used as substrates, while offering conformal encapsulation of electronic sensing elements in medical devices. Inorganic Si materials with open-mesh and pre-strained mechanical designs have also been attractive in flexible hybrid electronics, due to their superior electrical properties, along with their engineered stretchability. Most recently, a new category of biodegradable materials has developed to fabricate transient bioelectronics that work as temporary electrodes or supporting encapsulants. The main advantage of these materials is their natural dissolution in biofluids, while their byproducts are safely absorbed in the body. Thus, they have been widely studied to design new platforms of medical implants.

The current challenges of the transient materials are on the control of the materials' degradation rates and the programming of dissolution triggering. Possible solutions include the addition of external stimulation (either chemical or electrical) units to the system or the integration of the lifecycle-programmed materials in the system. In addition, the underlying physics and chemistry of dissolution behaviors still need to be investigated further so that the devices can disappear after a desired lifetime. Another area to further study is the development of biodegradable integrated circuits (ICs) and chip components for high performance electronics. Replacing the existing materials and material processing steps in the circuit and device fabrication will be very challenging and require an extensive study of mechanics, surface chemistry, physical interactions, and electronics. Collectively, a comprehensive study of engineered transient materials and the integration of high-performance ICs promises next-generation, low-profile, implantable bioelectronics.

Acknowledgments: Youngjae Chun acknowledges Central Research Development Fund at the University of Pittsburgh. Woon-Hong Yeo acknowledges a seed grant from the Institute for Electronics and Nanotechnology (IEN) at Georgia Institute of Technology, a research grant from the Fundamental Research Program (PNK5061) of Korea Institute of Materials Science (KIMS), and funds from the Marcus Foundation, The Georgia Research Alliance, and the Georgia Tech Foundation through their support of the Marcus Center for Therapeutic Cell Characterization and Manufacturing (MC3M) at Georgia Institute of Technology.

Author Contributions: Youngjae Chun and Woon-Hong Yeo conceived and designed the materials in the paper; All authors conducted reviews of the materials used in implantable biomedical devices and they wrote the paper together.

References

1. Joung, Y.-H. Development of implantable medical devices: From an engineering perspective. *Int. Neurourol. J.* **2013**, *17*, 98–106. [CrossRef] [PubMed]
2. Fiandra, O. The first pacemaker implant in america. *Pacing Clin. Electrophysiol.* **1988**, *11*, 1234–1238. [CrossRef] [PubMed]
3. Wilson, B.S.; Dorman, M.F. Cochlear implants: A remarkable past and a brilliant future. *Hear. Res.* **2008**, *242*, 3–21. [CrossRef] [PubMed]
4. DiMarco, J.P. Implantable cardioverter—Defibrillators. *N. Engl. J. Med.* **2003**, *349*, 1836–1847. [CrossRef] [PubMed]

5. Rijkhoff, N.; Wijkstra, H.; Van Kerrebroeck, P.; Debruyne, F. Urinary bladder control by electrical stimulation: Review of electrical stimulation techniques in spinal cord injury. *Neurourol. Urodyn.* **1997**, *16*, 39–53. [CrossRef]

6. Howe, C.; Lee, Y.; Chen, Y.; Chun, Y.; Yeo, W.-H. An implantable, stretchable microflow sensor integrated with a thin-film nitinol stent. In Proceedings of the IEEE 66th Electronic Components and Technology Conference (ECTC), Las Vegas, NV, USA, 31 May–3 June 2016; pp. 1638–1643.

7. Onuki, Y.; Bhardwaj, U.; Papadimitrakopoulos, F.; Burgess, D.J. A Review of the Biocompatibility of Implantable Devices: Current Challenges to Overcome Foreign Body Response. *J. Diabetes Sci. Technol.* **2008**, *2*, 1003–1015. [CrossRef] [PubMed]

8. Helmus, M.N.; Gibbons, D.F.; Cebon, D. Biocompatibility: Meeting a key functional requirement of next-generation medical devices. *Toxicol. Pathol.* **2008**, *36*, 70–80. [CrossRef] [PubMed]

9. Herbert, R.; Kim, J.-H.; Kim, Y.S.; Lee, H.M.; Yeo, W.-H. Soft material-enabled, flexible hybrid electronics for medicine, healthcare, and human-machine interfaces. *Materials* **2018**, *11*, 187. [CrossRef] [PubMed]

10. Bäcklund, Y.; Rosengren, L.; Hök, B.; Svedbergh, B. Passive silicon transensor intended for biomedical, remote pressure monitoring. *Sens. Actuators A Phys.* **1990**, *21*, 58–61. [CrossRef]

11. Rosengren, L.; Rangsten, P.; Bäcklund, Y.; Hök, B.; Svedbergh, B.; Selén, G. A system for passive implantable pressure sensors. *Sens. Actuators A Phys.* **1994**, *43*, 55–58. [CrossRef]

12. Walter, P.; Schnakenberg, U.; vom Bögel, G.; Ruokonen, P.; Krüger, C.; Dinslage, S.; Handjery, H.C.L.; Richter, H.; Mokwa, W.; Diestelhorst, M.; et al. Development of a completely encapsulated intraocular pressure sensor. *Ophthalmic Res.* **2000**, *32*, 278–284. [CrossRef] [PubMed]

13. Stangel, K.; Kolnsberg, S.; Hammerschmidt, D.; Hosticka, B.; Trieu, H.; Mokwa, W. A programmable intraocular CMOS pressure sensor system implant. *IEEE J. Solid-State Circ.* **2001**, *36*, 1094–1100. [CrossRef]

14. Ganji, B.A.; Shahiri-Tabarestani, M. A novel high sensitive mems intraocular capacitive pressure sensor. *Microsyst. Technol.* **2013**, *19*, 187–194. [CrossRef]

15. Chitnis, G.; Maleki, T.; Samuels, B.; Cantor, L.B.; Ziaie, B. A minimally invasive implantable wireless pressure sensor for continuous IOP monitoring. *IEEE Trans. Biomed. Eng.* **2013**, *60*, 250–256. [CrossRef] [PubMed]

16. Chatzandroulis, S.; Tsoukalas, D.; Neukomm, P.A. A miniature pressure system with a capacitive sensor and a passive telemetry link for use in implantable applications. *J. Microelectromech. Syst.* **2000**, *9*, 18–23. [CrossRef]

17. Yoon, H.J.; Jung, J.M.; Jeong, J.S.; Yang, S.S. Micro devices for a cerebrospinal fluid (CSF) shunt system. *Sens. Actuators A Phys.* **2004**, *110*, 68–76. [CrossRef]

18. Murphy, O.H.; Bahmanyar, M.R.; Borghi, A.; McLeod, C.N.; Navaratnarajah, M.; Yacoub, M.H.; Toumazou, C. Continuous in vivo blood pressure measurements using a fully implantable wireless saw sensor. *Biomed. Microdevices* **2013**, *15*, 737–749. [CrossRef] [PubMed]

19. Ye, X.; Fang, L.; Liang, B.; Wang, Q.; Wang, X.; He, L.; Bei, W.; Ko, W.H. Studies of a high-sensitive surface acoustic wave sensor for passive wireless blood pressure measurement. *Sens. Actuators A Phys.* **2011**, *169*, 74–82. [CrossRef]

20. Liang, B.; Fang, L.; Tu, C.; Zhou, C.; Wang, X.; Wang, Q.; Wang, P.; Ye, X. A novel implantable saw sensor for blood pressure monitoring. In Proceedings of the 2011 16th International Solid-State Sensors, Actuators and Microsystems Conference (TRANSDUCERS), Beijing, China, 5–9 June 2011; pp. 2184–2187.

21. Bal, B.S.; Rahaman, M. Orthopedic applications of silicon nitride ceramics. *Acta Biomater.* **2012**, *8*, 2889–2898. [CrossRef]

22. Melik, R.; Perkgoz, N.K.; Unal, E.; Puttlitz, C.; Demir, H.V. Bio-implantable passive on-chip RF-mems strain sensing resonators for orthopaedic applications. *J. Micromech. Microeng.* **2008**, *18*, 115017. [CrossRef]

23. McGilvray, K.C.; Unal, E.; Troyer, K.L.; Santoni, B.G.; Palmer, R.H.; Easley, J.T.; Demir, H.V.; Puttlitz, C.M. Implantable microelectromechanical sensors for diagnostic monitoring and post-surgical prediction of bone fracture healing. *J. Orthop. Res.* **2015**, *33*, 1439–1446. [CrossRef] [PubMed]

24. Raj, R.; Lakshmanan, S.; Apigo, D.; Kanwal, A.; Liu, S.; Russell, T.; Madsen, J.R.; Thomas, G.A.; Farrow, R.C. Demonstration that a new flow sensor can operate in the clinical range for cerebrospinal fluid flow. *Sens. Actuators A Phys.* **2015**, *234*, 223–231. [CrossRef] [PubMed]

25. Apigo, D.J.; Bartholomew, P.L.; Russell, T.; Kanwal, A.; Farrow, R.C.; Thomas, G.A. An angstrom-sensitive, differential mems capacitor for monitoring the milliliter dynamics of fluids. *Sens. Actuators A Phys.* **2016**, *251*, 234–240. [CrossRef] [PubMed]

26. Apigo, D.J.; Bartholomew, P.L.; Russell, T.; Kanwal, A.; Farrow, R.C.; Thomas, G.A. Evidence of an application of a variable mems capacitive sensor for detecting shunt occlusions. *Sci. Rep.* **2017**, *7*, 46039. [CrossRef] [PubMed]

27. Chen, X.; Brox, D.; Assadsangabi, B.; Hsiang, Y.; Takahata, K. Intelligent telemetric stent for wireless monitoring of intravascular pressure and its in vivo testing. *Biomed. Microdevices* **2014**, *16*, 745–759. [CrossRef] [PubMed]

28. Chen, X.; Brox, D.; Assadsangabi, B.; Ali, M.S.M.; Takahata, K. A stainless-steel-based implantable pressure sensor chip and its integration by microwelding. *Sens. Actuators A Phys.* **2017**, *257*, 134–144. [CrossRef]

29. Scholten, K.; Meng, E. Materials for microfabricated implantable devices: A review. *Lab Chip* **2015**, *15*, 4256–4272. [CrossRef] [PubMed]

30. Kalvesten, E.; Smith, L.; Tenerz, L.; Stemme, G. The first surface micromachined pressure sensor for cardiovascular pressure measurements. In Proceedings of the 1998 Eleventh Annual International Workshop on Micro Electro Mechanical Systems (MEMS 98), Heidelberg, Germany, 25–29 January 1998; pp. 574–579.

31. Ziaie, B.; Najafi, K. An implantable microsystem for tonometric blood pressure measurement. *Biomed. Microdevices* **2001**, *3*, 285–292. [CrossRef]

32. Park, J.; Lee, N.-E.; Lee, J.; Park, J.; Park, H. Deep dry etching of borosilicate glass using sf6 and sf6/Ar inductively coupled plasmas. *Microelectron. Eng.* **2005**, *82*, 119–128. [CrossRef]

33. Kolari, K.; Saarela, V.; Franssila, S. Deep plasma etching of glass for fluidic devices with different mask materials. *J. Micromech. Microeng.* **2008**, *18*, 064010. [CrossRef]

34. Theodor, M.; Ruh, D.; Fiala, J.; Förster, K.; Heilmann, C.; Manoli, Y.; Beyersdorf, F.; Zappe, H.; Seifert, A. Subcutaneous blood pressure monitoring with an implantable optical sensor. *Biomed. Microdevices* **2013**, *15*, 811–820. [CrossRef] [PubMed]

35. Fiala, J.; Bingger, P.; Ruh, D.; Foerster, K.; Heilmann, C.; Beyersdorf, F.; Zappe, H.; Seifert, A. An implantable optical blood pressure sensor based on pulse transit time. *Biomed. Microdevices* **2013**, *15*, 73–81. [CrossRef] [PubMed]

36. Theodor, M.; Fiala, J.; Ruh, D.; Foerster, K.; Heilmann, C.; Beyersdorf, F.; Manoli, Y.; Zappe, H.; Seifert, A. Implantable accelerometer system for the determination of blood pressure using reflected wave transit time. *Sens. Actuators A Phys.* **2014**, *206*, 151–158. [CrossRef]

37. Grayson, A.C.R.; Shawgo, R.S.; Johnson, A.M.; Flynn, N.T.; Li, Y.; Cima, M.J.; Langer, R. A biomems review: Mems technology for physiologically integrated devices. *Proc. IEEE* **2004**, *92*, 6–21. [CrossRef]

38. Potkay, J.A. Long term, implantable blood pressure monitoring systems. *Biomed. Microdevices* **2008**, *10*, 379–392. [CrossRef] [PubMed]

39. Lee, H.; Bellamkonda, R.V.; Sun, W.; Levenston, M.E. Biomechanical analysis of silicon microelectrode-induced strain in the brain. *J. Neural Eng.* **2005**, *2*, 81–89. [CrossRef] [PubMed]

40. Mishra, S.; Norton, J.J.; Lee, Y.; Lee, D.S.; Agee, N.; Chen, Y.; Chun, Y.; Yeo, W.-H. Soft, conformal bioelectronics for a wireless human-wheelchair interface. *Biosens. Bioelectron.* **2017**, *91*, 796–803. [CrossRef] [PubMed]

41. Lee, Y.; Nicholls, B.; Lee, D.S.; Chen, Y.; Chun, Y.; Ang, C.S.; Yeo, W.-H. Soft electronics enabled ergonomic human-computer interaction for swallowing training. *Sci. Rep.* **2017**, *7*, 46697. [CrossRef] [PubMed]

42. Norton, J.J.; Lee, D.S.; Lee, J.W.; Lee, W.; Kwon, O.; Won, P.; Jung, S.-Y.; Cheng, H.; Jeong, J.-W.; Akce, A.; et al. Soft, curved electrode systems capable of integration on the auricle as a persistent brain–computer interface. *Proc. Natl. Acad. Sci. USA* **2015**, *112*, 3920–3925. [CrossRef] [PubMed]

43. Araci, I.E.; Su, B.; Quake, S.R.; Mandel, Y. An implantable microfluidic device for self-monitoring of intraocular pressure. *Nat. Med.* **2014**, *20*, 1074–1078. [CrossRef] [PubMed]

44. Jung, T.; Yang, S. Highly stable liquid metal-based pressure sensor integrated with a microfluidic channel. *Sensors* **2015**, *15*, 11823–11835. [CrossRef] [PubMed]

45. Koley, G.; Liu, J.; Nomani, M.W.; Yim, M.; Wen, X.; Hsia, T.-Y. Miniaturized implantable pressure and oxygen sensors based on polydimethylsiloxane thin films. *Mater. Sci. Eng. C* **2009**, *29*, 685–690. [CrossRef] [PubMed]

46. Chiang, C.-C.; Lin, C.-C.K.; Ju, M.-S. An implantable capacitive pressure sensor for biomedical applications. *Sens. Actuators A Phys.* **2007**, *134*, 382–388. [CrossRef]

47. Adrega, T.; Lacour, S. Stretchable gold conductors embedded in PDMS and patterned by photolithography: Fabrication and electromechanical characterization. *J. Micromech. Microeng.* **2010**, *20*, 055025. [CrossRef]

48. Lacour, S.P.; Benmerah, S.; Tarte, E.; FitzGerald, J.; Serra, J.; McMahon, S.; Fawcett, J.; Graudejus, O.; Yu, Z.; Morrison, B. Flexible and stretchable micro-electrodes for in vitro and in vivo neural interfaces. *Med. Biol. Eng. Comput.* **2010**, *48*, 945–954. [CrossRef] [PubMed]

49. Wu, W.-Y.; Zhong, X.; Wang, W.; Miao, Q.; Zhu, J.-J. Flexible PDMS-based three-electrode sensor. *Electrochem. Commun.* **2010**, *12*, 1600–1604. [CrossRef]

50. Aquilina, K.; Thoresen, M.; Chakkarapani, E.; Pople, I.K.; Coakham, H.B.; Edwards, R.J. Preliminary evaluation of a novel intraparenchymal capacitive intracranial pressure monitor. *J. Neurosurg.* **2011**, *115*, 561–569. [CrossRef] [PubMed]

51. Bingger, P.; Zens, M.; Woias, P. Highly flexible capacitive strain gauge for continuous long-term blood pressure monitoring. *Biomed. Microdevices* **2012**, *14*, 573–581. [CrossRef] [PubMed]

52. Chen, G.-Z.; Chan, I.-S.; Lam, D.C. Capacitive contact lens sensor for continuous non-invasive intraocular pressure monitoring. *Sens. Actuators A Phys.* **2013**, *203*, 112–118. [CrossRef]

53. Chen, G.-Z.; Chan, I.-S.; Leung, L.K.; Lam, D.C. Soft wearable contact lens sensor for continuous intraocular pressure monitoring. *Med. Eng. Phys.* **2014**, *36*, 1134–1139. [CrossRef] [PubMed]

54. Farandos, N.M.; Yetisen, A.K.; Monteiro, M.J.; Lowe, C.R.; Yun, S.H. Contact lens sensors in ocular diagnostics. *Adv. Healthc. Mater.* **2015**, *4*, 792–810. [CrossRef] [PubMed]

55. Chen, P.-J.; Rodger, D.C.; Agrawal, R.; Saati, S.; Meng, E.; Varma, R.; Humayun, M.S.; Tai, Y.-C. Implantable micromechanical parylene-based pressure sensors for unpowered intraocular pressure sensing. *J. Micromech. Microeng.* **2007**, *17*, 1931. [CrossRef]

56. Chen, P.-J.; Rodger, D.C.; Saati, S.; Humayun, M.S.; Tai, Y.-C. Microfabricated implantable parylene-based wireless passive intraocular pressure sensors. *J. Microelectromech. Syst.* **2008**, *17*, 1342–1351. [CrossRef]

57. Chen, P.-J.; Saati, S.; Varma, R.; Humayun, M.S.; Tai, Y.-C. Wireless intraocular pressure sensing using microfabricated minimally invasive flexible-coiled LC sensor implant. *J. Microelectromech. Syst.* **2010**, *19*, 721–734. [CrossRef]

58. Ha, D.; de Vries, W.N.; John, S.W.; Irazoqui, P.P.; Chappell, W.J. Polymer-based miniature flexible capacitive pressure sensor for intraocular pressure (IOP) monitoring inside a mouse eye. *Biomed. Microdevices* **2012**, *14*, 207–215. [CrossRef] [PubMed]

59. Kim, B.J.; Jin, W.; Baldwin, A.; Yu, L.; Christian, E.; Krieger, M.D.; McComb, J.G.; Meng, E. Parylene mems patency sensor for assessment of hydrocephalus shunt obstruction. *Biomed. Microdevices* **2016**, *18*, 87. [CrossRef] [PubMed]

60. Kim, B.J.; Kuo, J.T.; Hara, S.A.; Lee, C.D.; Yu, L.; Gutierrez, C.; Hoang, T.; Pikov, V.; Meng, E. 3d parylene sheath neural probe for chronic recordings. *J. Neural Eng.* **2013**, *10*, 045002. [CrossRef] [PubMed]

61. Kuo, J.T.; Kim, B.J.; Hara, S.A.; Lee, C.D.; Gutierrez, C.A.; Hoang, T.Q.; Meng, E. Novel flexible parylene neural probe with 3d sheath structure for enhancing tissue integration. *Lab Chip* **2013**, *13*, 554–561. [CrossRef] [PubMed]

62. Rodger, D.C.; Fong, A.J.; Li, W.; Ameri, H.; Ahuja, A.K.; Gutierrez, C.; Lavrov, I.; Zhong, H.; Menon, P.R.; Meng, E.; et al. Flexible parylene-based multielectrode array technology for high-density neural stimulation and recording. *Sens. Actuators B Chem.* **2008**, *132*, 449–460. [CrossRef]

63. Yu, L.; Gutierrez, C.A.; Meng, E. An electrochemical microbubble-based mems pressure sensor. *J. Microelectromech. Syst.* **2016**, *25*, 144–152. [CrossRef]

64. Zhao, Z.; Kim, E.; Luo, H.; Zhang, J.; Xu, Y. Flexible deep brain neural probes based on a parylene tube structure. *J. Micromech. Microeng.* **2017**, *28*, 015012. [CrossRef]

65. Chen, L.Y.; Tee, B.C.-K.; Chortos, A.L.; Schwartz, G.; Tse, V.; Lipomi, D.J.; Wong, H.-S.P.; McConnell, M.V.; Bao, Z. Continuous wireless pressure monitoring and mapping with ultra-small passive sensors for health monitoring and critical care. *Nat. Commun.* **2014**, *5*, 5028. [CrossRef] [PubMed]

66. Chen, Y.-Y.; Lai, H.-Y.; Lin, S.-H.; Cho, C.-W.; Chao, W.-H.; Liao, C.-H.; Tsang, S.; Chen, Y.-F.; Lin, S.-Y. Design and fabrication of a polyimide-based microelectrode array: Application in neural recording and repeatable electrolytic lesion in rat brain. *J. Neurosci. Methods* **2009**, *182*, 6–16. [CrossRef] [PubMed]

67. Kim, S.; Bhandari, R.; Klein, M.; Negi, S.; Rieth, L.; Tathireddy, P.; Toepper, M.; Oppermann, H.; Solzbacher, F. Integrated wireless neural interface based on the utah electrode array. *Biomed. Microdevices* **2009**, *11*, 453–466. [CrossRef] [PubMed]

68. Rousche, P.J.; Pellinen, D.S.; Pivin, D.P.; Williams, J.C.; Vetter, R.J.; Kipke, D.R. Flexible polyimide-based intracortical electrode arrays with bioactive capability. *IEEE Trans. Biomed. Eng.* **2001**, *48*, 361–371. [CrossRef] [PubMed]

69. Viventi, J.; Kim, D.-H.; Vigeland, L.; Frechette, E.S.; Blanco, J.A.; Kim, Y.-S.; Avrin, A.E.; Tiruvadi, V.R.; Hwang, S.-W.; Vanleer, A.C.; et al. Flexible, foldable, actively multiplexed, high-density electrode array for mapping brain activity in vivo. *Nat. Neurosci.* **2011**, *14*, 1599–1605. [CrossRef] [PubMed]

70. Kang, B.; Hwang, H.; Lee, S.H.; Kang, J.Y.; Park, J.-H.; Seo, C.; Park, C. A wireless intraocular pressure sensor with variable inductance using a ferrite material. *J. Semicond. Technol. Sci.* **2013**, *13*, 355–360. [CrossRef]

71. Shin, K.-S.; Jang, C.-I.; Kim, M.J.; Yun, K.-S.; Park, K.H.; Kang, J.Y.; Lee, S.H. Development of novel implantable intraocular pressure sensors to enhance the performance in in vivo tests. *J. Microelectromech. Syst.* **2015**, *24*, 1896–1905. [CrossRef]

72. Chiu, Y.-Y.; Lin, W.-Y.; Wang, H.-Y.; Huang, S.-B.; Wu, M.-H. Development of a piezoelectric polyvinylidene fluoride (PVDF) polymer-based sensor patch for simultaneous heartbeat and respiration monitoring. *Sens. Actuators A Phys.* **2013**, *189*, 328–334. [CrossRef]

73. Li, C.; Wu, P.-M.; Shutter, L.A.; Narayan, R.K. Dual-mode operation of flexible piezoelectric polymer diaphragm for intracranial pressure measurement. *Appl. Phys. Lett.* **2010**, *96*, 053502. [CrossRef]

74. Sharma, T.; Aroom, K.; Naik, S.; Gill, B.; Zhang, J.X. Flexible thin-film PVDF-TRFE based pressure sensor for smart catheter applications. *Ann. Biomed. Eng.* **2013**, *41*, 744–751. [CrossRef] [PubMed]

75. Sharma, T.; Je, S.-S.; Gill, B.; Zhang, J.X. Patterning piezoelectric thin film PVDF–TRFE based pressure sensor for catheter application. *Sens. Actuators A Phys.* **2012**, *177*, 87–92. [CrossRef]

76. Jeong, J.; Lee, S.W.; Min, K.S.; Kim, S.J. A novel multilayered planar coil based on biocompatible liquid crystal polymer for chronic implantation. *Sens. Actuators A Phys.* **2013**, *197*, 38–46. [CrossRef]

77. Jeong, J.; Lee, S.W.; Min, K.S.; Shin, S.; Jun, S.B.; Kim, S.J. Liquid crystal polymer (LCP), an attractive substrate for retinal implant. *Sens. Mater.* **2012**, *24*, 189–203.

78. Lee, S.Y.; Park, K.-I.; Huh, C.; Koo, M.; Yoo, H.G.; Kim, S.; Ah, C.S.; Sung, G.Y.; Lee, K.J. Water-resistant flexible gan led on a liquid crystal polymer substrate for implantable biomedical applications. *Nano Energy* **2012**, *1*, 145–151. [CrossRef]

79. Wang, K.; Liu, C.-C.; Durand, D.M. Flexible nerve stimulation electrode with iridium oxide sputtered on liquid crystal polymer. *IEEE Trans. Biomed. Eng.* **2009**, *56*, 6–14. [CrossRef] [PubMed]

80. Chow, E.Y.; Chlebowski, A.L.; Irazoqui, P.P. A miniature-implantable RF-wireless active glaucoma intraocular pressure monitor. *IEEE Trans. Biomed. Circ. Syst.* **2010**, *4*, 340–349. [CrossRef] [PubMed]

81. Lee, D.-W.; Choi, Y.-S. A novel pressure sensor with a PDMS diaphragm. *Microelectron. Eng.* **2008**, *85*, 1054–1058. [CrossRef]

82. Lötters, J.C.; Olthuis, W.; Veltink, P.; Bergveld, P. The mechanical properties of the rubber elastic polymer polydimethylsiloxane for sensor applications. *J. Micromech. Microeng.* **1997**, *7*, 145–147. [CrossRef]

83. Whitesides, G.M. The origins and the future of microfluidics. *Nature* **2006**, *442*, 368–373. [CrossRef] [PubMed]

84. Larmagnac, A.; Eggenberger, S.; Janossy, H.; Vörös, J. Stretchable electronics based on Ag-PDMS composites. *Sci. Rep.* **2014**, *4*, 7254. [CrossRef] [PubMed]

85. Varel, Ç.; Shih, Y.-C.; Otis, B.P.; Shen, T.S.; Böhringer, K.F. A wireless intraocular pressure monitoring device with a solder-filled microchannel antenna. *J. Micromech. Microeng.* **2014**, *24*, 045012. [CrossRef]

86. Pal, R.K.; Pradhan, S.; Narayanan, L.; Yadavalli, V.K. Micropatterned conductive polymer biosensors on flexible pdms films. *Sens. Actuators B Chem.* **2018**, *259*, 498–504. [CrossRef]

87. Choi, S.; Lee, H.; Ghaffari, R.; Hyeon, T.; Kim, D.H. Recent advances in flexible and stretchable bio-electronic devices integrated with nanomaterials. *Adv. Mater.* **2016**, *28*, 4203–4218. [CrossRef] [PubMed]

88. Carl Ward, T.; Timothy Perry, J. Dynamic mechanical properties of medical grade silicone elastomer stored in simulated body fluids. *J. Biomed. Mater. Res. Part A* **1981**, *15*, 511–525. [CrossRef] [PubMed]

89. Puskas, J.E.; Chen, Y. Biomedical application of commercial polymers and novel polyisobutylene-based thermoplastic elastomers for soft tissue replacement. *Biomacromolecules* **2004**, *5*, 1141–1154. [CrossRef] [PubMed]

90. Hassler, C.; von Metzen, R.P.; Ruther, P.; Stieglitz, T. Characterization of Parylene C as an encapsulation material for implanted neural prostheses. *J. Biomed. Mater. Res. Part B Appl. Biomater.* **2010**, *93*, 266–274. [CrossRef] [PubMed]

91. Lecomte, A.; Degache, A.; Descamps, E.; Dahan, L.; Bergaud, C. In vitro and in vivo biostability assessment of chronically-implanted Parylene C neural sensors. *Sens. Actuators B Chem.* **2017**, *251*, 1001–1008. [CrossRef]

92. Liaw, D.-J.; Wang, K.-L.; Huang, Y.-C.; Lee, K.-R.; Lai, J.-Y.; Ha, C.-S. Advanced polyimide materials: Syntheses, physical properties and applications. *Prog. Polym. Sci.* **2012**, *37*, 907–974. [CrossRef]

93. Starr, P.; Bartels, K.; Agrawal, C.M.; Bailey, S. A thin-film pressure transducer for implantable and intravascular blood pressure sensing. *Sens. Actuators A Phys.* **2016**, *248*, 38–45. [CrossRef]

94. Hasenkamp, W.; Forchelet, D.; Pataky, K.; Villard, J.; Van Lintel, H.; Bertsch, A.; Wang, Q.; Renaud, P. Polyimide/su-8 catheter-tip mems gauge pressure sensor. *Biomed. Microdevices* **2012**, *14*, 819–828. [CrossRef] [PubMed]

95. Zhang, Q.; Li, H.; Poh, M.; Xia, F.; Cheng, Z.-Y.; Xu, H.; Huang, C. An all-organic composite actuator material with a high dielectric constant. *Nature* **2002**, *419*, 284–287. [CrossRef] [PubMed]

96. Xu, H.; Cheng, Z.-Y.; Olson, D.; Mai, T.; Zhang, Q.; Kavarnos, G. Ferroelectric and electromechanical properties of poly (vinylidene-fluoride–trifluoroethylene–chlorotrifluoroethylene) terpolymer. *Appl. Phys. Lett.* **2001**, *78*, 2360–2362. [CrossRef]

97. Wang, X.; Engel, J.; Liu, C. Liquid crystal polymer (LCP) for mems: Processes and applications. *J. Micromech. Microeng.* **2003**, *13*, 628–633. [CrossRef]

98. Kim, D.H.; Xiao, J.; Song, J.; Huang, Y.; Rogers, J.A. Stretchable, curvilinear electronics based on inorganic materials. *Adv. Mater.* **2010**, *22*, 2108–2124. [CrossRef] [PubMed]

99. Zhou, H.; Seo, J.-H.; Paskiewicz, D.M.; Zhu, Y.; Celler, G.K.; Voyles, P.M.; Zhou, W.; Lagally, M.G.; Ma, Z. Fast flexible electronics with strained silicon nanomembranes. *Sci. Rep.* **2013**, *3*, 1291. [CrossRef] [PubMed]

100. Zhang, K.; Seo, J.-H.; Zhou, W.; Ma, Z. Fast flexible electronics using transferrable silicon nanomembranes. *J. Phys. D Appl. Phys.* **2012**, *45*, 143001. [CrossRef]

101. Kim, D.-H.; Ghaffari, R.; Lu, N.; Rogers, J.A. Flexible and stretchable electronics for biointegrated devices. *Annu. Rev. Biomed. Eng.* **2012**, *14*, 113–128. [CrossRef] [PubMed]

102. Rogers, J.; Lagally, M.; Nuzzo, R. Synthesis, assembly and applications of semiconductor nanomembranes. *Nature* **2011**, *477*, 45–53. [CrossRef] [PubMed]

103. Viventi, J.; Kim, D.-H.; Moss, J.D.; Kim, Y.-S.; Blanco, J.A.; Annetta, N.; Hicks, A.; Xiao, J.; Huang, Y.; Callans, D.J.; et al. A conformal, bio-interfaced class of silicon electronics for mapping cardiac electrophysiology. *Sci. Transl. Med.* **2010**, *2*, 24ra22. [CrossRef] [PubMed]

104. Cheng, T.; Zhang, Y.; Lai, W.Y.; Huang, W. Stretchable thin-film electrodes for flexible electronics with high deformability and stretchability. *Adv. Mater.* **2015**, *27*, 3349–3376. [CrossRef] [PubMed]

105. Chung, H.J.; Sulkin, M.S.; Kim, J.S.; Goudeseune, C.; Chao, H.Y.; Song, J.W.; Yang, S.Y.; Hsu, Y.Y.; Ghaffari, R.; Efimov, I.R.; et al. Stretchable, multiplexed ph sensors with demonstrations on rabbit and human hearts undergoing ischemia. *Adv. Healthc. Mater.* **2014**, *3*, 59–68. [CrossRef] [PubMed]

106. Tan, M.J.; Owh, C.; Chee, P.L.; Kyaw, A.K.K.; Kai, D.; Loh, X.J. Biodegradable electronics: Cornerstone for sustainable electronics and transient applications. *J. Mater. Chem. C* **2016**, *4*, 5531–5558. [CrossRef]

107. Yin, L.; Cheng, H.; Mao, S.; Haasch, R.; Liu, Y.; Xie, X.; Hwang, S.W.; Jain, H.; Kang, S.K.; Su, Y. Dissolvable metals for transient electronics. *Adv. Funct. Mater.* **2014**, *24*, 645–658. [CrossRef]

108. Wang, H.; Shi, Z. In vitro biodegradation behavior of magnesium and magnesium alloy. *J. Biomed. Mater. Res. Part B Appl. Biomater.* **2011**, *98*, 203–209. [CrossRef] [PubMed]

109. Song, G. Control of biodegradation of biocompatable magnesium alloys. *Corros. Sci.* **2007**, *49*, 1696–1701. [CrossRef]

110. Hwang, S.-W.; Tao, H.; Kim, D.-H.; Cheng, H.; Song, J.-K.; Rill, E.; Brenckle, M.A.; Panilaitis, B.; Won, S.M.; Kim, Y.-S.; et al. A physically transient form of silicon electronics. *Science* **2012**, *337*, 1640–1644. [CrossRef] [PubMed]

111. Luo, M.; Martinez, A.W.; Song, C.; Herrault, F.; Allen, M.G. A microfabricated wireless RF pressure sensor made completely of biodegradable materials. *J. Microelectromech. Syst.* **2014**, *23*, 4–13. [CrossRef]

112. Kang, S.K.; Hwang, S.W.; Yu, S.; Seo, J.H.; Corbin, E.A.; Shin, J.; Wie, D.S.; Bashir, R.; Ma, Z.; Rogers, J.A. Biodegradable thin metal foils and spin-on glass materials for transient electronics. *Adv. Funct. Mater.* **2015**, *25*, 1789–1797. [CrossRef]

113. Curry, E.J.; Ke, K.; Chorsi, M.T.; Wrobel, K.S.; Miller, A.N.; Patel, A.; Kim, I.; Feng, J.; Yue, L.; Wu, Q.; et al. Biodegradable piezoelectric force sensor. *Proc. Natl. Acad. Sci. USA* **2018**, *115*, 909–914. [CrossRef] [PubMed]

114. Kang, S.-K.; Murphy, R.K.; Hwang, S.-W.; Lee, S.M.; Harburg, D.V.; Krueger, N.A.; Shin, J.; Gamble, P.; Cheng, H.; Yu, S. Bioresorbable silicon electronic sensors for the brain. *Nature* **2016**, *530*, 71–76. [CrossRef] [PubMed]

115. Yu, K.J.; Kuzum, D.; Hwang, S.-W.; Kim, B.H.; Juul, H.; Kim, N.H.; Won, S.M.; Chiang, K.; Trumpis, M.; Richardson, A.G.; et al. Bioresorbable silicon electronics for transient spatiotemporal mapping of electrical activity from the cerebral cortex. *Nat. Mater.* **2016**, *15*, 782–791. [CrossRef] [PubMed]

116. Boutry, C.M.; Nguyen, A.; Lawal, Q.O.; Chortos, A.; Bao, Z. Fully biodegradable pressure sensor, viscoelastic behavior of PGS dielectric elastomer upon degradation. In Proceedings of the 2015 IEEE SENSORS, Busan, Korea, 1–4 November 2015; pp. 1–4.

117. Kim, D.-H.; Kim, Y.-S.; Amsden, J.; Panilaitis, B.; Kaplan, D.L.; Omenetto, F.G.; Zakin, M.R.; Rogers, J.A. Silicon electronics on silk as a path to bioresorbable, implantable devices. *Appl. Phys. Lett.* **2009**, *95*, 133701. [CrossRef]

118. Pal, R.K.; Farghaly, A.A.; Wang, C.; Collinson, M.M.; Kundu, S.C.; Yadavalli, V.K. Conducting polymer-silk biocomposites for flexible and biodegradable electrochemical sensors. *Biosens. Bioelectron.* **2016**, *81*, 294–302. [CrossRef] [PubMed]

119. Tao, H.; Kaplan, D.L.; Omenetto, F.G. Silk materials—A road to sustainable high technology. *Adv. Mater.* **2012**, *24*, 2824–2837. [CrossRef] [PubMed]

120. Makadia, H.K.; Siegel, S.J. Poly lactic-co-glycolic acid (PLGA) as biodegradable controlled drug delivery carrier. *Polymers* **2011**, *3*, 1377–1397. [CrossRef] [PubMed]

121. Gentile, P.; Chiono, V.; Carmagnola, I.; Hatton, P.V. An overview of poly (lactic-co-glycolic) acid (PLGA)-based biomaterials for bone tissue engineering. *Int. J. Mol. Sci.* **2014**, *15*, 3640–3659. [CrossRef] [PubMed]

122. Hwang, S.W.; Song, J.K.; Huang, X.; Cheng, H.; Kang, S.K.; Kim, B.H.; Kim, J.H.; Yu, S.; Huang, Y.; Rogers, J.A. High-performance biodegradable/transient electronics on biodegradable polymers. *Adv. Mater.* **2014**, *26*, 3905–3911. [CrossRef] [PubMed]

123. Wang, Y.; Ameer, G.A.; Sheppard, B.J.; Langer, R. A tough biodegradable elastomer. *Nat. Biotechnol.* **2002**, *20*, 602–606. [CrossRef] [PubMed]

124. Boutry, C.M.; Nguyen, A.; Lawal, Q.O.; Chortos, A.; Rondeau-Gagné, S.; Bao, Z. A sensitive and biodegradable pressure sensor array for cardiovascular monitoring. *Adv. Mater.* **2015**, *27*, 6954–6961. [CrossRef] [PubMed]

125. Lewitus, D.; Smith, K.L.; Shain, W.; Kohn, J. Ultrafast resorbing polymers for use as carriers for cortical neural probes. *Acta Biomater.* **2011**, *7*, 2483–2491. [CrossRef] [PubMed]

126. Kim, D.-H.; Viventi, J.; Amsden, J.J.; Xiao, J.; Vigeland, L.; Kim, Y.-S.; Blanco, J.A.; Panilaitis, B.; Frechette, E.S.; Contreras, D.; et al. Dissolvable films of silk fibroin for ultrathin conformal bio-integrated electronics. *Nat. Mater.* **2010**, *9*, 511–517. [CrossRef] [PubMed]

Review

Flexible and Stretchable Bio-Integrated Electronics Based on Carbon Nanotube and Graphene

Taemin Kim †, Myeongki Cho † and Ki Jun Yu *

School of Electrical Engineering, Yonsei University, Seoul 03722, Korea; taeminkim@yonsei.ac.kr (T.K.);
mango_cho@yonsei.ac.kr (M.C.)
* Correspondence: kijunyu@yonsei.ac.kr; Tel.: +82-2-2123-2769
† These authors contributed equally to this work.

Received: 31 May 2018; Accepted: 6 July 2018; Published: 8 July 2018

Abstract: Scientific and engineering progress associated with increased interest in healthcare monitoring, therapy, and human-machine interfaces has rapidly accelerated the development of bio-integrated multifunctional devices. Recently, compensation for the cons of existing materials on electronics for health care systems has been provided by carbon-based nanomaterials. Due to their excellent mechanical and electrical properties, these materials provide benefits such as improved flexibility and stretchability for conformal integration with the soft, curvilinear surfaces of human tissues or organs, while maintaining their own unique functions. This review summarizes the most recent advanced biomedical devices and technologies based on two most popular carbon based materials, carbon nanotubes (CNTs) and graphene. In the beginning, we discuss the biocompatibility of CNTs and graphene by examining their cytotoxicity and/or detrimental effects on the human body for application to bioelectronics. Then, we scrutinize the various types of flexible and/or stretchable substrates that are integrated with CNTs and graphene for the construction of high-quality active electrode arrays and sensors. The convergence of these carbon-based materials and bioelectronics ensures scalability and cooperativity in various fields. Finally, future works with challenges are presented in bio-integrated electronic applications with these carbon-based materials.

Keywords: flexible electronics; carbon-based nano-materials; bio-integrated electronics

---

1. Introduction

Over the past few decades, commercial nanoscale electronic devices, based on semiconductor wafers composed of single-crystal inorganic materials, have achieved high performance with a significant development of fabrication processes. Despite these technological advances, devices with bulky, rigid, and planar materials have limitations for use as flexible or stretchable electronics [1–3] because of their poor mechanical properties, such as high Young's modulus and mechanical rigidity [4], which can easily induce device fracture when it comes under conditions of large mechanical deformation. To resolve this problem, the establishment of soft–hard integrated materials [5,6] or the integration of ultra-thin semiconducting materials with soft substrates are strongly suggested for flexible and stretchable electronics [7–10]. Recently, studies on flexible and stretchable electronic devices have demonstrated various electronic applications, such as passive/active circuit elements (e.g., resistor, diode, and transistor), sensors (e.g., strain, temperature, and electrochemical sensor), wireless radio-frequency (RF) communication components (e.g., capacitor, inductor, oscillator, and antenna), and light-emitting elements (e.g., organic light-emitting diode) on substrates with various materials [11–14]. Furthermore, the flexible or stretchable mechanical properties of these devices offer potentially unique opportunities in bio-electronic applications for health monitoring and treating diseases [15–21], mainly because of their ultrathin profiles that eliminate the mechanical mismatch between the tissue or organs and electronics. The tissues of the human body have soft, curvilinear surfaces and even have dynamic movements, such as the swelling and contraction of heart [22,23]. Flexible or stretchable electronics on polymer-based substrates [24–28] (e.g., parylene, SU8, and polyimide), which reduce the mismatch between the device and the organs, provide the possibility for conformal, intimate integration with tissues for advanced healthcare from the measurement of electrophysiological signals [29], delivery of drugs or stimulations [30], to well-established human-machine interfaces [21,31]. Recently, carbon-based nanomaterials, such as graphene and carbon nanotubes (CNTs), have been applied in the field of bioelectronics because of their extraordinary mechanical properties, and unique electrical and optical properties based on their inherent geometric structures [32–35]. Both graphene and CNTs show similar properties, except for their physical structures. Particularly, the extraordinary mechanical properties of these materials, such as an approximately 1 TPa Young's modulus and 25% fracture strain [36,37], along with their subnanometer thicknesses are highly advantageous when manufacturing flexible and stretchable bioelectronics for conformal contact with tissues. Furthermore, graphene and CNTs are suitable materials for use in bioelectronics, because they have high stability and low biofouling in biological environments as well as high reproducibility with direct growth [38–42]. In this review, we introduce some of the key flexible electronics with carbon-based materials that can be utilized not only for intrinsically conformal contact with tissues, but also for biocompatibility in physiological environments for applications in bio-integrated electronics.

2. Biocompatibility

Using solely biocompatible materials is one of the most significant undertakings in biomedical systems, especially for epidermal electronics [5,11] and implantable electronics [15,17,20,43,44]. Before soft materials are integrated into existing or new novel bioelectronics, toxicity, such as cytotoxicity, and the immune system response to these materials must be precisely examined. Previous studies about the toxicity of carbon nanomaterials in the human body have been reported [45–48]. However, the effects on the long-term usage of these materials in biological environments has not been clearly carried out yet. Therefore, we are still required to understand the potential toxicity of carbon nanomaterials for long-term implants.

Over the last few decades, carbon-based materials such as pyrolytic carbon (PC) and diamond-like carbon (DLC) have been used in the biomedical field. PC has been used in biomedical implants and coating materials, especially in the production of heart valve prostheses [49]. Most PCs are anisotropic in nature. By advancing the manufacturing using chemical vapor deposition (CVD) processes, isotropic PC coatings on heart valves have been achieved [50]. PC heart valves show excellent biocompatibility, including good blood compatibility, good adhesion to endothelial cells, and minimal adhesion to inactive platelets [51–53]. More recently, DLC has been developed more than PC as a biomedical carbon-based material. Via postprocessing DLC coating, properties such as hardness, surface lubrication, and insolubility can be obtained through carbon-based nanomaterials, along with biocompatibility [54–57]. Conventional carbon-based biomedical materials have been successfully commercialized because of their superior biocompatibility. Herein, we discuss the specific biocompatibility of graphene and carbon nanotubes, which can be extensively used as devices in biomedical engineering or human healthcare systems.

CNTs have always been controversial in terms of biocompatibility. These conflicting arguments are influenced by the effects of the impurities in CNTs, the length of CNTs, surface chemistry, dispersion interaction between various factors, and experimental variables, including different doses of CNTs, cell population, and evaluation systems [58]. CNTs are a type of nanoparticle that is detrimental to the body when inserted into the body, because of their high surface area and surface toxicity. Recent studies on CNT nanoparticles associated with biocompatible materials address the toxicity in lungs, skin irritation, and cytotoxicity. Asbestosis, lung cancer, and pleural mesothelioma can be triggered by CNT inhalation [59–61]. As the CNT resembles asbestos, which has the fibrous structural characteristics, CNTs are regionally deposited in the lung and cause the diseases [62,63]. Brown et al.

reported that the release of the pro-inflammatory cytokine tumor necrosis factor (TNF-α) and the reactive oxygen species is increased in the monocytic cells exposed to nanotubes, and frustrated phagocytosis is observed, which could inhibit the elimination of nanotubes from the lungs by macrophage [64]. However, the pathway of CNT nanoparticles into the lungs without inhalation has not been clearly eluded. There are previous studies on skin irritation caused by CNT [65,66], however, Ema et al. evaluated dermal and eye irritation and skin sensitization experiments in rabbits and guinea pigs and demonstrated Nikkiso-single-walled carbon nanotubes (SWCNTs), super-growth SWCNTs, and the Mitsui product of multi-walled carbon nanotubes (MWCNTs) were not irritants to the skin or eyes [67]. The cytotoxicity of CNT has been more broadly researched. In the past, the cytotoxicity of CNTs in mammalian cells has been examined in various types of in vitro cells [65,66,68,69]. In contrast, some reports have determined noncytotoxic CNT in mammalian cells [70–73]. Depending on the degrees of aggregation of the CNTs, the physical shapes and surface areas of CNTs can be formed differently, which leads to different effects of cytotoxicity. Belyanskaya et al. investigated the effects of differently agglomerated SWCNTs on neural cells from the central and peripheral nervous system of the chicken embryo. The results showed that the more agglomerated SWCNTs are the more toxic, by decreasing the overall DNA content [74]. Wick et al. demonstrated that well-dispersed CNTs are less cytotoxic, and rope-like agglomerated CNTs are more cytotoxic than asbestos on human MSTO-211H cells [75]. On the other hand, however, Mutlu et al. reported that highly dispersed SWCNTs show nontoxicity in vitro or in vivo [76]. Also, Kolosnjaj-Tabi et al. demonstrated no granuloma formation when the length of the CNTs, which are intraperitoneal injections, of mice was under 10μm [77]. On the side of the CNTs over the substrate, not the particles or fiber, there are many reports about CNT-based biomedical materials with their acceptable biocompatibility. Elias et al. examined carbon nanofiber compacts for biomaterials because of their unique adhesion with bone-forming osteoblast cells, and determined the biocompatibility for their usage as orthopedic or dental implantable materials [78]. Similarly, Lobo et al. demonstrated the biocompatibility of vertically aligned MWCNT films by culturing L929 mouse fibroblast cells [79]. The cell adhesion and morphology of the L929 mouse fibroblast cells were observed after seven days using scanning electron microscopy. The cells from the first layer spread out and survived for seven days by clearly showing a good cytocompatibility of aligned MWCNT to the fibroblast cells. In neuronal growth studies, chemically functionalized multi-walled carbon nanotubes (MWCNTs), especially neurons grown on MWNTs with positive charge, showed the excellent neurite outgrowth and good biocompatibility [80].

The chemical composition and crystalline structure of both graphene and CNTs are similar. However, some properties of these two materials are different, because the two-dimensional (2D) structure of graphene is fundamentally different from the cylindrical structure of CNT. These materials have different mechanisms of interaction with cell systems, resulting in definite differences in the toxicity of living cells between graphene and CNT. Therefore, the biocompatibility of graphene should be carefully examined separately from CNTs. Recent studies on dispersed graphene flakes using a liquid-phase exfoliation (LPE) [81] of graphite to obtain biocompatible graphene flakes, have been reported [82–84]. Castagnola et al. used a complete serum to disperse graphene for the exploration of biological interactions and resolved the protein composition on the dispersed graphene nanoflakes [85]. The results showed that most of the proteins, except for apolipoprotein B100, in a complete serum have good affinity, which proves the biological identity of the graphene nanoflakes. However, in a real complex biological environment, the evaluation of an interaction between the dispersed graphene flakes and proteins is complicated and should be considered more carefully. On the other hand, some research repeatedly reported that the graphene sheet interacts with protein molecules and causes a disruption to the structure and function of proteins. The graphene sheet disrupted the α-helical structures of the peptides by interacting with the villin headpiece (HP35) protein [86,87]. The distortion in the villin headpiece (HP35) proteins proceeds with almost losing the α-helical elements, which are adsorbed to the graphene surface, because of the π–π stacking interaction between the graphene lattice and the aromatic residues. Zhao et al. demonstrated that graphene also disrupted the double stranded DNA segments [88]. In an aqueous solution, the self-assembly phenomenon between the DNA segments with graphene were observed. Some of the DNA are vertically standing while others are horizontally laying on the graphene surface. The latter case of the DNA segments, which were damaged because of several mutations, leads to DNA toxicity. There are studies of graphene with regards to biocompatibility. Sahni et al. presented a great biocompatibility of graphene by culturing rat neuronal cells grown on graphene-coated surfaces [89]. The results verified the biocompatibility of graphene surfaces with direct neuronal interfaces. As another example, Kalbacova et al. demonstrated that human osteoblasts and mesenchymal stromal cells can be cultured on monolayer graphene films [90]. The results showed that, compared to that of cells that were cultured on a silicon oxide substrate, the adherence and proliferation of the cells cultured on the graphene were better, and no cytotoxicity for human osteoblasts and mesenchymal stromal cells was demonstrated. Furthermore, Li et al. also successfully cultured neural stem cells on three-dimensional graphene foams (3D-GFs) [91]. The cytotoxicity of the 3D-GFs was estimated by calcein acetoxymethyl ester (calcein-AM) and EthD-I staining assays, with a 2D graphene film as the control. There was no difference in the cell viability between the 3D-GFs and 2D graphene film five days after culturing without abnormal cell apoptosis on 3D-GFs. In addition, Bendali et al. demonstrated the excellent cytocompatibility of single-layer graphene by directly growing primary adult retinal ganglion cells on a graphene surface without any peptide coating [92]. This work differs from the existing studies in its use of purified adult and differentiated retinal neurons instead of stem cells.

3. CNT/Graphene Based Microelectrodes for Neural Interface

3.1. CNT-Based Microelectrodes

CNTs are considered promising materials for flexible neural interfaces because of their high conductivity and mechanical strength. The electrophysiological signals such as electroencephalogram (EEG) or electrocorticogram (ECoG) are directly affected by the impedance of the electrodes, which lead to the extremely low current injected into the electrodes [93]. High impedance gives rise to more interference of the thermal noise that, as a result, decreases the sensitivity towards the signal. Hence, lowering the impedance by increasing the effective contact area of the electrodes has been prime importance in order to measure the more reliable action potentials with minimal noise. However, the low impedance and high spatial resolution of neural electrodes are in a tradeoff relationship. As the ultimate goal of the neural interfaces is to selectively measure the action potential of every single neuron, increasing the size of the electrodes is contrary to this goal. Therefore, it is attracting great attention as a method to increase the sensitivity while maintaining a high selectivity, by replacing conventional planar electrodes with a variety of nanomaterials or by making nanostructured surfaces using such nanomaterials [94–96] to achieve a higher contact areal-density in a unit area.

CNT-based microelectrodes have much higher surface areas than those of flat surface electrodes in the same planar surface as a result of the unique one-dimensional (1D) geometry of CNTs (Figure 1), which gives three-dimensional (3D) space to the surface. This high surface area combined with the inherently outstanding conductivity and stability [97–99] makes CNTs promising as active nanomaterials for highly sensitive neural interfaces. Jan et al. coated three PtIr electrodes with MWCNT, synthesized by layer-by-layer (LBL) deposition, poly(3,4-ethylenedioxythiophene) (PEDOT), and IrOx, respectively, and compared their electrochemical properties to electrochemical impedance spectroscopy (EIS) and cyclic voltammetry (CV) [100]. As shown in the SEM image and the EIS graph, in Figure 1a,b, the roughness of the surface decreases in the order of MWCNT, PEDOT, and IrOx, and the impedance increases in the same order. The thickness of the MWCNT LBL coating increases by 7 nm with the addition of bilayers, resulting in 700 nm at 100 LBL. At a 700 nm coating thicknesses among the various electrodes, the MWCNT LBL electrodes show 30% and 60% lower impedances than those of the PEDOT and IrOx electrodes, respectively, at 1 kHz.

Similar to CNTs, PEDOT, well known as a representative conducting polymer, has also been regarded as a biocompatible sensing material that provides improved selectivity and enables further surface modification [101]. The electrochemical properties of PEDOT-coated electrodes can be further enhanced when fabricated with CNTs, while maintaining high mechanical flexibility. Gerwig et al. demonstrated that the surface roughness of the PEDOT-CNT composites was higher than that of the pure PEDOT, because of the porous morphology (Figure 1c) [102]. This increased surface roughness, as well as the CNTs acting as a conducting network, provides improved impedance spectra. Figure 1d shows an additional example of a CNT coating, where the CNTs are mixed with conventional metal particles rather than soft materials, so as to increase the effective surface area [103]. Compared with the bare Au contacts, the CNT-Au composite contacts offered lower electrochemical impedance over a wide band of the frequency range from 10 Hz to 100 kHz; in particular, the impedance was 18 times lower at 1 kHz. The ultrathin neural probe with these CNT-Au composite electrodes successfully recorded the neural activity from the rat's hippocampus CA1 area, with a high signal-to-noise ratio (SNR).

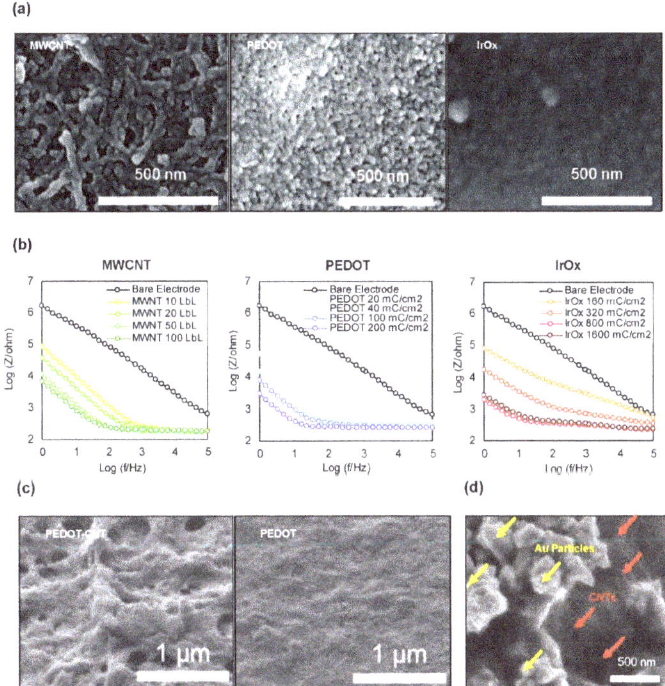

Figure 1. Surface texturing by coating with nanomaterials to lower the electrochemical impedance: (**a**) SEM images of PtIr electrodes coated with multi-walled carbon nanotubes (MWCNTs) (left), poly(3,4-ethylenedioxythiophene) (PEDOT) (middle), and IrOx (right); (**b**) electrochemical impedance spectroscopy (EIS) of three surface-modified electrodes with the increasing film thicknesses over a frequency range of 1–105 Hz. Reproduced with permission from reference [100], Copyright 2009, American Chemical Society; (**c**) SEM images to show the surface morphology of the PEDOT-carbon nanotubes (CNTs) composite (left) and pure PEDOT (right) coatings. Reproduced with permission from reference Gerwig et al. [102], Copyright 2012, Frontiers; (**d**) SEM image of an electrodeposited Au-CNT composite network. Reproduced with permission from reference Xiang et al. [103], copyright 2014, IOP Publishing Ltd.

Electrophysiological signals recorded using novel microelectrodes provide an excellent temporal resolution, but the optical imaging of neural circuits must be followed to achieve both high temporal and spatial resolution [104–106]. Furthermore, unlike electrical stimulation, the advent of optogenetics, which can simultaneously perform electrophysiological readout and stimulation, has led to additional demand for transparent devices [107,108]. However, conventional opaque electrodes and interconnects preclude the optical imaging and stimulation of tissues just below the site, where those components are located, and produce light-induced artifacts that contaminate the electrical recording [109]. These challenges limit the use of metal electrodes, such as Pt or stainless steel, which exhibit higher electrical conductivity than that of CNTs and graphene. Even though indium-tin oxide (ITO) is widely used as a transparent conductor because of its high transmittance in the visible light region, it is difficult to use in bio-integrated electronics, because of its poor mechanical flexibility, which induces fractures under large deformations in the neural surface.

Here, we introduce recent advances in transparent and flexible microelectrode arrays, based on CNTs and graphene for simultaneous electrical and optical interaction with brain tissues. Figure 2a shows an example of the transparent and stretchable microelectrode array from CNT thin films fabricated by Zhang et al. for optogenetic stimulations with simultaneous ECoG recording [110]. The stretchable transparent electrode array with these films integrates an elastic PDMS substrate, CNT electrodes and interconnects, and SU-8 insulation layer, retaining a high optical transmittance across a broad wavelength range of 400 nm to 2.5 μm. The CNT web-like thin film was produced through CVD, while the solvent-induced condensation process created a void area between the tubes, contributing to the high optical transparency under mechanical strain (Figure 2b) [111]. Furthermore, as the percolation network structures maintain high electrical conductivity even under large mechanical strain, the device shows considerably less impedance change than that of the graphene electrode, even at higher strains (Figure 2c). The cyclic deformation test demonstrates the possibility of long-term electrophysiological recording, because there is no significant change in impedance with up to 6000 cycles of the mechanical fatigue test under 20% tensile or compressive strain. (Figure 2d). With these excellent optical, electrical, and mechanical properties, the in vivo optogenetic stimulation and corresponding neural recordings from Thy1-ChR2-YFP mice were concurrently taken using the transparent electrode arrays. Figure 2e shows the larger amplitude of optical evoked action potentials, with a higher intensity and duration of stimulus by blue laser illumination. The light-induced artifact comparison of the CNT electrode and Au control electrode in saline demonstrates that the potentials induced by the laser stimuli are not light-induced artifacts but evoked potentials by depolarization (Figure 2f).

3.2. Graphene-Based Microelectrodes

Graphene is another candidate for a transparent flexible electrodes arrays because of its ultrathin geometry with a unique two-dimensional structure and mechanical flexibility. Kuzum et al. presented a transparent, flexible neural interface for simultaneous ECoG and calcium imaging, by transferring CVD-grown graphene electrodes on a polyimide substrate (Figure 3a) [38]. The transparent graphene electrode does not act as an image artifact, blocking the fluorescence of the dentate gyrus in a calcium indicator-stained hippocampal slice when it is excited by a confocal microscope. As a result, the calcium transient mapping for the six randomly selected cells within an electrode site shows a remarkable coincidence with the recorded interictal-like event from the electrode, as shown in Figure 3b–d. The advantage of this single-cell resolution of calcium imaging, combined with the advantage of the temporal resolution of graphene microelectrodes, provides extensive information on neural activities.

The development of transparent graphene electrode arrays enables the optogenetic stimulation of the underlying brain tissue, as in the case of CNTs, as previously described. Park et al. developed a fully transparent and flexible graphene-based carbon-layered electrode array (CLEAR) by coating the graphene electrodes with Parylene C films (Figure 3e) [112]. The fabrication of CLEAR involves the deposition of Au/Cr traces, in addition to the four monolayers graphene. However, the use of these metals is only for the interconnection between the graphene electrodes and the printed circuit board, so it does not affect the transparency of the electrode sites. The transparent CLEAR device can be implanted on the cerebral cortex of a Thy1:ChR2 mouse for optogenetic stimulation (Figure 3f). As shown in Figure 3g, the optically evoked potentials show the increase of amplitudes with an increase in intensity of the laser stimuli, and are divided into an initial peak and a following longer peak. The control experiment, in which mice were killed after the original illumination experiment, revealed that the initial peak was a light-induced artifact and that the second peak was an evoked action potential.

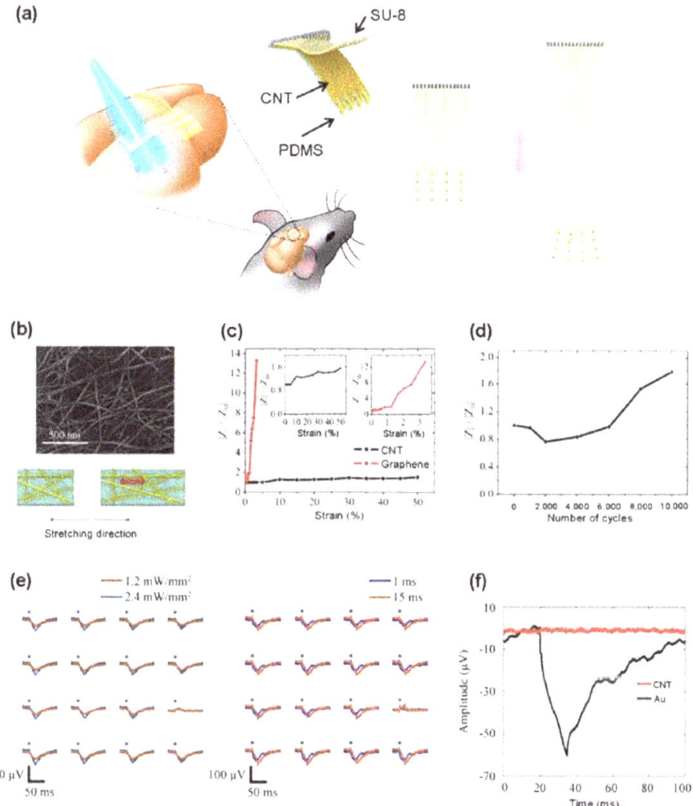

Figure 2. CNT-based transparent electrode arrays for simultaneous electrical recording and optical stimulation: (**a**) schematic illustrations of the stretchable transparent CNT electrode array and its application for optogenetics; (**b**) an SEM image of the CNT network thin film (top) and a schematic illustration of the CNT network thin film under stretching (bottom); (**c**) the strain-dependent impedance change at 1 kHz of the CNT and graphene electrodes under the applied tensile strain; (**d**) impedance change at 1 kHz of a CNT electrode during 10,000 stretching cycles with 20% strain; (**e**) the mapping of evoked potentials after optical stimulations under different stimulus intensity (left) and duration (right). Blue dots indicate the stimulus time points; and (**f**) the light-induced artifacts in CNT and Au electrodes under a photostimulus (2.4 mW/mm², 15 ms). The gray box indicates the duration of stimulation pulses. Reproduced with permission from Zhang et al. [110], copyright 2018, American Chemical Society.

Microelectrodes are capable of not only recording the neural signals, but also exciting or inhibiting the action potentials by applying an electrical stimulation to neurons [113]. As the electrical current is injected into the neural tissue from the electrodes, the polarity of the neuron membrane changes, which evokes action potentials. There are two mechanisms by which electrical charges are transferred from electrodes to tissue, the capacitive mechanism and the faradaic mechanism [114]. The capacitive mechanism is safer than the faradaic injection is, in that it stimulates by charging and discharging the surface of the electrode without directly injecting charge into the tissue. Park et al. developed another CLEAR transparent electrodes that allow the electrical stimulation current to be delivered to the cortex with a capacitive mechanism, and simultaneously enables optical in vivo monitoring (Figure 3h,i) [115]. The authors compared the capabilities of the CLEAR device and the platinum electrode array during electrical stimulation using transgenic mice with the novel calcium indicator GCaMP6f [89,116]. In the case of CLEAR device, the fluorescence evoked by the electrical stimulation is clearly visible, whereas in the platinum arrays, it is obstructed by the opaque electrodes and interconnects (Figure 3j,k). Figure 3l,m shows the graphical representations of the fluorescence intensity corresponding to the CLEAR device and platinum array, respectively. As the peak response of fluorescence occurs at the electrode site where the electrical pulse is transmitted, no peak response can be observed at the platinum fluorescence intensity.

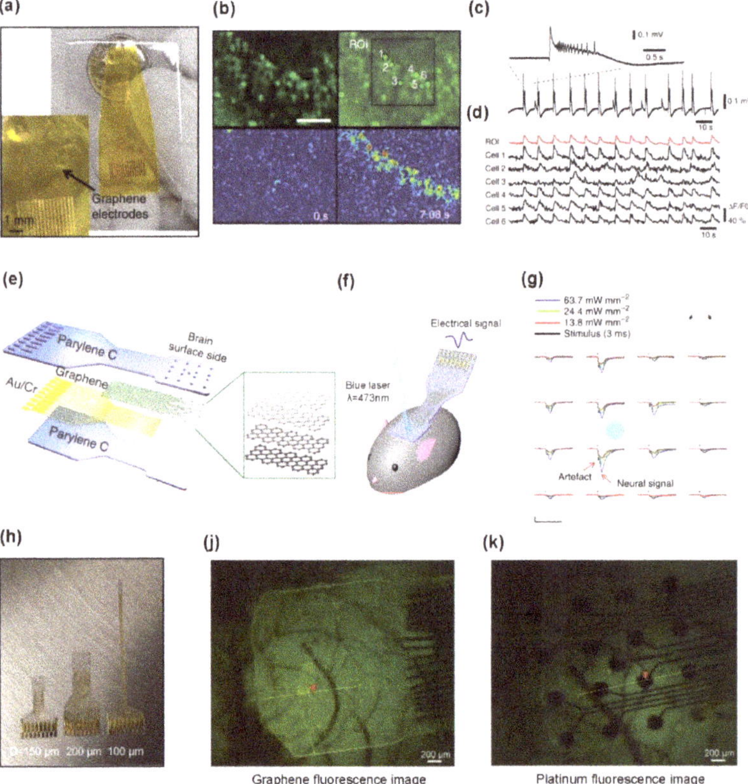

Figure 3. *Cont.*

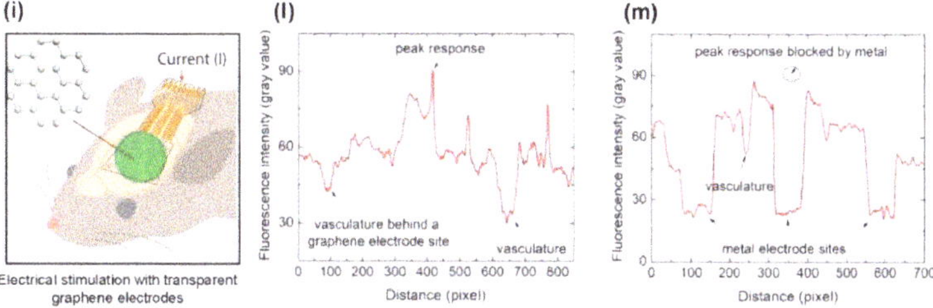

Figure 3. Graphene-based transparent devices for optical imaging and optogenetics: (**a**) photograph of a 16-electrode transparent electrocorticogram (ECoG) array. The electrode size is 300 × 300 μm^2 each; (**b**) top left: a steady-state fluorescence image of the dentate gyrus in an OBG-1 AM-stained hippocampal slice, where the blue laser and fluorescence emission penetrated through the transparent graphene electrode. Top right: region of interest (ROI) and six random cells within the electrode. Bottom: color-coded images of normalized fluorescence change (ΔF/F0) at the baseline (0 s, left) and at peak response (7.08 s, right); (**c**) time-dependent electrophysiological recording from graphene electrode demonstrates interictal-like activity; (**d**) time-dependent calcium transient (ΔF/F0) for the six cells within the ROI in b. Increases in calcium are generally consistent with interictal-like events in c. Reproduced with permission from reference Kuzum et al. [38], copyright 2014, Macmillan Publishers Limited; (**e**) schematic diagram of carbon-layered electrode array (CLEAR) device consisting of the layered structures; (**f**) schematic illustration of optogenetic testing, where the CLEAR device was implanted on the cortex of a Thyl:ChR2 mouse, with an optical fiber delivering blue laser stimuli to the neural cells; (**g**) average optical evoked responses recorded by the CLEAR device. The x-scale bar, 50 ms; y-scale bar, 100 μV. Reproduced with permission from reference Park et al. [112], copyright 2014, Macmillan Publishers Limited; (**h**) optical camera image of three types of graphene electrode arrays of different diameters (100, 150, and 200 μm); (**i**) schematic drawing of graphene μECoG electrodes implantation and electrical stimulation in GCaMP6f mice. Fluorescence images after the electrical stimulation delivered to the cortex through a single graphene electrode (**j**) and a single platinum electrode (**k**); the stimulation site is marked with a red triangle. Fluorescence intensity over (**l**) graphene electrodes (data from yellow line in j) and (**m**) platinum electrodes (data from the yellow line in k). Reproduced with permission from reference Park et al. [115], copyright 2018, American Chemical Society.

4. CNT/Graphene Based Sensors

4.1. Field-Effect Transistors (FET) for Biosensors

4.1.1. CNT Field-Effect Transistors

Since the first demonstration by the Dekker group in 1998 [117], semiconducting single-walled carbon nanotubes (SWCNTs) have been broadly used as channel layers for field effect transistors (FETs). The SWCNT channel has mobility tunable by varying the arrangement of each tube. Kocabas et al. showed that the mobility of a randomly scattered SWCNT network (~10 cm^2/V·s) is much lower than that of a perfectly aligned SWCNT in one direction (~1000 cm^2/V·s) [118,119]. Although several postsynthesis strategies aligning SWCNTs in one direction with electric fields [120], magnetic fields [121], or mechanical force [122] have been introduced, the alignment during the growth of the CNTs through the CVD process is the most efficient in terms of fabrication simplicity and high quality of the product [123].

The SWCNT FETs have been used extensively for label-free biosensors because the small diameters of SWCNTs match well with the size of biomolecules, such as protein, nucleic acid, antigen, and bacteria [124,125]. A great deal of research has been done on SWCNTs as biosensors. In general, they are proposed to sense biomolecules by two major mechanisms, electrostatic gating and Schottky barrier effects [126–128]. However, the fact that these two mechanisms have different gate-voltage dependencies makes it difficult to choose the appropriate gate-voltage. Thus, the Dekker group presented the metal passivation strategy, which inhibits the Schottky barrier effect, making electrostatic gating the dominant mechanism. A more detailed introduction to the mechanisms of biosensing with SWCNT FETs can be found in a report from the Dekker group [129].

Although the hydrophobic nature of the SWCNT surface induces the nonspecific bonding of biomolecules, surface modification through biofunctionalization can lead to selectively immobilizing biomolecules for detection [130]. For example, Kim et al. employed the SWCNT FETs functionalized with various ratios of linkers to spacers for the detection of a prostate cancer marker (PSA-ACT complex) [131]. In this way, they successfully lowered the detection limit to 1.0 ng/mL, without labeling the marker protein. In addition to detecting proteins, Star et al. developed a covalently attached complementary DNA strand (cDNA) on the surface of SWCNT FETs to bind the target single-stranded DNA, resulting in DNA adsorption and hybridization [132]. More information on SWCNT FET biosensors can be found in other reviews [133–135].

4.1.2. Graphene Field-Effect Transistors

Among the exceptional properties of graphene, low mechanical stiffness, high optical transparency, and high electrical conductivity are the most noticeable. As the electrical properties of the CNT, in the previous section, rely on the number of layers in the tube, those of graphene also show a similar dependency. The conductivity of graphene decreases gradually with the increasing number of layers by forming graphite. [136]. Bolotin et al. showed that the carrier mobility of a freely suspended monolayer of graphene exceeded 200,000 $cm^2/V \cdot s$ [137]. In graphene, carriers move as though they are massless fermions. This phenomenon results in a high mobility of graphene that allows significant interest for the applications in next-generation high-speed field-effect transistors [138]. However, although graphene has a high mobility that outperforms other semiconducting materials, the absence of a bandgap in the pristine graphene layer hinders its applications as a semiconductor. In the energy band structure of graphene, there is a region where the valence band and conduction band overlap, but as density of state is zero at that point, the undoped graphene is classified as a semimetal rather than a semiconductor [139]. To this end, various strategies for controlling the bandgap of graphene have been studied and can be found in several review papers [140–142].

Like SWCNTs, graphene is very promising material for an active biosensor to detect target biomolecules, because of its combination of properties, including a high conductivity and large surface area. Ohno et al. introduced an FET that uses a nonfunctionalized single layer of graphene as a channel for detecting electrolyte pH and protein concentrations [143]. The conductance of their graphene-based FET (GFET) shows a linear pH dependency and an increasing inclination towards protein adsorption, up to the several hundred picomolar level. Surface functionalization methods involving covalent or noncovalent interactions for SWCNT sensors can also be applied to graphene sensors for the selective adsorption of biomolecules [144]. Mohanty and Berry reported a label-free ssDNA detector using chemically modified graphene with controllable functionality [145]. They suggested that by manipulating the surface functionalization, the sensitivity and polarity of the GFET device can be modified. Another CVD-grown n-doped GFET was fabricated by Dong et al., to detect ssDNA with a high sensitivity of 0.1 nM [146]. Moreover, the outstanding mechanical flexibility of graphene offers potential applications in wearable or implantable electronics. The CVD-grown GFET produced by Kwak et al. was implemented as a flexible glucose sensor by integrating GFET on polyethylene terephthalate (PET) as a substrate [147]. The glucose detection range of the fabricated GFET is 3.3–10.9 mM, which is comparable to that of the conventional screen test used to diagnose diabetes. A perfectly flexible FET sensor based on reduced graphene oxide (rGO) was developed by He et al. to detect fibronectin at a concentration as low as 0.5 mM [148].

4.2. Flexible Sensors for Wearable Devices

The CNT or graphene thin films can be assembled with a soft elastomeric substrate, such as PDMS or Ecoflex, to form stretchable and transparent electronics with various functions [149,150]. These carbon-based stretchable devices can be utilized as wearable or implantable bio-integrated sensors, because they can provide conformal contact to the curvilinear shapes of human body. The most important performance factor of a wearable sensor is its sensitivity. Particularly, in the case of a strain sensor, the gauge factor (GF) determines the sensitivity. A variety of elastomeric composites of metals, semiconductors, and carbon nanostructures showed an improved GF value (>15) and stretchability [151–153]. However, optical transparency is another important feature of wearable sensors, to enhance the aesthetic effect. For this reason, CNTs and graphene-based wearable sensors with high GF, stretchability, and considerable transparency are attracting attentions. Roh et al. fabricated a transparent and patchable strain sensor by stacking a nanohybrid film of SWCNT and an elastomeric composite of polyurethane (PU)-poly(3,4-ethylenedioxythiophene) polystyrenesulfonate (PEDOT:PSS) on a PDMS substrate (Figure 4a) [154]. Their sandwich-like piezoresistive strain sensor, in which SWCNT nanofillers are confined in a conductive elastomeric matrix, have a gauge factor of 62, which outperforms conventional metallic or carbon-based strain sensors [155–157]. These highly sensitive strain sensors can be conformably attached to several parts of the face, sensing the change of facial expression due to muscle movement or wrinkling of the skin (Figure 4b). The strain sensors onto the forehead and skin near the mouth clearly distinguish between the subject's laughing and crying. Normally, when a person laughs, the movement of the mouth is larger than that of the other facial regions, while the movement of the forehead becomes remarkable when crying. Figure 4c–f demonstrates the quantified results from the resistance change of the strain sensors attached on forehead and near the mouth.

Capacitive touch sensors are sometimes preferred over resistive sensors, because of their capabilities of multitouch sensing. Kang et al. presented a graphene-based stretchable touch sensor capable of multitouch sensing and three-dimensional sensing for the recognition of 3D shapes [158]. These touch sensors consist of graphene top and bottom electrodes transferred onto an ultrathin polyethylene terephthalate (PET) substrate, an intermediate dielectric layer, and a monolayer graphene for bottom grounding (Figure 5a). The completed sensor array has a transmittance of more than 80% in the visible range, and its thin structure and stretchability allow conformal contact to the curved surface (Figure 5b). The stretchability is further improved when the auxetic mesh-like structure is applied to the device [159]. The touch sensor array of the auxetic structure was attached to the palm with conformal contacts and used to remote control the movement of a toy car through simple finger motions (Figure 5c,d).

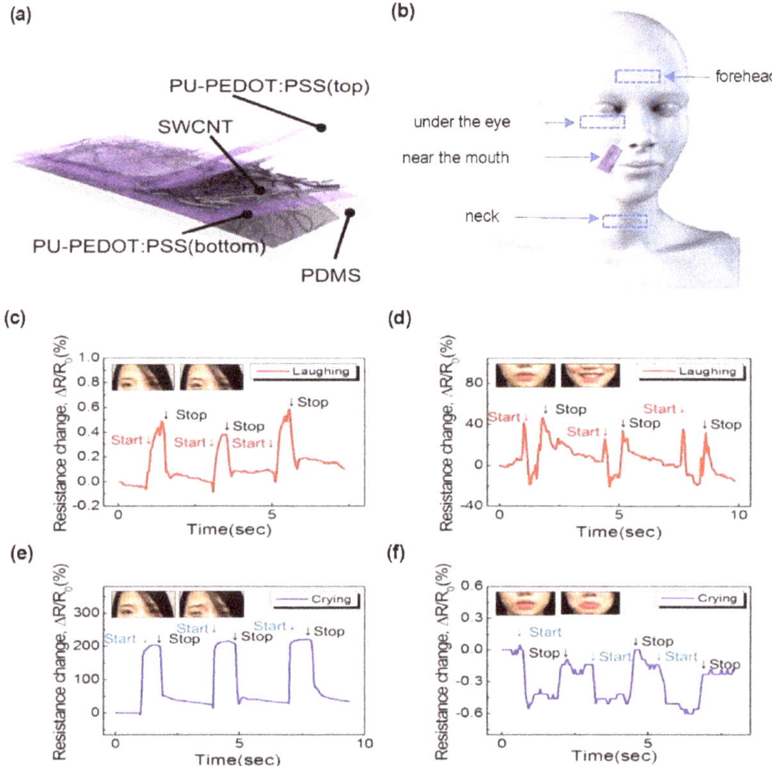

Figure 4. CNT-based strain sensor for facial expression recognition: (**a**) schematic illustration of the cross-section of the strain sensor showing the three-layer stacked structure of the polyurethane poly(3,4-ethylenedioxythiophene) polystyrenesulfonate (PU-PEDOT:PSS)/single-walled carbon nanotubes (SWCNTs)/PU-PEDOT:PSS composite elastomer on a PDMS substrate; (**b**) schematic illustration of stretchable transparent strain sensors attached to four different facial regions to sense skin strains by muscle movements during the expression of emotions; (**c**–**f**) time-dependent ΔR/R0 responses of the sensor attached to (**c**) forehead and (**d**) skin near the mouth with the subject laughing, and of the sensor attached on the (**e**) forehead and (**f**) skin near the mouth with the subject crying. Reproduced with permission from Roh et al. [154], copyright 2015, American Chemical Society.

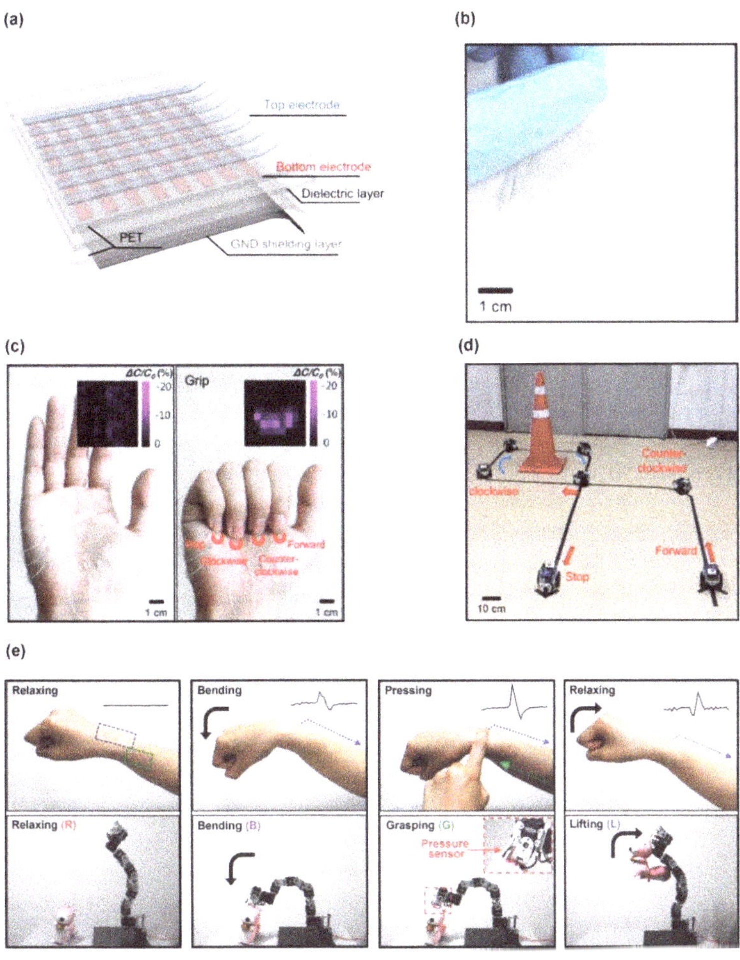

Figure 5. Graphene-based wearable sensors for a human-machine interface: (**a**) schematic illustration of a graphene-based capacitive sensor consisting of two graphene electrodes layers and an intermediate dielectric layer; (**b**) optical camera image of 64-channel touch sensor array; (**c**) images of stretchable touch sensor mounted on a palm for remote control application. Inset: capacitance changes for spread (left) and grip (right) status of the palm; and (**d**) images of a toy car operated by the stretchable remote control in c. Reproduced with permission from Kang et al. [158], copyright 2017, American Chemical Society. (**e**) The transparent motion sensor and the electrotactile stimulator attached on the wrist and forearm, respectively. The bending, pressing, and relaxing of the wrist lead to the bending, grasping, and lifting of the robot arm, respectively. Reproduced with permission from Lim et al. [160], copyright 2014, WILEY-VCH.

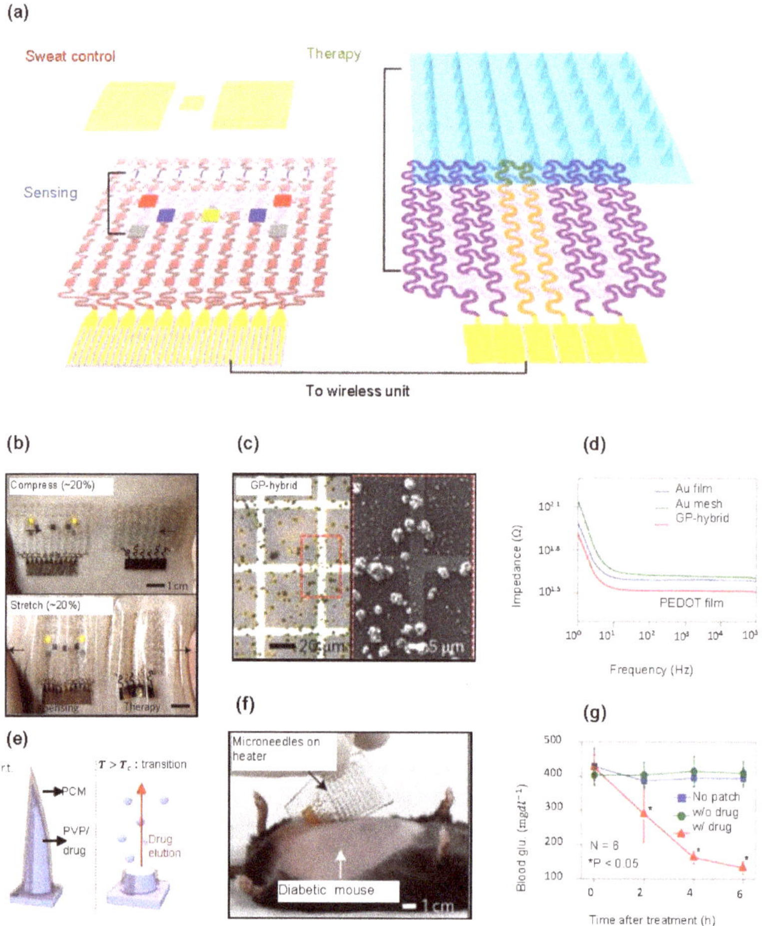

Figure 6. Graphene-based electrochemical sensors and microneedles for diabetes monitoring and therapy: (**a**) schematic illustration of the diabetes patch system consisting of sweat control, sensing (humidity, glucose, pH, and tremor sensor) and therapy components (heater, temperature sensor, and microneedles with drugs) on a silicone substrate; (**b**) photographs of the diabetes patch attached on the human skin under 20% applied compressive (top) and tensile (bottom) strain; (**c**) optical microscopy (left) and SEM (right) images of the graphene-hybrid with the PEDOT electrodeposition; (**d**) bode plots of the three electrodes in PBS after PEDOT electrodeposition; (**e**) schematic illustrations of the thermally active bioresorbable microneedles; (**f**) optical camera image of the therapeutic patch with microneedles on heater laminated on the skin near the abdomen of the diabetic (db/db) mouse; and (**g**) blood glucose concentrations of the db/db mice for the experimental group and control groups. The error bars and asterisks indicate the standard deviation in each group and significant difference ($p < 0.05$) among the groups at each time point relatively. Reproduced with permission from Lee et al. [161], copyright 2016, Macmillan Publishers Limited.

Another example of a carbon-based transparent and stretchable motion sensor was introduced by Lim et al. [160]. Their stretchable and wearable piezoelectric motion sensors and electrotactile stimulators based on a graphene heterostructure are utilized as an interactive human–machine interface (iHMI) that bidirectionally connects human and machines. In the iHMI system, motion sensors convert human motion into electrical signals that control the machine, and the electrotactile stimulators deliver feedback signals from the machine back to the human body. The graphene heterostructure can be used in both sensors and stimulators of the iHMI system. The motion sensor consists of a heterostructure with a piezoelectric polymer/SWCNT composite film sandwiched between two layers of CVD-grown graphene and polymethylmethacrylate (PMMA) insulating layers. The SWCNTs are embedded in polylactic acid (PLA), which is a piezoelectric polymer [162], to enhance SNR. The electrotactile stimulator consists of a heterostructure containing silver nanowire (AgNW) networks embedded in graphene layers, epoxy encapsulation layers, and a PDMS substrate. The stretchable iHMI system attached to the human arm can interactively control the robot arm. Through a data acquisition board and a special software program, the robot arm identified several movements of the human arm and performed the corresponding actions (Figure 5e). Furthermore, electrical stimulation can provide feedback to the user to prevent excessive operation of the robot arm.

Lee et al. have developed a patchable graphene-based device that can simultaneously diagnose and treat diabetes by combining various electrochemical sensors, including humidity, glucose, pH, tremor, and temperature, with thermoresponsive microneedles for drug delivery on stretchable silicone substrate (Figure 6a) [161]. Their multifunctional electrochemical device integrated on soft materials and serpentine design provides extremely conformal contact to the skin even under 20% tensile or compressive strains (Figure 6b). In addition, the GP-hybrids, consisting of a bilayer of Au mesh and gold-doped graphene, facilitate surface functionalization via electrodeposition, providing better electrochemical properties (Figure 6c,d). The high glucose concentration detected by the sensing part activates the heater in the therapy part, and the heat generated thereby dissolves the microneedles containing the drug to perform drug delivery into the blood (Figure 6e). As shown in Figure 6f,g, lamination of the therapeutic patch onto the abdomen of diabetic rats resulted in significantly reduced blood glucose levels, compared to those of the control groups.

5. Conclusions

This review covers some of the latest bioelectronics based on carbon nanotubes and graphene, the most representative carbon nanomaterials. In addition to their high electrical conductivity and optical transparency, excellent mechanical properties that are not found in conventional metal or semiconducting materials have allowed them to be extensively studied as wearable or implantable electronics. However, even though those physical properties of CNTs and graphene are outstanding, they cannot be integrated into the human body if their definite biocompatibility is not established. CNTs and graphene show varying biocompatibility depending on their concentration, method of synthesis, functionalization, and the type of cells to which they are applied. Although most studies using carbon nanomaterials in mammalian cells have shown biocompatibility, careful consideration of the application types is needed, as there are some studies that show toxicity to certain tissues, such as lung and skin. We have introduced biocompatible and flexible CNT/graphene-based bioelectronics that can be applied in epidermal or implantable electronics in two categories, according to their application, namely, microelectrodes and sensors.

The surface coating of traditional microelectrodes with CNTs or composites of CNTs and other conductive materials, including PEDOT and gold, remarkably increases the surface roughness, thereby resulting in a high SNR. Furthermore, the high optical transmittance of graphene and CNT thin films enables optical imaging and optogenetic stimulation, at the same time as electrical recording, offering high temporal and spatial resolution. CNTs and graphene with tunable bandgaps are suitable for sensors that detect biomolecules, such as proteins and nucleic acids, with a high sensitivity due to their high mobility and surface area. Recent studies have developed a novel human–machine interface that converts human facial expressions and movements into electrical signals using carbon-based piezoresistive, capacitive, and piezoelectric strain sensors. The integration of various electrochemical sensors and microneedles for drug delivery presents a new scheme that enables the diagnosis and treatment of hyperglycemia with a thin and transparent graphene-based patch.

These technological advances in carbon-based materials have contributed greatly to the realization of neural mapping, diagnosis and treatment of various diseases, and to the human–machine interface. Nevertheless, there are still unsolved issues that must be addressed to make these applications even better. Although CNTs and graphene have electrochemical properties that surpass other conventional electrodes for neural interfaces, there are still size-related problems in measuring the activity of all single neurons. To be integrated with a highly stretchable substrate for a more conformal interface with the human body, improved manufacturing technologies, such as low-temperature processes or transfer processes, are required. Furthermore, although many examples of implantable devices have been introduced, there is still a lack of research on devices for long-term usage in the biological environments. For example, CNTs and graphene can be oxidized and peeled off from the electrode surfaces under the electrical stimulation [163]. Establishing complete biocompatibility and creating fully biodegradable electronics, combined with the proper encapsulation layers is another challenge requiring solutions in the future [164]. Continued research on carbon-based materials is expected to overcome these challenges and make an important impact on human healthcare.

Author Contributions: T.K., M.C., and K.J.Y. collected the data, contributed to the scientific discussions, and co-wrote the manuscript. Conceptualization, T.K., M.C., and K.J.Y.; Resources, T.K., M.C., and K.J.Y.; Writing-Original Draft Preparation, T.K., M.C., and K.J.Y; Writing-Review & Editing, T.K., M.C., and K.J.Y.; Supervision, K.J.Y;; Funding Acquisition, K.J.Y., T.K. and M.C. contributed equally to this work.

Funding: National Research Foundation of Korea: NRF-2017R1C1B5017728, Yonsei University Future-Leading Research Initiative of 2017: RMS22017-22-00

Acknowledgments: T.K., M.C., and K.J.Y. acknowledge the support from the National Research Foundation of Korea (Grant No.NRF-2017R1C1B5017728) and the Yonsei University Future-Leading Research Initiative of 2018 (RMS22018-22-0028).

Conflicts of Interest: The authors declare no competing financial interests.

References

1. Moore, G.E. Cramming more components onto integrated circuits, reprinted from electronics, volume 38, number 8, April 19, 1965, pp.114 ff. *IEEE Solid-State Circuits Soc. Newsl.* **2006**, *11*, 33–35. [CrossRef]
2. Arns, R.G. The other transistor: Early history of the metal-oxide semiconductor field-effect transistor. *Eng. Sci. Educ. J.* **1998**, *7*, 233–240. [CrossRef]
3. Vogel, E. Technology and metrology of new electronic materials and devices. *Nat. Nanotechnol.* **2007**, *2*, 25–32. [CrossRef] [PubMed]
4. Rogers, J.; Lagally, M.; Nuzzo, R. Synthesis, assembly and applications of semiconductor nanomembranes. *Nature* **2011**, *477*, 45. [CrossRef] [PubMed]
5. Xu, S.; Zhang, Y.; Jia, L.; Mathewson, K.E.; Jang, K.-I.; Kim, J.; Fu, H.; Huang, X.; Chava, P.; Wang, R. Soft microfluidic assemblies of sensors, circuits, and radios for the skin. *Science* **2014**, *344*, 70–74. [CrossRef] [PubMed]
6. Jang, K.-I.; Li, K.; Chung, H.U.; Xu, S.; Jung, H.N.; Yang, Y.; Kwak, J.W.; Jung, H.H.; Song, J.; Yang, C. Self-assembled three dimensional network designs for soft electronics. *Nat. Commun.* **2017**, *8*, 15894. [CrossRef] [PubMed]
7. Lewis, J. Material challenge for flexible organic devices. *Mater. Today* **2006**, *9*, 38–45. [CrossRef]
8. Jeong, J.W.; Shin, G.; Park, S.I.; Yu, K.J.; Xu, L.; Rogers, J.A. Soft materials in neuroengineering for hard problems in neuroscience. *Neuron* **2015**, *86*, 175–186. [CrossRef] [PubMed]
9. Yu, K.J.; Yan, Z.; Han, M.; Rogers, J.A. Inorganic semiconducting materials for flexible and stretchable electronics. *Npj Flex. Electron.* **2017**, *1*, 4. [CrossRef]

10. Kim, T.-I.; Kim, M.J.; Jung, Y.H.; Jang, H.; Dagdeviren, C.; Pao, H.A.; Cho, S.J.; Carlson, A.; Yu, K.J.; Ameen, A. Thin film receiver materials for deterministic assembly by transfer printing. *Chem. Mater.* **2014**, *26*, 3502–3507. [CrossRef]

11. Kim, D.-H.; Lu, N.; Ma, R.; Kim, Y.-S.; Kim, R.-H.; Wang, S.; Wu, J.; Won, S.M.; Tao, H.; Islam, A.; et al. Epidermal electronics. *Science* **2011**, *333*, 838–843. [CrossRef] [PubMed]

12. Liu, Z.; Xu, J.; Chen, D.; Shen, G. Flexible electronics based on inorganic nanowires. *Chem. Soc. Rev.* **2015**, *44*, 161–192. [CrossRef] [PubMed]

13. Su, Y.; Ping, X.; Yu, K.J.; Lee, J.W.; Fan, J.A.; Wang, B.; Li, M.; Li, R.; Harburg, D.V.; Huang, Y. In-plane deformation mechanics for highly stretchable electronics. *Adv. Mater.* **2017**, *29*, 1604989. [CrossRef] [PubMed]

14. Kang, K.; Cho, Y.; Yu, K.J. Novel nano-materials and nano-fabrication techniques for flexible electronic systems. *Micromachines* **2018**, *9*, 263. [CrossRef]

15. Kang, S.-K.; Murphy, R.K.; Hwang, S.-W.; Lee, S.M.; Harburg, D.V.; Krueger, N.A.; Shin, J.; Gamble, P.; Cheng, H.; Yu, S. Bioresorbable silicon electronic sensors for the brain. *Nature* **2016**, *530*, 71–76. [CrossRef] [PubMed]

16. Chen, X.; Park, Y.J.; Kang, M.; Kang, S.-K.; Koo, J.; Shinde, S.M.; Shin, J.; Jeon, S.; Park, G.; Yan, Y. Cvd-grown monolayer mos2 in bioabsorbable electronics and biosensors. *Nat. Commun.* **2018**, *9*, 1690. [CrossRef] [PubMed]

17. Park, S.I.; Brenner, D.S.; Shin, G.; Morgan, C.D.; Copits, B.A.; Chung, H.U.; Pullen, M.Y.; Noh, K.N.; Davidson, S.; Oh, S.J. Soft, stretchable, fully implantable miniaturized optoelectronic systems for wireless optogenetics. *Nat. Biotechnol.* **2015**, *33*, 1280. [CrossRef] [PubMed]

18. Dagdeviren, C.; Shi, Y.; Joe, P.; Ghaffari, R.; Balooch, G.; Usgaonkar, K.; Gur, O.; Tran, P.L.; Crosby, J.R.; Meyer, M. Conformal piezoelectric systems for clinical and experimental characterization of soft tissue biomechanics. *Nat. Mater.* **2015**, *14*, 728. [CrossRef] [PubMed]

19. Kim, J.; Lee, M.; Shim, H.J.; Ghaffari, R.; Cho, H.R.; Son, D.; Jung, Y.H.; Soh, M.; Choi, C.; Jung, S. Stretchable silicon nanoribbon electronics for skin prosthesis. *Nat. Commun.* **2014**, *5*, 5747. [CrossRef] [PubMed]

20. Fang, H.; Yu, K.J.; Gloschat, C.; Yang, Z.; Song, E.; Chiang, C.-H.; Zhao, J.; Won, S.M.; Xu, S.; Trumpis, M. Capacitively coupled arrays of multiplexed flexible silicon transistors for long-term cardiac electrophysiology. *Nat. Biomed. Eng.* **2017**, *1*, 0038. [CrossRef] [PubMed]

21. Yu, K.J.; Kuzum, D.; Hwang, S.-W.; Kim, B.H.; Juul, H.; Kim, N.H.; Won, S.M.; Chiang, K.; Trumpis, M.; Richardson, A.G.; et al. Bioresorbable silicon electronics for transient spatiotemporal mapping of electrical activity from the cerebral cortex. *Nat. Mater.* **2016**, *15*, 782. [CrossRef] [PubMed]

22. Kim, D.-H.; Ghaffari, R.; Lu, N.; Rogers, J.A. Flexible and stretchable electronics for biointegrated devices. *Annu. Rev. Biomed. Eng.* **2012**, *14*, 113–128. [CrossRef] [PubMed]

23. Pernot, M.; Fujikura, K.; Fung-Kee-Fung, S.D.; Konofagou, E.E. Ecg-gated, mechanical and electromechanical wave imaging of cardiovascular tissues in vivo. *Ultrasound Med. Biol.* **2007**, *33*, 1075–1085. [CrossRef] [PubMed]

24. Takeuchi, S.; Ziegler, D.; Yoshida, Y.; Mabuchi, K.; Suzuki, T. Parylene flexible neural probes integrated with microfluidic channels. *Lab Chip* **2005**, *5*, 519–523. [CrossRef] [PubMed]

25. Castagnola, V.; Descamps, E.; Lecestre, A.; Dahan, L.; Remaud, J.; Nowak, L.G.; Bergaud, C. Parylene-based flexible neural probes with pedot coated surface for brain stimulation and recording. *Biosens. Bioelectron.* **2015**, *67*, 450–457. [CrossRef] [PubMed]

26. Zhao, Z.; Kim, E.; Luo, H.; Zhang, J.; Xu, Y. Flexible deep brain neural probes based on a parylene tube structure. *J. Micromech. Microeng.* **2017**, *28*, 015012. [CrossRef]

27. Lo, M.-C.; Wang, S.; Singh, S.; Damodaran, V.B.; Ahmed, I.; Coffey, K.; Barker, D.; Saste, K.; Kals, K.; Kaplan, H.M. Evaluating the in vivo glial response to miniaturized parylene cortical probes coated with an ultra-fast degrading polymer to aid insertion. *J. Neural Eng.* **2018**, *15*, 036002. [CrossRef] [PubMed]

28. Luan, L.; Wei, X.; Zhao, Z.; Siegel, J.J.; Potnis, O.; Tuppen, C.A.; Lin, S.; Kazmi, S.; Fowler, R.A.; Holloway, S. Ultraflexible nanoelectronic probes form reliable, glial scar–free neural integration. *Sci. Adv.* **2017**, *3*, e1601966. [CrossRef] [PubMed]

29. Miller, K.J.; Weaver, K.E.; Ojemann, J.G. Direct electrophysiological measurement of human default network areas. *Proc. Natl. Acad. Sci. USA* **2009**, *106*, 12174–12177. [CrossRef] [PubMed]

30. Lu, C.; Froriep, U.P.; Koppes, R.A.; Canales, A.; Caggiano, V.; Selvidge, J.; Bizzi, E.; Anikeeva, P. Polymer fiber probes enable optical control of spinal cord and muscle function in vivo. *Adv. Funct. Mater.* **2014**, *24*, 6594–6600. [CrossRef]

31. Nicolelis, M.A.L.; Lebedev, M.A. Principles of neural ensemble physiology underlying the operation of brain–machine interfaces. *Nat. Rev. Neurosci.* **2009**, *10*, 530–540. [CrossRef] [PubMed]

32. Allen, M.J.; Tung, V.C.; Kaner, R.B. Honeycomb carbon: A review of graphene. *Chem. Rev.* **2010**, *110*, 132–145. [CrossRef] [PubMed]

33. Bandaru, P.R. Electrical properties and applications of carbon nanotube structures. *J. Nanosci. Nanotechnol.* **2007**, *7*, 1239–1267. [CrossRef] [PubMed]

34. Nair, R.R.; Blake, P.; Grigorenko, A.N.; Novoselov, K.S.; Booth, T.J.; Stauber, T.; Peres, N.M.R.; Geim, A.K. Fine structure constant defines visual transparency of graphene. *Science* **2008**, *320*, 1308. [CrossRef] [PubMed]

35. Kaempgen, M.; Duesberg, G.S.; Roth, S. Transparent carbon nanotube coatings. *Appl. Surf. Sci.* **2005**, *252*, 425–429. [CrossRef]

36. Lee, C.; Wei, X.; Kysar, J.W.; Hone, J. Measurement of the elastic properties and intrinsic strength of monolayer graphene. *Science* **2008**, *321*, 385–388. [CrossRef] [PubMed]

37. Transport properties of carbon nanotubes. *Physical Properties of Carbon Nanotubes*; Imperial College Press: London, UK, 2011; pp. 137–162.

38. Kuzum, D.; Takano, H.; Shim, E.; Reed, J.C.; Juul, H.; Richardson, A.G.; De Vries, J.; Bink, H.; Dichter, M.A.; Lucas, T.H. Transparent and flexible low noise graphene electrodes for simultaneous electrophysiology and neuroimaging. *Nat. Commun.* **2014**, *5*, 5259. [CrossRef] [PubMed]

39. Nayagam, D.A.; Williams, R.A.; Chen, J.; Magee, K.A.; Irwin, J.; Tan, J.; Innis, P.; Leung, R.T.; Finch, S.; Williams, C.E. Biocompatibility of immobilized aligned carbon nanotubes. *Small* **2011**, *7*, 1035–1042. [CrossRef] [PubMed]

40. Ping, J.; Wu, J.; Wang, Y.; Ying, Y. Simultaneous determination of ascorbic acid, dopamine and uric acid using high-performance screen-printed graphene electrode. *Biosens. Bioelectron.* **2012**, *34*, 70–76. [CrossRef] [PubMed]

41. Choi, W.; Lahiri, I.; Seelaboyina, R.; Kang, Y.S. Synthesis of graphene and its applications: A review. *Crit. Rev. Solid State Mater. Sci.* **2010**, *35*, 52–71. [CrossRef]

42. Szunerits, S.; Boukherroub, R.; Downard, A.; Zhu, J.-J. *Nanocarbons for Electroanalysis*; John Wiley & Sons: Hoboken, NJ, USA, 2017.

43. Kim, T.-I.; McCall, J.G.; Jung, Y.H.; Huang, X.; Siuda, E.R.; Li, Y.; Song, J.; Song, Y.M.; Pao, H.A.; Kim, R.-H. Injectable, cellular-scale optoelectronics with applications for wireless optogenetics. *Science* **2013**, *340*, 211–216. [CrossRef] [PubMed]

44. Curry, E.J.; Ke, K.; Chorsi, M.T.; Wrobel, K.S.; Miller, A.N.; Patel, A.; Kim, I.; Feng, J.; Yue, L.; Wu, Q. Biodegradable piezoelectric force sensor. *Proc. Natl. Acad. Sci. USA* **2018**, *115*, 909–914. [CrossRef] [PubMed]

45. Lam, C.-W.; James, J.T.; McCluskey, R.; Arepalli, S.; Hunter, R.L. A review of carbon nanotube toxicity and assessment of potential occupational and environmental health risks. *Crit. Rev. Toxicol.* **2006**, *36*, 189–217. [CrossRef] [PubMed]

46. Johnston, H.J.; Hutchison, G.R.; Christensen, F.M.; Peters, S.; Hankin, S.; Aschberger, K.; Stone, V. A critical review of the biological mechanisms underlying the in vivo and in vitro toxicity of carbon nanotubes: The contribution of physico-chemical characteristics. *Nanotoxicology* **2010**, *4*, 207–246. [CrossRef] [PubMed]

47. Sanchez, V.C.; Jachak, A.; Hurt, R.H.; Kane, A.B. Biological interactions of graphene-family nanomaterials: An interdisciplinary review. *Chem. Res. Toxicol.* **2011**, *25*, 15–34. [CrossRef] [PubMed]

48. Seabra, A.B.; Paula, A.J.; de Lima, R.; Alves, O.L.; Duran, N. Nanotoxicity of graphene and graphene oxide. *Chem. Res. Toxicol.* **2014**, *27*, 159–168. [CrossRef] [PubMed]

49. Bokros, J.C. Carbon biomedical devices. *Carbon* **1977**, *15*, 353–371. [CrossRef]

50. Kaae, J.L. The mechanism of the deposition of pyrolytic carbons. *Carbon* **1985**, *23*, 665–673. [CrossRef]

51. Cenni, E.; Granchi, D.; Arciola, C.R.; Ciapetti, G.; Savarino, L.; Stea, S.; Cavedagna, D.; Di Leo, A.; Pizzoferrato, A. Adhesive protein expression on endothelial cells after contact *in vitro* with polyethylene terephthalate coated with pyrolytic carbon. *Biomaterials* **1995**, *16*, 1223–1227. [CrossRef]

52. Ling, M.; George, S. Fatigue behavior of a pyrolytic carbon. *J. Biomed. Mater. Res.* **2000**, *51*, 61–68.

53. Haubold, A. Blood/carbon interactions. *ASAIO J.* **1983**, *29*, 88.

54. Cui, F.Z.; Li, D.J. A review of investigations on biocompatibility of diamond-like carbon and carbon nitride films. *Surf. Coat. Technol.* **2000**, *131*, 481–487. [CrossRef]

55. Grill, A. Diamond-like carbon coatings as biocompatible materials—An overview. *Diam. Relat. Mater.* **2003**, *12*, 166–170. [CrossRef]

56. Sheeja, D.; Tay, B.K.; Nung, L.N. Feasibility of diamond-like carbon coatings for orthopaedic applications. *Diam. Relat. Mater.* **2004**, *13*, 184–190. [CrossRef]

57. Lifshitz, Y. Diamond-like carbon—Present status. *Diam. Relat. Mater.* **1999**, *8*, 1659–1676. [CrossRef]

58. Cui, H.-F.; Vashist, S.K.; Al-Rubeaan, K.; Luong, J.H.T.; Sheu, F.-S. Interfacing carbon nanotubes with living mammalian cells and cytotoxicity issues. *Chem. Res. Toxicol.* **2010**, *23*, 1131–1147. [CrossRef] [PubMed]

59. LaDou, J. The asbestos cancer epidemic. *Environ. Health Perspect.* **2004**, *112*, 285–290. [CrossRef] [PubMed]

60. Fujita, K.; Fukuda, M.; Endoh, S.; Maru, J.; Kato, H.; Nakamura, A.; Shinohara, N.; Uchino, K.; Honda, K. Size effects of single-walled carbon nanotubes on in vivo and in vitro pulmonary toxicity. *Inhal. Toxicol.* **2015**, *27*, 207–223. [CrossRef] [PubMed]

61. Hamilton, R.F.; Wu, Z.; Mitra, S.; Shaw, P.K.; Holian, A. Effect of mwcnt size, carboxylation, and purification on in vitro and in vivo toxicity, inflammation and lung pathology. *Part. Fibre Toxicol.* **2013**, *10*, 57. [CrossRef] [PubMed]

62. Donaldson, K.; Aitken, R.; Tran, L.; Stone, V.; Duffin, R.; Forrest, G.; Alexander, A. Carbon nanotubes: A review of their properties in relation to pulmonary toxicology and workplace safety. *Toxicol. Sci.* **2006**, *92*, 5–22. [CrossRef] [PubMed]

63. Huczko, A.; Lange, H.; Całko, E.; Grubek-Jaworska, H.; Droszcz, P. Physiological testing of carbon nanotubes: Are they asbestos-like? *Fuller. Sci. Technol.* **2001**, *9*, 251–254. [CrossRef]

64. Brown, D.; Kinloch, I.; Bangert, U.; Windle, A.; Walter, D.; Walker, G.; Scotchford, C.; Donaldson, K.; Stone, V. An in vitro study of the potential of carbon nanotubes and nanofibres to induce inflammatory mediators and frustrated phagocytosis. *Carbon* **2007**, *45*, 1743–1756. [CrossRef]

65. Monteiro-Riviere, N.A.; Nemanich, R.J.; Inman, A.O.; Wang, Y.Y.; Riviere, J.E. Multi-walled carbon nanotube interactions with human epidermal keratinocytes. *Toxicol. Lett.* **2005**, *155*, 377–384. [CrossRef] [PubMed]

66. Shvedova, A.; Castranova, V.; Kisin, E.; Schwegler-Berry, D.; Murray, A.; Gandelsman, V.; Maynard, A.; Baron, P. Exposure to carbon nanotube material: Assessment of nanotube cytotoxicity using human keratinocyte cells. *J. Toxicol. Environ. Health Part A* **2003**, *66*, 1909–1926. [CrossRef] [PubMed]

67. Ema, M.; Matsuda, A.; Kobayashi, N.; Naya, M.; Nakanishi, J. Evaluation of dermal and eye irritation and skin sensitization due to carbon nanotubes. *Regul. Toxicol. Pharmacol.* **2011**, *61*, 276–281. [CrossRef] [PubMed]

68. Jia, G.; Wang, H.; Yan, L.; Wang, X.; Pei, R.; Yan, T.; Zhao, Y.; Guo, X. Cytotoxicity of carbon nanomaterials: Single-wall nanotube, multi-wall nanotube, and fullerene. *Environ. Sci. Technol.* **2005**, *39*, 1378–1383. [CrossRef] [PubMed]

69. Cui, D.; Tian, F.; Ozkan, C.S.; Wang, M.; Gao, H. Effect of single wall carbon nanotubes on human hek293 cells. *Toxicol. Lett.* **2005**, *155*, 73–85. [CrossRef] [PubMed]

70. Chen, X.; Tam, U.C.; Czlapinski, J.L.; Lee, G.S.; Rabuka, D.; Zettl, A.; Bertozzi, C.R. Interfacing carbon nanotubes with living cells. *J. Am. Chem. Soc.* **2006**, *128*, 6292–6293. [CrossRef] [PubMed]

71. Dumortier, H.; Lacotte, S.; Pastorin, G.; Marega, R.; Wu, W.; Bonifazi, D.; Briand, J.-P.; Prato, M.; Muller, S.; Bianco, A. Functionalized carbon nanotubes are non-cytotoxic and preserve the functionality of primary immune cells. *Nano Lett.* **2006**, *6*, 1522–1528. [CrossRef] [PubMed]

72. Flahaut, E.; Durrieu, M.C.; Remy-Zolghadri, M.; Bareille, R.; Baquey, C. Investigation of the cytotoxicity of ccvd carbon nanotubes towards human umbilical vein endothelial cells. *Carbon* **2006**, *44*, 1093–1099. [CrossRef]

73. Silvano, G.; Claudio, B.; Valter, B.; Giorgio, G.; Claudio, N. Carbon nanotube biocompatibility with cardiac muscle cells. *Nanotechnology* **2006**, *17*, 391.

74. Belyanskaya, L.; Weigel, S.; Hirsch, C.; Tobler, U.; Krug, H.F.; Wick, P. Effects of carbon nanotubes on primary neurons and glial cells. *Neurotoxicology* **2009**, *30*, 702–711. [CrossRef] [PubMed]

75. Wick, P.; Manser, P.; Limbach, L.K.; Dettlaff-Weglikowska, U.; Krumeich, F.; Roth, S.; Stark, W.J.; Bruinink, A. The degree and kind of agglomeration affect carbon nanotube cytotoxicity. *Toxicol. Lett.* **2007**, *168*, 121–131. [CrossRef] [PubMed]

76. Mutlu, G.M.; Budinger, G.S.; Green, A.A.; Urich, D.; Soberanes, S.; Chiarella, S.E.; Alheid, G.F.; McCrimmon, D.R.; Szleifer, I.; Hersam, M.C. Biocompatible nanoscale dispersion of single-walled carbon nanotubes minimizes in vivo pulmonary toxicity. *Nano Lett.* **2010**, *10*, 1664–1670. [CrossRef] [PubMed]

77. Kolosnjaj-Tabi, J.; Hartman, K.B.; Boudjemaa, S.; Ananta, J.S.; Morgant, G.; Szwarc, H.; Wilson, L.J.; Moussa, F. In vivo behavior of large doses of ultrashort and full-length single-walled carbon nanotubes after oral and intraperitoneal administration to swiss mice. *ACS Nano* **2010**, *4*, 1481–1492. [CrossRef] [PubMed]

78. Elias, K.L.; Price, R.L.; Webster, T.J. Enhanced functions of osteoblasts on nanometer diameter carbon fibers. *Biomaterials* **2002**, *23*, 3279–3287. [CrossRef]

79. Lobo, A.; Corat, M.; Antunes, E.; Palma, M.; Pacheco-Soares, C.; Garcia, E.; Corat, E. An evaluation of cell proliferation and adhesion on vertically-aligned multi-walled carbon nanotube films. *Carbon* **2010**, *48*, 245–254. [CrossRef]

80. Hu, H.; Ni, Y.; Montana, V.; Haddon, R.C.; Parpura, V. Chemically functionalized carbon nanotubes as substrates for neuronal growth. *Nano Lett.* **2004**, *4*, 507–511. [CrossRef] [PubMed]

81. Hernandez, Y.; Nicolosi, V.; Lotya, M.; Blighe, F.M.; Sun, Z.; De, S.; McGovern, I.; Holland, B.; Byrne, M.; Gun'Ko, Y.K. High-yield production of graphene by liquid-phase exfoliation of graphite. *Nat. Nanotechnol.* **2008**, *3*, 563. [CrossRef] [PubMed]

82. Ahadian, S.; Estili, M.; Surya, V.J.; Ramón-Azcón, J.; Liang, X.; Shiku, H.; Ramalingam, M.; Matsue, T.; Sakka, Y.; Bae, H. Facile and green production of aqueous graphene dispersions for biomedical applications. *Nanoscale* **2015**, *7*, 6436–6443. [CrossRef] [PubMed]

83. Laaksonen, P.; Kainlauri, M.; Laaksonen, T.; Shchepetov, A.; Jiang, H.; Ahopelto, J.; Linder, M.B. Interfacial engineering by proteins: Exfoliation and functionalization of graphene by hydrophobins. *Angew. Chem. Int. Ed.* **2010**, *49*, 4946–4949. [CrossRef] [PubMed]

84. Gravagnuolo, A.M.; Morales-Narváez, E.; Longobardi, S.; da Silva, E.T.; Giardina, P.; Merkoçi, A. In situ production of biofunctionalized few-layer defect-free microsheets of graphene. *Adv. Funct. Mater.* **2015**, *25*, 2771–2779. [CrossRef]

85. Castagnola, V.; Zhao, W.; Boselli, L.; Giudice, M.L.; Meder, F.; Polo, E.; Paton, K.; Backes, C.; Coleman, J.; Dawson, K. Biological recognition of graphene nanoflakes. *Nat. Commun.* **2018**, *9*, 1577. [CrossRef] [PubMed]

86. Ou, L.; Luo, Y.; Wei, G. Atomic-level study of adsorption, conformational change, and dimerization of an α-helical peptide at graphene surface. *J. Phys. Chem. B* **2011**, *115*, 9813–9822. [CrossRef] [PubMed]

87. Zuo, G.; Zhou, X.; Huang, Q.; Fang, H.; Zhou, R. Adsorption of villin headpiece onto graphene, carbon nanotube, and c60: Effect of contacting surface curvatures on binding affinity. *J. Phys. Chem. C* **2011**, *115*, 23323–23328. [CrossRef]

88. Zhao, X. Self-assembly of DNA segments on graphene and carbon nanotube arrays in aqueous solution: A molecular simulation study. *J. Phys. Chem. C* **2011**, *115*, 6181–6189. [CrossRef]

89. Sahni, D.; Jea, A.; Mata, J.A.; Marcano, D.C.; Sivaganesan, A.; Berlin, J.M.; Tatsui, C.E.; Sun, Z.; Luerssen, T.G.; Meng, S.; et al. Biocompatibility of pristine graphene for neuronal interface. *J. Neurosurg. Pediatr.* **2013**, *11*, 575–583. [CrossRef] [PubMed]

90. Kalbacova, M.; Broz, A.; Kong, J.; Kalbac, M. Graphene substrates promote adherence of human osteoblasts and mesenchymal stromal cells. *Carbon* **2010**, *48*, 4323–4329. [CrossRef]

91. Li, N.; Zhang, Q.; Gao, S.; Song, Q.; Huang, R.; Wang, L.; Liu, L.; Dai, J.; Tang, M.; Cheng, G. Three-dimensional graphene foam as a biocompatible and conductive scaffold for neural stem cells. *Sci. Rep.* **2013**, *3*, 1604. [CrossRef] [PubMed]

92. Bendali, A.; Hess, L.H.; Seifert, M.; Forster, V.; Stephan, A.F.; Garrido, J.A.; Picaud, S. Purified neurons can survive on peptide-free graphene layers. *Adv. Healthc. Mater.* **2013**, *2*, 929–933. [CrossRef] [PubMed]

93. Seker, E.; Berdichevsky, Y.; Begley, M.R.; Reed, M.L.; Staley, K.J.; Yarmush, M.L. The fabrication of low-impedance nanoporous gold multiple-electrode arrays for neural electrophysiology studies. *Nanotechnology* **2010**, *21*, 125504. [CrossRef] [PubMed]

94. Green, R.A.; Lovell, N.H.; Wallace, G.G.; Poole-Warren, L.A. Conducting polymers for neural interfaces: Challenges in developing an effective long-term implant. *Biomaterials* **2008**, *29*, 3393–3399. [CrossRef] [PubMed]

95. Suyatin, D.B.; Wallman, L.; Thelin, J.; Prinz, C.N.; Jörntell, H.; Samuelson, L.; Montelius, L.; Schouenborg, J. Nanowire-based electrode for acute in vivo neural recordings in the brain. *PLoS ONE* **2013**, *8*, e56673. [CrossRef] [PubMed]

96. Robinson, J.T.; Jorgolli, M.; Shalek, A.K.; Yoon, M.-H.; Gertner, R.S.; Park, H. Vertical nanowire electrode arrays as a scalable platform for intracellular interfacing to neuronal circuits. *Nat. Nanotechnol.* **2012**, *7*, 180. [CrossRef] [PubMed]

97. Ansaldo, A.; Castagnola, E.; Maggiolini, E.; Fadiga, L.; Ricci, D. Superior electrochemical performance of carbon nanotubes directly grown on sharp microelectrodes. *ACS Nano* **2011**, *5*, 2206–2214. [CrossRef] [PubMed]

98. Baranauskas, G.; Maggiolini, E.; Castagnola, E.; Ansaldo, A.; Mazzoni, A.; Angotzi, G.N.; Vato, A.; Ricci, D.; Panzeri, S.; Fadiga, L. Carbon nanotube composite coating of neural microelectrodes preferentially improves the multiunit signal-to-noise ratio. *J. Neural Eng.* **2011**, *8*, 066013. [CrossRef] [PubMed]

99. Eleftheriou, C.G.; Zimmermann, J.B.; Kjeldsen, H.D.; David-Pur, M.; Hanein, Y.; Sernagor, E. Carbon nanotube electrodes for retinal implants: A study of structural and functional integration over time. *Biomaterials* **2017**, *112*, 108–121. [CrossRef] [PubMed]

100. Jan, E.; Hendricks, J.L.; Husaini, V.; Richardson-Burns, S.M.; Sereno, A.; Martin, D.C.; Kotov, N.A. Layered carbon nanotube-polyelectrolyte electrodes outperform traditional neural interface materials. *Nano Lett.* **2009**, *9*, 4012–4018. [CrossRef] [PubMed]

101. Balamurugan, A.; Chen, S.-M. Poly (3, 4-ethylenedioxythiophene-co-(5-amino-2-naphthalenesulfonic acid))(pedot-pans) film modified glassy carbon electrode for selective detection of dopamine in the presence of ascorbic acid and uric acid. *Anal. Chim. Acta* **2007**, *596*, 92–98. [CrossRef] [PubMed]

102. Gerwig, R.; Fuchsberger, K.; Schroeppel, B.; Link, G.S.; Heusel, G.; Kraushaar, U.; Schuhmann, W.; Stett, A.; Stelzle, M. Pedot–cnt composite microelectrodes for recording and electrostimulation applications: Fabrication, morphology, and electrical properties. *Front. Neuroeng.* **2012**, *5*, 8. [CrossRef] [PubMed]

103. Xiang, Z.; Yen, S.-C.; Xue, N.; Sun, T.; Tsang, W.M.; Zhang, S.; Liao, L.-D.; Thakor, N.V.; Lee, C. Ultra-thin flexible polyimide neural probe embedded in a dissolvable maltose-coated microneedle. *J. Micromech. Microeng.* **2014**, *24*, 065015. [CrossRef]

104. Stosiek, C.; Garaschuk, O.; Holthoff, K.; Konnerth, A. In vivo two-photon calcium imaging of neuronal networks. *Proc. Natl. Acad. Sci. USA* **2003**, *100*, 7319–7324. [CrossRef] [PubMed]

105. Dombeck, D.A.; Harvey, C.D.; Tian, L.; Looger, L.L.; Tank, D.W. Functional imaging of hippocampal place cells at cellular resolution during virtual navigation. *Nat. Neurosci.* **2010**, *13*, 1433. [CrossRef] [PubMed]

106. Wedeen, V.J. The geometric structure of the brain fiber pathways (vol 335, pg 1628, 2012). *Science* **2012**, *336*, 670.

107. Anikeeva, P.; Andalman, A.S.; Witten, I.; Warden, M.; Goshen, I.; Grosenick, L.; Gunaydin, L.A.; Frank, L.M.; Deisseroth, K. Optetrode: A multichannel readout for optogenetic control in freely moving mice. *Nat. Neurosci.* **2012**, *15*, 163. [CrossRef] [PubMed]

108. Park, S.; Guo, Y.; Jia, X.; Choe, H.K.; Grena, B.; Kang, J.; Park, J.; Lu, C.; Canales, A.; Chen, R. One-step optogenetics with multifunctional flexible polymer fibers. *Nat. Neurosci.* **2017**, *20*, 612. [CrossRef] [PubMed]

109. Cardin, J.A.; Carlén, M.; Meletis, K.; Knoblich, U.; Zhang, F.; Deisseroth, K.; Tsai, L.-H.; Moore, C.I. Targeted optogenetic stimulation and recording of neurons in vivo using cell-type-specific expression of channelrhodopsin-2. *Nature protocols* **2010**, *5*, 247. [CrossRef] [PubMed]

110. Zhang, J.; Liu, X.; Xu, W.; Luo, W.; Li, M.; Chu, F.; Xu, L.; Cao, A.; Guan, J.; Tang, S. Stretchable transparent electrode arrays for simultaneous electrical and optical interrogation of neural circuits in vivo. *Nano Lett.* **2018**, *18*, 2903–2911. [CrossRef] [PubMed]

111. Li, Z.; Jia, Y.; Wei, J.; Wang, K.; Shu, Q.; Gui, X.; Zhu, H.; Cao, A.; Wu, D. Large area, highly transparent carbon nanotube spiderwebs for energy harvesting. *J. Mater. Chem.* **2010**, *20*, 7236–7240. [CrossRef]

112. Park, D.-W.; Schendel, A.A.; Mikael, S.; Brodnick, S.K.; Richner, T.J.; Ness, J.P.; Hayat, M.R.; Atry, F.; Frye, S.T.; Pashaie, R. Graphene-based carbon-layered electrode array technology for neural imaging and optogenetic applications. *Nat. Commun.* **2014**, *5*, ncomms6258. [CrossRef] [PubMed]

113. Tehovnik, E.J. Electrical stimulation of neural tissue to evoke behavioral responses. *J. Neurosci. Methods* **1996**, *65*, 1–17. [CrossRef]

114. Dymond, A.M. Characteristics of the metal-tissue interface of stimulation electrodes. *IEEE Trans. Biomed. Eng.* **1976**, 274–280. [CrossRef]

115. Park, D.-W.; Ness, J.P.; Brodnick, S.K.; Esquibel, C.; Novello, J.; Atry, F.; Baek, D.-H.; Kim, H.; Bong, J.; Swanson, K.I. Electrical neural stimulation and simultaneous in vivo monitoring with transparent graphene electrode arrays implanted in gcamp6f mice. *ACS Nano* **2018**, *12*, 148–157. [CrossRef] [PubMed]

116. Chen, T.-W.; Wardill, T.J.; Sun, Y.; Pulver, S.R.; Renninger, S.L.; Baohan, A.; Schreiter, E.R.; Kerr, R.A.; Orger, M.B.; Jayaraman, V. Ultrasensitive fluorescent proteins for imaging neuronal activity. *Nature* **2013**, *499*, 295. [CrossRef] [PubMed]

117. Tans, S.J.; Verschueren, A.R.; Dekker, C. Room-temperature transistor based on a single carbon nanotube. *Nature* **1998**, *393*, 49. [CrossRef]

118. Kocabas, C.; Hur, S.H.; Gaur, A.; Meitl, M.A.; Shim, M.; Rogers, J.A. Guided growth of large-scale, horizontally aligned arrays of single-walled carbon nanotubes and their use in thin-film transistors. *Small* **2005**, *1*, 1110–1116. [CrossRef] [PubMed]

119. Kang, S.J.; Kocabas, C.; Ozel, T.; Shim, M.; Pimparkar, N.; Alam, M.A.; Rotkin, S.V.; Rogers, J.A. High-performance electronics using dense, perfectly aligned arrays of single-walled carbon nanotubes. *Nat. Nanotechnol.* **2007**, *2*, 230. [CrossRef] [PubMed]

120. Chen, X.; Saito, T.; Yamada, H.; Matsushige, K. Aligning single-wall carbon nanotubes with an alternating-current electric field. *Appl. Phys. Lett.* **2001**, *78*, 3714–3716. [CrossRef]

121. Kimura, T.; Ago, H.; Tobita, M.; Ohshima, S.; Kyotani, M.; Yumura, M. Polymer composites of carbon nanotubes aligned by a magnetic field. *Adv. Mater.* **2002**, *14*, 1380–1383. [CrossRef]

122. Lu, L.; Chen, W. Large-scale aligned carbon nanotubes from their purified, highly concentrated suspension. *ACS Nano* **2010**, *4*, 1042–1048. [CrossRef] [PubMed]

123. Huang, S.; Maynor, B.; Cai, X.; Liu, J. Ultralong, well-aligned single-walled carbon nanotube architectureson surfaces. *Adv. Mater.* **2003**, *15*, 1651–1655. [CrossRef]

124. Tang, X.; Bansaruntip, S.; Nakayama, N.; Yenilmez, E.; Chang, Y.-L.; Wang, Q. Carbon nanotube DNA sensor and sensing mechanism. *Nano Lett.* **2006**, *6*, 1632–1636. [CrossRef] [PubMed]

125. Maehashi, K.; Matsumoto, K. Label-free electrical detection using carbon nanotube-based biosensors. *Sensors* **2009**, *9*, 5368–5378. [CrossRef] [PubMed]

126. Hu, P.; Fasoli, A.; Park, J.; Choi, Y.; Estrela, P.; Maeng, S.; Milne, W.; Ferrari, A. Self-assembled nanotube field-effect transistors for label-free protein biosensors. *J. Appl. Phys.* **2008**, *104*, 074310. [CrossRef]

127. Gui, E.L.; Li, L.-J.; Zhang, K.; Xu, Y.; Dong, X.; Ho, X.; Lee, P.S.; Kasim, J.; Shen, Z.; Rogers, J.A. DNA sensing by field-effect transistors based on networks of carbon nanotubes. *J. Am. Chem. Soc.* **2007**, *129*, 14427–14432. [CrossRef] [PubMed]

128. Star, A.; Gabriel, J.-C.P.; Bradley, K.; Grüner, G. Electronic detection of specific protein binding using nanotube fet devices. *Nano Lett.* **2003**, *3*, 459–463. [CrossRef]

129. Heller, I.; Janssens, A.M.; Männik, J.; Minot, E.D.; Lemay, S.G.; Dekker, C. Identifying the mechanism of biosensing with carbon nanotube transistors. *Nano Lett.* **2008**, *8*, 591–595. [CrossRef] [PubMed]

130. Karousis, N.; Tagmatarchis, N.; Tasis, D. Current progress on the chemical modification of carbon nanotubes. *Chem. Rev.* **2010**, *110*, 5366–5397. [CrossRef] [PubMed]

131. Kim, J.P.; Lee, B.Y.; Lee, J.; Hong, S.; Sim, S.J. Enhancement of sensitivity and specificity by surface modification of carbon nanotubes in diagnosis of prostate cancer based on carbon nanotube field effect transistors. *Biosens. Bioelectron.* **2009**, *24*, 3372–3378. [CrossRef] [PubMed]

132. Star, A.; Tu, E.; Niemann, J.; Gabriel, J.-C.P.; Joiner, C.S.; Valcke, C. Label-free detection of DNA hybridization using carbon nanotube network field-effect transistors. *Proc. Natl. Acad. Sci. USA* **2006**, *103*, 921–926. [CrossRef] [PubMed]

133. Allen, B.L.; Kichambare, P.D.; Star, A. Carbon nanotube field-effect-transistor-based biosensors. *Adv. Mater.* **2007**, *19*, 1439–1451. [CrossRef]

134. Yáñez-Sedeño, P.; Pingarrón, J.M.; Riu, J.; Rius, F.X. Electrochemical sensing based on carbon nanotubes. *TrAC Trends Anal. Chem.* **2010**, *29*, 939–953. [CrossRef]

135. Yang, N.; Chen, X.; Ren, T.; Zhang, P.; Yang, D. Carbon nanotube based biosensors. *Sens. Actuators B Chem.* **2015**, *207*, 690–715. [CrossRef]

136. Nirmalraj, P.N.; Lutz, T.; Kumar, S.; Duesberg, G.S.; Boland, J.J. Nanoscale mapping of electrical resistivity and connectivity in graphene strips and networks. *Nano Lett.* **2010**, *11*, 16–22. [CrossRef] [PubMed]

137. Bolotin, K.I.; Sikes, K.; Jiang, Z.; Klima, M.; Fudenberg, G.; Hone, J.; Kim, P.; Stormer, H. Ultrahigh electron mobility in suspended graphene. *Solid State Commun.* **2008**, *146*, 351–355. [CrossRef]

138. Du, X.; Skachko, I.; Barker, A.; Andrei, E.Y. Approaching ballistic transport in suspended graphene. *Nat. Nanotechnol.* **2008**, *3*, 491. [CrossRef] [PubMed]

139. Ohta, T.; Bostwick, A.; Seyller, T.; Horn, K.; Rotenberg, E. Controlling the electronic structure of bilayer graphene. *Science* **2006**, *313*, 951–954. [CrossRef] [PubMed]

140. Lu, G.; Yu, K.; Wen, Z.; Chen, J. Semiconducting graphene: Converting graphene from semimetal to semiconductor. *Nanoscale* **2013**, *5*, 1353–1368. [CrossRef] [PubMed]

141. Guo, B.; Fang, L.; Zhang, B.; Gong, J.R. Graphene doping: A review. *Insci. J.* **2011**, *1*, 80–89. [CrossRef]

142. Zhang, Y.; Tang, T.-T.; Girit, C.; Hao, Z.; Martin, M.C.; Zettl, A.; Crommie, M.F.; Shen, Y.R.; Wang, F. Direct observation of a widely tunable bandgap in bilayer graphene. *Nature* **2009**, *459*, 820. [CrossRef] [PubMed]

143. Ohno, Y.; Maehashi, K.; Yamashiro, Y.; Matsumoto, K. Electrolyte-gated graphene field-effect transistors for detecting ph and protein adsorption. *Nano Lett.* **2009**, *9*, 3318–3322. [CrossRef] [PubMed]

144. Liu, Y.; Dong, X.; Chen, P. Biological and chemical sensors based on graphene materials. *Chem. Soc. Rev.* **2012**, *41*, 2283–2307. [CrossRef] [PubMed]

145. Mohanty, N.; Berry, V. Graphene-based single-bacterium resolution biodevice and DNA transistor: Interfacing graphene derivatives with nanoscale and microscale biocomponents. *Nano Lett.* **2008**, *8*, 4469–4476. [CrossRef] [PubMed]

146. Dong, X.; Shi, Y.; Huang, W.; Chen, P.; Li, L.J. Electrical detection of DNA hybridization with single-base specificity using transistors based on cvd-grown graphene sheets. *Adv. Mater.* **2010**, *22*, 1649–1653. [CrossRef] [PubMed]

147. Kwak, Y.H.; Choi, D.S.; Kim, Y.N.; Kim, H.; Yoon, D.H.; Ahn, S.-S.; Yang, J.-W.; Yang, W.S.; Seo, S. Flexible glucose sensor using cvd-grown graphene-based field effect transistor. *Biosens. Bioelectron.* **2012**, *37*, 82–87. [CrossRef] [PubMed]

148. He, Q.; Wu, S.; Gao, S.; Cao, X.; Yin, Z.; Li, H.; Chen, P.; Zhang, H. Transparent, flexible, all-reduced graphene oxide thin film transistors. *ACS Nano* **2011**, *5*, 5038–5044. [CrossRef] [PubMed]

149. Yamada, T.; Hayamizu, Y.; Yamamoto, Y.; Yomogida, Y.; Izadi-Najafabadi, A.; Futaba, D.N.; Hata, K. A stretchable carbon nanotube strain sensor for human-motion detection. *Nat. Nanotechnol.* **2011**, *6*, 296. [CrossRef] [PubMed]

150. Bandodkar, A.J.; Jeerapan, I.; You, J.-M.; Nuñez-Flores, R.; Wang, J. Highly stretchable fully-printed cnt-based electrochemical sensors and biofuel cells: Combining intrinsic and design-induced stretchability. *Nano Lett.* **2015**, *16*, 721–727. [CrossRef] [PubMed]

151. Amjadi, M.; Pichitpajongkit, A.; Lee, S.; Ryu, S.; Park, I. Highly stretchable and sensitive strain sensor based on silver nanowire–elastomer nanocomposite. *ACS Nano* **2014**, *8*, 5154–5163. [CrossRef] [PubMed]

152. Xiao, X.; Yuan, L.; Zhong, J.; Ding, T.; Liu, Y.; Cai, Z.; Rong, Y.; Han, H.; Zhou, J.; Wang, Z.L. High-strain sensors based on zno nanowire/polystyrene hybridized flexible films. *Adv. Mater.* **2011**, *23*, 5440–5444. [CrossRef] [PubMed]

153. Boland, C.S.; Khan, U.; Backes, C.; O'Neill, A.; McCauley, J.; Duane, S.; Shanker, R.; Liu, Y.; Jurewicz, I.; Dalton, A.B. Sensitive, high-strain, high-rate bodily motion sensors based on graphene–rubber composites. *ACS Nano* **2014**, *8*, 8819–8830. [CrossRef] [PubMed]

154. Roh, E.; Hwang, B.-U.; Kim, D.; Kim, B.-Y.; Lee, N.-E. Stretchable, transparent, ultrasensitive, and patchable strain sensor for human–machine interfaces comprising a nanohybrid of carbon nanotubes and conductive elastomers. *ACS Nano* **2015**, *9*, 6252–6261. [CrossRef] [PubMed]

155. Lee, J.; Kim, S.; Lee, J.; Yang, D.; Park, B.C.; Ryu, S.; Park, I. A stretchable strain sensor based on a metal nanoparticle thin film for human motion detection. *Nanoscale* **2014**, *6*, 11932–11939. [CrossRef] [PubMed]

156. Cai, L.; Song, L.; Luan, P.; Zhang, Q.; Zhang, N.; Gao, Q.; Zhao, D.; Zhang, X.; Tu, M.; Yang, F. Super-stretchable, transparent carbon nanotube-based capacitive strain sensors for human motion detection. *Sci. Rep.* **2013**, *3*, 3048. [CrossRef] [PubMed]

157. Wang, Y.; Yang, R.; Shi, Z.; Zhang, L.; Shi, D.; Wang, E.; Zhang, G. Super-elastic graphene ripples for flexible strain sensors. *ACS Nano* **2011**, *5*, 3645–3650. [CrossRef] [PubMed]

158. Kang, M.; Kim, J.; Jang, B.; Chae, Y.; Kim, J.-H.; Ahn, J.-H. Graphene-based three-dimensional capacitive touch sensor for wearable electronics. *ACS Nano* **2017**, *11*, 7950–7957. [CrossRef] [PubMed]

159. Wang, K.; Chang, Y.-H.; Chen, Y.; Zhang, C.; Wang, B. Designable dual-material auxetic metamaterials using three-dimensional printing. *Mater. Des.* **2015**, *67*, 159–164. [CrossRef]

160. Lim, S.; Son, D.; Kim, J.; Lee, Y.B.; Song, J.K.; Choi, S.; Lee, D.J.; Kim, J.H.; Lee, M.; Hyeon, T. Transparent and stretchable interactive human machine interface based on patterned graphene heterostructures. *Adv. Funct. Mater.* **2015**, *25*, 375–383. [CrossRef]

161. Lee, H.; Choi, T.K.; Lee, Y.B.; Cho, H.R.; Ghaffari, R.; Wang, L.; Choi, H.J.; Chung, T.D.; Lu, N.; Hyeon, T. A graphene-based electrochemical device with thermoresponsive microneedles for diabetes monitoring and therapy. *Nat. Nanotechnol.* **2016**, *11*, 566. [CrossRef] [PubMed]

162. Yoshida, T.; Imoto, K.; Nakai, T.; Uwami, R.; Kataoka, T.; Inoue, M.; Fukumoto, T.; Kamimura, Y.; Kato, A.; Tajitsu, Y. Piezoelectric motion of multilayer film with alternate rows of optical isomers of chiral polymer film. *Jpn. J. Appl. Phys.* **2011**, *50*, 09ND13. [CrossRef]

163. Vomero, M.; Castagnola, E.; Ciarpella, F.; Maggiolini, E.; Goshi, N.; Zucchini, E.; Carli, S.; Fadiga, L.; Kassegne, S.; Ricci, D. Highly stable glassy carbon interfaces for long-term neural stimulation and low-noise recording of brain activity. *Sci. Rep.* **2017**, *7*, 40332. [CrossRef] [PubMed]

164. Fang, H.; Zhao, J.; Yu, K.J.; Song, E.; Farimani, A.B.; Chiang, C.-H.; Jin, X.; Xue, Y.; Xu, D.; Du, W. Ultrathin, transferred layers of thermally grown silicon dioxide as biofluid barriers for biointegrated flexible electronic systems. *Proc. Natl. Acad. Sci. USA* **2016**, *113*, 11682–11687. [CrossRef] [PubMed]

 *materials*

Review

Biocompatible and Implantable Optical Fibers and Waveguides for Biomedicine

Roya Nazempour [1], Qianyi Zhang [2], Ruxing Fu [2] and Xing Sheng [1,*]

[1] Department of Electronic Engineering, Beijing National Research Center for Information science and Technology, Tsinghua University, Beijing 100084, China; rui-y16@mails.tsinghua.edu.cn
[2] School of Materials Science and Engineering, Tsinghua University, Beijing 100084, China; qy-zhang15@mails.tsinghua.edu.cn (Q.Z.); furx14@mails.tsinghua.edu.cn (R.F.)
* Correspondence: xingsheng@tsinghua.edu.cn; Tel.: +86-10-6278-2515

Received: 24 May 2018; Accepted: 21 July 2018; Published: 25 July 2018

Abstract: Optical fibers and waveguides in general effectively control and modulate light propagation, and these tools have been extensively used in communication, lighting and sensing. Recently, they have received increasing attention in biomedical applications. By delivering light into deep tissue via these devices, novel applications including biological sensing, stimulation and therapy can be realized. Therefore, implantable fibers and waveguides in biocompatible formats with versatile functionalities are highly desirable. In this review, we provide an overview of recent progress in the exploration of advanced optical fibers and waveguides for biomedical applications. Specifically, we highlight novel materials design and fabrication strategies to form implantable fibers and waveguides. Furthermore, their applications in various biomedical fields such as light therapy, optogenetics, fluorescence sensing and imaging are discussed. We believe that these newly developed fiber and waveguide based devices play a crucial role in advanced optical biointerfaces.

Keywords: implantable devices; optical waveguides; optical fibers; biocompatible; biodegradable

1. Introduction

Optical fibers and waveguides are widely used in fiber-optic telecommunication, remote sensing and on-chip devices for a long time. With recent development of optical techniques towards medical applications, the interaction of light and living matter has always been favorable in a wide variety of medical purposes, such as laser surgery, phototherapy, biosensing and imaging [1–10]. However, the penetration depth of light in biological tissues at visible and near-infrared wavelengths is limited (<3 mm), due to absorption and scattering characteristics of tissues [11,12]. Implantable light sources, such as light emitting diodes and cell-based lasers [13–15], have been demonstrated for biomedical applications. Besides, injectable upconversion optoelectronic devices [16] or nanoparticles [17] have been developed to expand the optical penetration depth. However, these active implantable light sources still demand for further optimization to realize practical and clinical uses considering their sophisticated fabrication methods and biocompatibility issues. Alternatively, implantable fibers and waveguides provide an accessible way to deliver therapeutic or sensing light into deep tissues to overcome the penetration limit. Besides, optical fibers and waveguides are also serving the rapidly emerging photonic and optoelectronic implants to transmit optical signals and power [2,18]. In recent years, integrated optical fibers and waveguides have begun to emerge particularly in biomedical applications including optogenetics [19–21], fluorescence detection [22,23] and sensing [24,25]. Considering the impacts that these implanted devices might have on their host, the development of novel optical fibers and waveguides requires not only ideal optical properties, but also desirable biocompatibilities such as mechanical properties matching biological tissues, low cytotoxicity, minimally invasive injection, etc. In addition, optical fibers and waveguides made from biodegradable materials, which can physically disappear after use and eliminate the risk of further retraction, have attracted particular interests in the field.

Deriving from telecommunication, silica quartz based optical fibers have been used in biomedical field over decades [26,27]. The core-cladding structure in conventional silica fibers can be formed by the mature thermal drawing method and has the potential to deliver light energy into the tissue with low optical losses. However, their rigid and stiff nature might cause large tissue lesions, resulting in poor compatibilities with biological systems [28–30]. To resolve these challenges, implantable fibers and waveguides made from biocompatible and biodegradable materials have been actively explored.

In this article, we provide an overview on the recent advances of implantable optical fibers and waveguides for biomedical applications. In addition to inorganic fibers, progresses in flexible, stretchable, biocompatible and even biodegradable fibers and waveguides are discussed, associated with related materials, manufacturing strategies, device performance and functionalities. In particular, Table 1 summarizes representative materials and methods to form these fibers and waveguides, which will be explained in detail in Section 2. In Section 3, we highlight their applications in fields including optogenetics, phototherapy, sensing and imaging.

Table 1. A summary of representative materials and methods to form biocompatible fibers and waveguides.

Category	Materials	Fabrication Process
Inorganic materials	Silica, phosphate, silicon oxynitride	Thermal drawing, Lithography
Natural materials	Silk, cellulose, bacterial cells	Thermal drawing, Printing, Molding
Hydrogel	Agarose gel, PEG, alginate	Molding
Synthetic polymers	PLLA, PLA, PLGA	Thermal drawing, Molding
Elastomers	PDMS, POC-POMC	Molding
Multifunctional	COC, PC, CPE	Thermal drawing

2. Materials and Synthesis

Figure 1 provides an overview of representative approaches to synthesize and fabricate implantable fiber or waveguides. Thermal drawing (Figure 1a) is commonly used for conventional optical fibers made of silica or other inorganic materials. Employed with custom-designed drawing towers [31,32], multifunctional polymer fibers with delicate structures can also be drawn from designed preforms [33]. Fibers fabricated with such drawing processes will be discussed in detail in Sections 2.1 and 2.6. Three-dimensional (3D) printing (e.g., direct ink writing, Figure 1b) is a commonly used technique to develop implantable fibers and waveguides derived from organic materials. The inks flow through the printer nozzle at a controlled speed and then experience rapid solidification, forming optical waveguides and other photonic structures with desired shapes. As examples, Section 2.2 discusses biocompatible silk optical waveguides [34] developed by printing. Alternatively, lithographic techniques (Figure 1c) can be applied to fabricate waveguides, fluidic channels [35], etc. with features at micro- and even nanoscale. Molding (Figure 1d) is also an effective way to manufacturing optical fibers at low costs. In particular, Section 2.3 will discuss the development of hydrogel fibers [36], by ultraviolet (UV) curing optical materials in transparent tubes along with a dip-coating process to form claddings after extrusion. Molding techniques can also be applied to form fibers with other materials including synthetic polymers and elastomers, which will be respectively discussed in Sections 2.4 and 2.5.

Typical characteristics of various optical fibers and waveguides include their structural (e.g., porosity, crystallinity), optical (e.g., refractive index, loss), mechanical (e.g., Young's modulus, bending stiffness) and thermal properties (e.g., glass transition and melting temperatures), which play critical roles in their versatile applications for different fields. These properties can be evaluated by various approaches, for example, using electron microscopy and X-ray diffraction for structures, ellipsometry and spectroscopy for optics, dynamic mechanical analysis and thermal analysis.

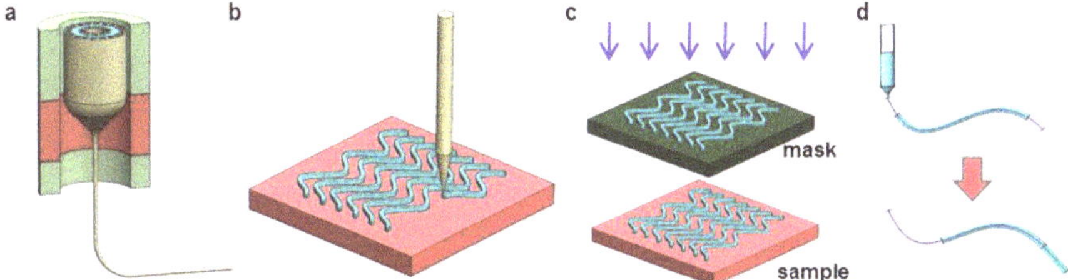

Figure 1. Schematic overview of representative approaches for fiber/waveguide fabrication. (**a**) Thermal drawing. (**b**) Printing. (**c**) Lithography. (**d**) Molding.

2.1. Inorganic Materials

Inorganic materials such as glasses (i.e., fused silica) are the base components for optical fibers (Figure 2a) and essential materials in the field of optics. Silica-based materials exhibit high optical transparency in a wide range of wavelengths from visible to near-infrared (near-IR), rendering optical fibers with extremely low propagation losses (as low as 0.2 dB/km around 1550 nm) [37]. Based on these glass materials, fibers with core-cladding structures can be formed at high temperatures using the thermal drawing method. The step or gradient refractive index profiles ensure the total internal reflection within the fibers and optical confinement in the core region.

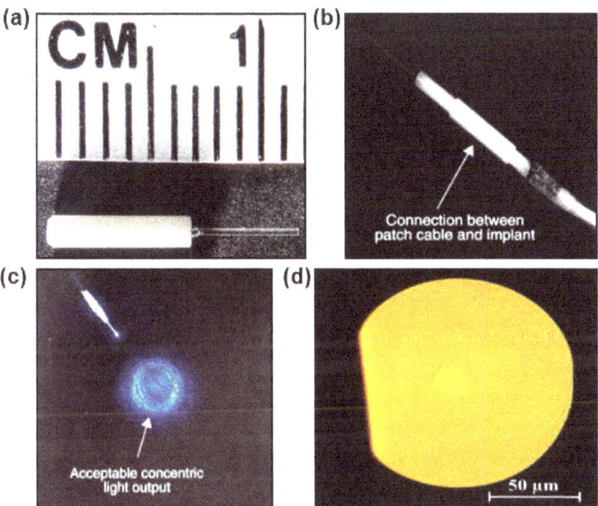

Figure 2. Inorganic fibers. (**a**) A silica fiber for tissue implant. (**b**) A silica fiber connected with a patch cable. Reproduced with permission [27]. Copyright 2011, Nature Publishing Group. (**c**) Transmission of light through a silica fiber. Reproduced with permission [27]. Copyright 2011, Nature Publishing Group. (**d**) The cross-sectional view of a calcium-phosphate glass based optical fiber. Reproduced with permission [38]. Copyright 2016, OSA.

Given that refractive indices of various biological tissues are generally ranged from 1.33 to 1.51 [11], the core-cladding structure confines photons into silica fibers with total internal reflection and guide light into deep tissues with negligible losses. The step-index structure has also been applied to the design of novel implantable optical fibers, which will be discussed hereinafter. Another advantage of silica-based implantable optical fibers lies in their chemical inertness, which ensures their long-term stability during operation. Standard silica optical fibers, inserted into a ceramic ferrule, can be connected with external light sources including LEDs and lasers by a cannula (Figure 2b), and serve as standard tools for optical neural interfaces [27]. As shown in Figure 2c, the output light profile from a silica fiber generally exhibits a uniform and centric power distribution.

Alternatively, highly transparent, degradable calcium-phosphate glasses (PGs) are considered as a novel category of materials for implantable photonic devices [38,39]. A new optical fiber made of this material recently has been manufactured using the thermal drawing technique [38]. Figure 2d presents a cross-section view of such a fiber. The fiber's diameter is 120 μm and the core diameter is 12 μm. With different fractions of calcium oxide (CaO) and magnesium oxide (MgO), the refractive index of the PGs can be adjusted and then proper PGs are chosen to form a core-cladding structure with a step-index profile. Based on such structures, single-mode fibers can be formed and the measured optical loss of the fiber is as low as 1.86 dB/m at the wavelength of 1300 nm. The propagation loss at visible wavelengths is slightly higher and reaches 4.67 dB/m at 633 nm. The biodegradation of PGs fibers core is evaluated in physiological conditions. It demonstrates a decrease of fiber diameters along with the weight losses. In addition, the fiber degradation rate varies with PGs' compositions, and higher CaO:MgO ratio leads to higher degradation rate.

Silicate or phosphate glasses provide ideal optical properties for low-loss fibers and waveguides. PGs based fibers also demonstrate their utility as fully bioresorbable implants. However, their rigidness and fragile characteristics render undesirable biocompatibilities with biological systems and then restrict their practical biomedical uses. Organic materials with better biocompatibilities have become promising candidates for optical implants, which will be discussed subsequently.

2.2. Natural Materials

Naturally derived materials with ideal biocompatibility and biodegradability have been widely investigated for a variety of medical uses, such as drug delivery [40,41], tissue engineering [42,43], sensing and imaging [44,45]. To date, various optical waveguides have been fabricated using natural materials such as silk [34], cellulose [46] and bacteria cells [47]. For example, bio-derived cellulose polymers can be thermally drawn to form a core-cladding fiber structure, with some cellulose powders in between to form a hollow channel for potential drug delivery [46]. The propagation loss is measured to be ~1 dB/cm. These cellulose fibers are fully dissolved after 1-day immersion in aqueous solutions and during the dissolution, the light transmittance increases with water intake. Optical waveguides can also be formed from bacterial cells (e.g., Escherichia coli, propagation loss ~0.23 dB/μm [47]), and their optical losses are needed to be further reduced for practical uses.

In recent years, silk fibers produced by *Bombyx mori* worm or spiders have been studied and their application in the medical field are extensively explored [48,49]. To enable the silk's capability in device applications, it is firstly converted to silk fibroin in a thin-film form through reverse engineering of the natural fiber generation process. The desirable biocompatibility and biodegradability of silk-based materials make them suitable for implantable devices that can be left in the body and are gradually resorbed by biological systems. Furthermore, silk materials exhibit mechanical properties similar to those of tissues and also offer distinctive optical properties. The combination of all these perfect properties makes silk an ideal material for a wide range of optical and photonic devices in medical applications [50].

Optical waveguides made from high-index (n = 1.54) regenerated *Bombyx mori* silk fibroin (Figure 3a) were fabricated on quartz (n = 1.52) via a printing based method [34]. The fabrication process of printed silk optical waveguides is schematically illustrated in Figure 1b. A position-adjustable condenser is applied to couple light inside the waveguide so that the transverse face is imaged and analyzed (Figure 3b). The average propagation loss of these waveguides is ~0.5 dB/cm. Recent investigations have demonstrated a biocompatible step-index optical waveguide made of silk fibroin [51,52]. This waveguide is fabricated with a high-index silk film (n = 1.54) for the core and a low-index silk fibroin hydrogel (n = 1.34) as the cladding layer, using injection-molding process.

These silk waveguides are able to guide light through tissue with a propagation loss of ~2 dB/cm, which is mainly attributed to the rough edge of the silk film. An example of light guiding through chicken breast tissues via a silk optical waveguide coupled to a single mode glass optical fiber and a green laser source is shown in Figure 3e [51]. Silkworm gut fibers are fabricated by extracting and stretching the silk glands, which can serve as light-diffusion waveguides with a rough surface leading to massive light scattering and have been demonstrated for an optical stimulation on cell proliferation [53].

Unlike the natural and regenerative silkworm silk that have been widely studied through years for developing waveguides, less attention has been paid to the spider silk due to challenges of manufacturing at a large scale. Recently, a new class of optical fibers made from native spider silk (n = 1.50) through a molding process was presented [54]. These fibers have a diameter of ~5 μm, which are much thinner compared to conventional ones. The electron microscopy image shows a smooth surface of these fibers (Figure 3c). Figure 3d illustrates light guidance through the fiber, with a propagation loss of ~10.5 dB/cm. Native spider silk fibers are also able to deliver light in physiological liquid and in an integrated photonic chip. Through genetic engineering, producing spider silk proteins at a large scale becomes possible [55]. Recently, researchers have fabricated optical waveguides by exploiting genetically engineered spider silk proteins [55]. The refractive index (n = 1.70) of these fibers is much higher than that of biological tissues (n = 1.33–1.51). In addition, with a low propagation loss of ~0.8 dB/cm lower than that of regenerative silkworm silk waveguides, these waveguides are capable of guiding and delivering light and energy into deep tissues. As represented in Figure 3f, after inserting the recombinant spider silk optical waveguide, the penetration length of the light into muscle increases from less than 1 cm to 3 cm [55].

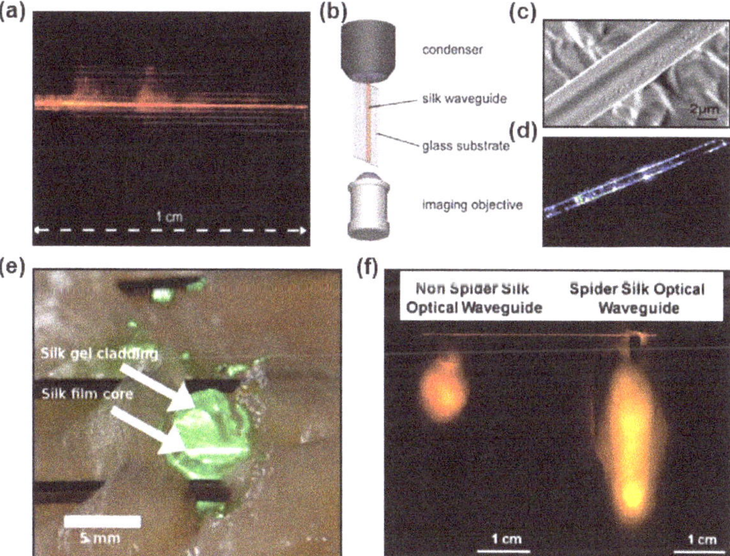

Figure 3. Waveguides made of natural materials. (**a**) Silk optical waveguide. Reproduced with permission [34]. Copyright 2010, Wiley-VCH. (**b**) Tools to analyze transverse faces of waveguides. Reproduced with permission [34]. Copyright 2010, Wiley-VCH. (**c**) Scanning electron microscopic (SEM) image of a silk fiber. Reproduced with permission. [54] Copyright 2013, AIP. (**d**) Micro-beam profile of a spider silk fiber. Reproduced with permission. [54] Copyright 2013, AIP. (**e**) An implanted silk optical fiber in tissue. Reproduced with permission. [51] Copyright 2015, OSA. (**f**) Comparison of light penetration in different waveguides. Reproduced with permission [55]. Copyright 2017, ACS.

2.3. Hydrogels

Hydrogels can be easily integrated within biological systems, because of their mechanical properties similar to those of biological tissues, as well as their high water content and porous structures [56,57]. These attributes allow them to encapsulate living cells and make them ideal components of bio-scaffolds in tissue engineering [57]. Hydrogels with optimized chemical and physical compositions act as favorable materials for light guiding [2]. Based on their capability of cell encapsulation, various cell-based optical sensing and phototherapy can be carried out in vitro or in vivo [58]. In addition, hydrogel based optical waveguides have potentials in various applications including drug delivery [59], optogenetics [58] and so on.

As an example, hydrogel-based planar optical waveguides are manufactured out of agarose and gelatin via spin coating [60]. The core region is made by cross-linked gelatin with a refractive index of 1.536, while the cladding layer is made by agarose with a refractive index of 1.497. Both agarose and gelatin are bio-derived polymers that meet the need of biocompatibility and biodegradability. Via spin-coating or direct molding, the size of planar waveguides can vary from a few μm to mm, with the potential for implantable applications. Agarose-based hydrogels are also exploited to form an optical waveguide integrated with a fluidic channel by soft lithography [35]. Materials of two different gel concentrations with a refractive index difference are applied, forming a two-layer structure to guide light effectively. A microfluidic channel is built on top of the waveguide for on-chip biosensing applications.

Shown in Figure 4a, hydrogels based on polyethylene glycol diacrylate (PEGDA) are utilized to fabricate slab waveguides using mold-injection and ultraviolet (UV) cross-linking processes, with particular applications in optogenetic stimulation and cell-based toxicity sensing [58]. Specifically, living cells can be encapsulated into the waveguide structure and perform sensing or therapeutic operations. In addition, excellent transparency allows PEGDA waveguides to simultaneously deliver excitation light and collect fluorescence signals by coupling to standard silica fibers. The optical properties of PEGDA hydrogels with different concentrations and molecular weights are characterized. Other properties, including stiffness, cell viability and swelling ratio, are taken into account. Hydrogels with a molecular weight of 5 kDa and a 10% w/v concentration are chosen to form implantable waveguides with an average optical loss of 0.23 dB/cm in the blue-green spectrum (from 450 nm to 550 nm). Figure 4b compares light guiding properties within the tissue with or without a hydrogel waveguide coupled to an external light source. By implanting the slab waveguide, the illumination area is expanded by 40 times.

To improve the light-guiding efficiency, step-index optical waveguides based on hydrogels are developed by molding and dip coating [36]. The difference of refractive indices between the polyethylene glycol (PEG) based core and the calcium alginate based cladding is critical for light confinement in waveguides. PEG has refractive index of about 1.46, while the low concentration of sodium alginate as the second hydrogel has a lower refractive index that is close to water (about 1.34). Figure 1d illustrates the molding process to form the core of such step-index hydrogel-based optical fibers. In the core, a platinum-cured silicone tube is used as a mold and filled with a hydrogel precursor solution, which is subsequently cross-linked through UV irradiation. Then the dip coating method is used to encapsulate the core with Ca^{2+} crosslinked hydrogels. In this process, the diameter of optical fibers could be adjusted by the number of dipping times. Figure 4c demonstrates the light guidance of such step-index optical fibers in the air and biological tissue, respectively. With a low optical loss (0.3 dB/cm in the air and 0.49 dB/cm in tissue) in visible ranges, these fibers have been implanted into live mice to justify the capability in sensing and photomedicine.

Fragile characteristic of hydrogel photonic devices is the result of weakness of common synthetic hydrogels which possess a brittle nature. These poor mechanical properties would cause difficulty for these hydrogels to be applied in implantable devices because during body movements, they might cause damage to the tissue. Therefore, improving the mechanical properties and stability of hydrogel fibers has become critically important [61], especially for biosensing applications. A novel hydrogel material with enhanced toughness and stretchability has been applied for optical fibers [61]. Specifically, a hybrid polymer network that contains both ionic and covalent bonds is introduced into hydrogels, improving the material robustness. Following this principle, highly stretchable hydrogel fibers are developed for strain sensing. Combining Ca^{2+} crosslinked alginate and covalently crosslinked polyacrylamide (PAAm), tough hydrogels are synthesized via UV polymerization (for the core) and dip coating (for the cladding). A 1.1 mm-diameter fiber can be stretched up to 730% and repeated 100 times (with 300% strain) without apparent plastic deformation. The optical absorption of dye-loaded fibers is proportional to the applied strain based on which the strain sensing function has been demonstrated. Such stretchable waveguides are envisioned to be used as optical strain sensors in wearable devices. In addition, another class of optical fibers has been recently developed for glucose sensing [62]. In this work, a hydrogel fiber is demonstrated, which is based on poly(acrylamide-co-poly(ethylene glycol) diacrylate) p(AM-co-PEGDA) for the core region and cladded with Ca^{2+} alginate and functionalized with phenylboronic acid. Figure 4d,e illustrate these hydrogels based optical fibers as well as their capability of guiding light through porcine tissues. Refractive indices of the cladding and the core are measured to be 1.34 and 1.46, respectively. The low optical loss measured in tissue phantoms is about 0.28 dB/cm. Their unique mechanical properties attract lots of attention in medical applications such as phototherapy and photomedicine [63]. Researchers have also manufactured a fluorescence slab waveguide made of hydrogels and doped with carbon dots to detect heavy metal ions, such as Hg^{2+} ions [64]. These waveguides exhibit smooth surface and obtain desirable transparency in the range of 400–800 nm, with an average propagation loss of less than 1.25 dB/cm.

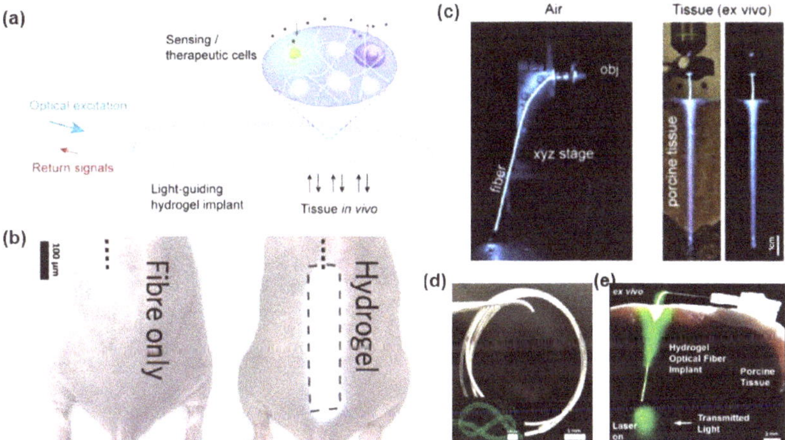

Figure 4. Hydrogel based waveguides. (**a**) Illustrating scheme of a light-guiding hydrogel with encapsulated cells. Reproduced with permission [58]. Copyright 2013, Nature Publishing Group. (**b**) Comparison of light scattering profiles with (**left**) and without (**right**) hydrogel implants. Reproduced with permission [58]. Copyright 2013, Nature Publishing Group. (**c**) Hydrogel fiber guiding light in air (**left**) and porcine slices (**right**). Reproduced with permission. [36] Copyright 2015, Wiley-VCH. (**d**) Image of a hydrogel optical fiber. Reproduced with permission [62]. Copyright 2017, Wiley-VCH. (**e**) Insertion of a fabricated hydrogel fiber in tissue. Reproduced with permission [62]. Copyright 2017, Wiley-VCH.

2.4. Synthetic Polymers

In the past, biodegradable synthetic polymers like poly (lactic acid) (PLA), poly (lactic-co-glycolic acid) (PLGA), and polycaprolactone (PCL) have been extensively explored for use as structural materials for bio-interfaces [65,66]. Degradation rates of these synthetic polymers are controllable and vary with molecular weights, chemical compositions (e.g., lactide/glycolide ratio) and aqueous environments. They have been widely used as biomedical implants like suture, stents and various injectable products. Although these polymers are mostly utilized as implantable structural materials in opaque forms, amorphous states of these polymers are capable of guiding light in tissue because of their low extinction and high refractive indices (~1.46), which are well-suited for optical waveguide devices. PLA and PLGA based implantable waveguides are fabricated by melt pressing and laser cutting, as shown in Figure 5a,b [67,68]. These waveguides are coupled to external light sources to direct light into the deep biological tissue for treatment. Photochemical tissue bonding (PTB) treatment of a full-thickness skin incision is successfully demonstrated. The modulation of the waveguide surface pattern can help optimize the output light profile by inducing bending loss and an optimal waveguide is demonstrated with a uniform light distribution for photobleaching (Figure 5c). The propagation losses of these bioresorbable waveguides are measured in different environment and presented in Figure 5d.

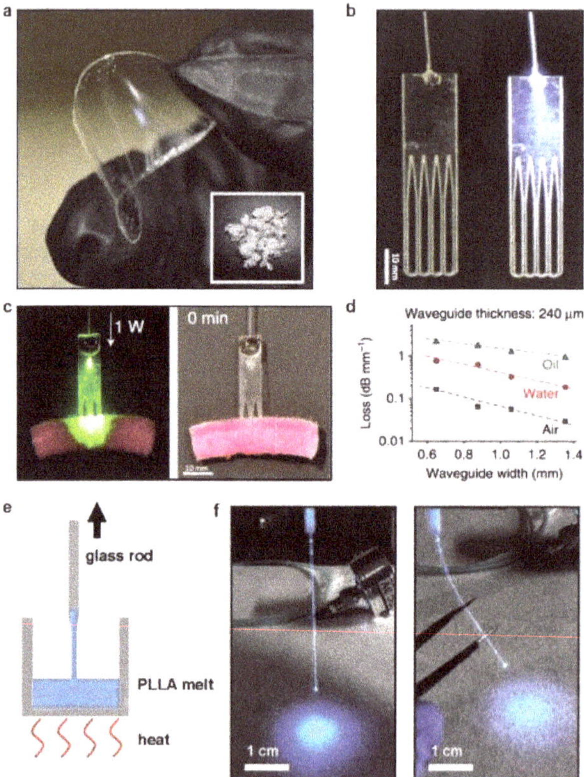

Figure 5. Fibers and waveguides made of biodegradable synthetic polymers. (**a**) A transparent, flexible PLLA film. Reproduced with permission [68]. Copyright 2016, Nature Publishing Group. (**b**) A comb-shaped PLLA waveguide. Reproduced with permission [68]. Copyright 2016, Nature Publishing Group. (**c**) Light delivery into deep tissues via the waveguide. Reproduced with permission [68]. Copyright 2016, Nature Publishing Group. (**d**) Measured optical loss of the waveguide in different media. Reproduced with permission [68]. Copyright 2016, Nature Publishing Group. (**e**) Process of forming thermally drawn PLLA fibers. Reproduced with permission [69]. Copyright 2018, Wiley-VCH. (**f**) A cylindrical PLLA fiber connected to a blue LED. Reproduced with permission [69]. Copyright 2018, Wiley-VCH.

Recently, poly(L-lactic acid) (PLLA) based optical fibers are fabricated via thermal drawing process [69]. Figure 5e schematically illustrates the fabrication process. In this method, PLLA powders are melted at 220 °C and fibers are formed through a crystalline-to-amorphous phase transition. Fiber diameters are manipulated by controlling the drawing speed. Figure 5f shows a formed PLLA fiber with a diameter of around 200 μm coupled to a blue LED light source. Due to the transparency of PLLA in the visible ranges, these optical fibers with cylindrical structures are able to deliver light in the air with a propagation loss of 1.64 dB/cm. The higher refractive index of PLLA (n = 1.47) than that of tissue makes PLLA fibers guide light into deep tissues effectively. These implantable fibers are applied as neural interfaces including optogenetics and fluorescence detection. With experiments in living mice, in vivo brain function modulations including neural sensing and interrogation are realized, along with a gradual biodegradation.

2.5. Elastomers

Optical waveguides made of polydimethylsiloxane (PDMS) based elastomers (n = 1.42) are demonstrated via molding process. The specific feature of this waveguide is the ability to carry adequate amounts of light uniformly by a tapered structure, contributing to the scleral cross linking treatment [70]. Recently, a fully biodegradable step-index optical fiber made from elastomers are developed [71]. For this biodegradable fiber, the core is made of poly(octamethylene maleate citrate) (POMC) while the cladding layer is formed with poly(octamethylene citrate) (POC). Citrate-based elastomers have been studied as biodegradable implants for tissue engineering, drug delivery, etc. and the monomer, citric acid, is an intermediate of metabolism. The elastomer fibers exhibit high flexibility (initial modulus ~3.39 MPa) and stretchability (elongation ~61.49%). With a measured propagation loss of ~0.4 dB/cm, they are capable of carrying sufficient amount of light deep to tissue. The degradation time (ranging from few weeks to years) of these fibers depends on the synthesis and structure of citrate-based elastomers. These developed step-index fibers are applied for in vivo fluorescence detection to study deep inside tissue. Besides, the image transmission function has also been displayed.

2.6. Hybrid Materials

Biomedical implants with versatile functions can help to realize a multimodal sensing and (or) stimulations, for instance, a real-time interrogation and monitoring of neural circuits using optical, electrical and chemical signals [33,72]. Such multifunctional waveguides can be fabricated from hybrid materials by integrating conductive wires (or coatings) and hollow channels with optical fibers. Most of the devices that had been used for recording the neural activities were mostly fabricated out of metals, semiconductors or glass which can easily cause harm to the surrounding tissues during animal body's movement because of their rigidity or stiffness characteristics. Polymer materials are employed for multifunctional fibers, considering their flexibility and excellent transparency which makes them suitable for the purpose of implantation as neural interfaces. To develop the delicate structure of these multifunctional fibers, thermal drawing process is used based on a well-designed preform as a starting structure so that different functions can be integrated in one fiber probe [31]. These hybrid materials should have similar mechanical and thermal properties to form the deterministic fiber structures. Moreover, in order to guide light, the internal and external layers of such optical fibers are required to have different refractive indices. Conductive polymers or silver nanowire coatings are integrated to record the electrophysiological signals. Meanwhile, hollow channels are developed for drug delivery and other fluidic injection. By controlling drawing temperatures and speeds, researchers can achieve the geometry of the device to fit specific applications.

On the basis of this idea, following materials are selected to fabricate hybrid-materials fibers for the use of optical neuromodulation: polycarbonate (PC; refractive index n = 1.58; glass transition temperature T_g = 145 °C; Young's modulus E = 2.38 GPa) and cyclic olefin copolymer (COC; n = 1.52, T_g = 158 °C, E = 3.0 GPa) [33,73]. Figure 6 schematically illustrates the thermal drawing method and steps of the fabrication to produce a multifunctional fiber. The structure of the fiber consists of an optical waveguide made of PC as the core and COC for the cladding, six electrodes made of conductive polyethylene with 5% graphite (gCPE), two hollow microfluidic channels and another PC based cladding layer for protection. The propagation loss is measured to be 1.6–2.6 dB/cm. The impedance at 1 kHz of the fiber electrode is approximately 1 MΩ under bending or flat conditions. The device can be implanted into animal brains and operated for a long time (about two months) considering their flexibility and small dimensions (Figure 6d). The fiber is able to simultaneously deliver drugs into the neuron cells, stimulate optical signals and record electrophysiological activities in one step. Specifically, viral injection is performed through the hollow channel of the multifunctional fibers and then colocalized protein expression, illumination and recording can be achieved in living animals [72].

With similar fabrication process, another kind of hybrid fibers is developed with stretchable materials, aiming to probe the spinal cord circuits of freely moving mice (Figure 6b). Similar materials as fibers shown in Figure 6a are selected for the core and the cladding, combining with polydimethylsiloxane (PDMS). A mesh of transparent silver nanowires (AgNWs) with a wire diameter of 70 nm and a length of 40 mm is dip-coated alongside the fiber and encapsulated with PDMS (Figure 6c), producing a conductive layer to collect electrophysiological signals. The electrode impedance is around 50 kΩ and can be tuned by choosing different concentrations of AgNWs solutions. Full elastomer-based AgNWs-coated fibers are formed based on COC elastomer as a core and PDMS as a cladding. The fiber can be stretched to 100% strain by integration 3 layers of AgNWs while maintain its electrical conductivity. The stretchable fibers are implanted into the lumbar region of the spinal cord (Figure 6e) and both spontaneous and optical stimulated neural signals can be recorded [74].

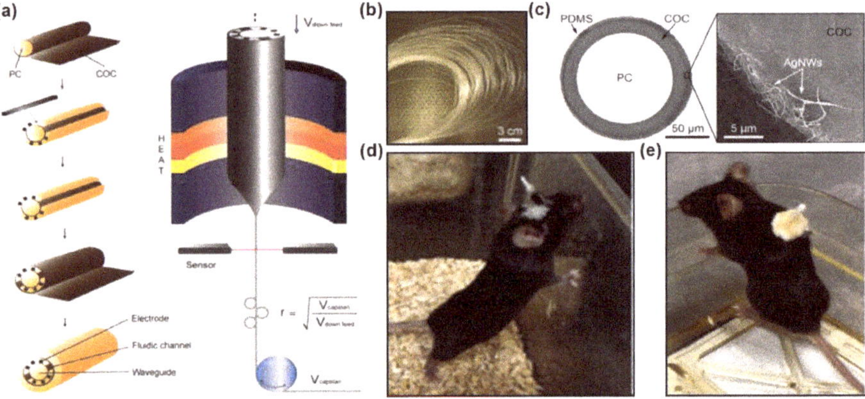

Figure 6. Hybrid, multifunctional fibers. (**a**) Steps involved in the fabrication of multifunctional fiber device. Reproduced with permission [72]. Copyright 2017, Nature Publishing Group. (**b**) Multifunctional fiber spools. Reproduced with permission [74]. Copyright 2017. AAAS. (**c**) Cross-sectional view of the fiber (right panel) and SEM image of the AgNW electrodes (left panel). Reproduced with permission [74]. Copyright 2017, AAAS. (**d**,**e**) Living mice with the implanted fiber probes in the brain and spiral cord, respectively. Reproduced with permission [72,74]. Copyright 2017, Nature Publishing Group and AAAS.

3. Biomedical Applications

In the previous section, we discussed the materials and fabrication processes for biocompatible and implantable optical fibers and waveguides. Fibers and other photonic structures based on biocompatible materials such as ceramics, hydrogels, synthetic polymers and even natural materials have been exploited in biomedical applications ranging from optogenetic stimulation, fluorescence photometry, surgery, phototherapy, to biochemical sensing and imaging. In this section, we provide examples of their wide applications in the development of optogenetics, surgery and phototherapy, optical sensing and bio-imaging. The cartoons in Figure 7 demonstrate three representative biomedical applications: optogenetics (Figure 7a), laser surgery (Figure 7b) and fluorescence sensing (Figure 7c), in which implantable fibers or waveguides play an important role.

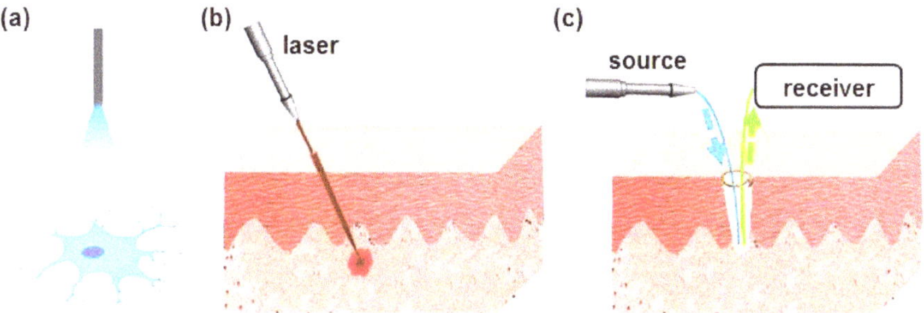

Figure 7. Schematically illustrated examples of applications for implantable fiber/waveguides. (**a**) Optogenetics. (**b**) Laser surgery. (**c**) Fluorescence sensing.

3.1. Optogenetics

Optogenetics, the combination of optical and genetic methods to activate or deactivate certain events in specific cells, is employed today to study activities of neurons and their related behavior. Over the decades, it has enabled acquisition of invaluable insight into a wide range of fields in physiology, pathology, behavior and even psychiatry. Optical fibers and waveguides, along with microbial opsins and vectors, are most commonly core features in optogenetics. Specific opsin gene expression, carried to well-defined cells by viral vector, encodes a protein that causes electrical current across cell membranes when illuminated by light. With excitation light delivered by optical fibers and waveguides, targeted cells exhibit events of interest. Optogenetic methods have become standard tools for studying neural circuits underpinning behavior in freely behaving animals (Figure 8a). Here, we provide studies exploring implantable, biocompatible optical fibers and waveguides with various functions for optogenetics [75].

Optical stimulation is often compared with electrical one in causing or inhibiting activities of neurons in brain regions. Although electrical stimulation has been employed in both experimental and clinical level to probe and control neural activities in discrete brain areas, it fails to control cells specifically, as shown in Figure 8b [20]. Optical stimulation, by contrast, targets only neuron type expressing microbial opsins (ChR2, NpHR, etc.). This characteristic also sheds light on single-cell control and cellular signal traveling.

Recently, much attention has been drawn to the development of optical fibers that can realize several functions with one fiber. Stretchable probes are demonstrated, which are made of polymer fiber coated with conductive meshes of silver nanowires [74]. These probes have the ability to simultaneously stimulate neurons and record electrophysiological activity. With flexible and stretchable characteristics, such fibers can be tailored for stimulating and monitoring electrophysiological activities in spinal cord. Park et al. made further attempt by integrating all steps of optogenetics in a single biocompatible platform [72]. These fibers contain micro channels, electrodes and waveguides core to enable viral vector injection, optical stimulation and simultaneous electrophysiological recording, all within the general dimensions of fibers used in optogenetic studies. Figure 8c shows the probe equipped by an optical ferrule, electrical connector, and an injection tube to fit its functions. Microfluidic channels within fibers deliver liquid into deep brain tissue efficiently during the tests in Figure 8d. The device maintains reliable optical stimulation capabilities for months, with high signal to noise ratio (Figure 8e). Such technology allows for one-step surgery, providing minimally invasive alternatives to the current practice in optogenetics.

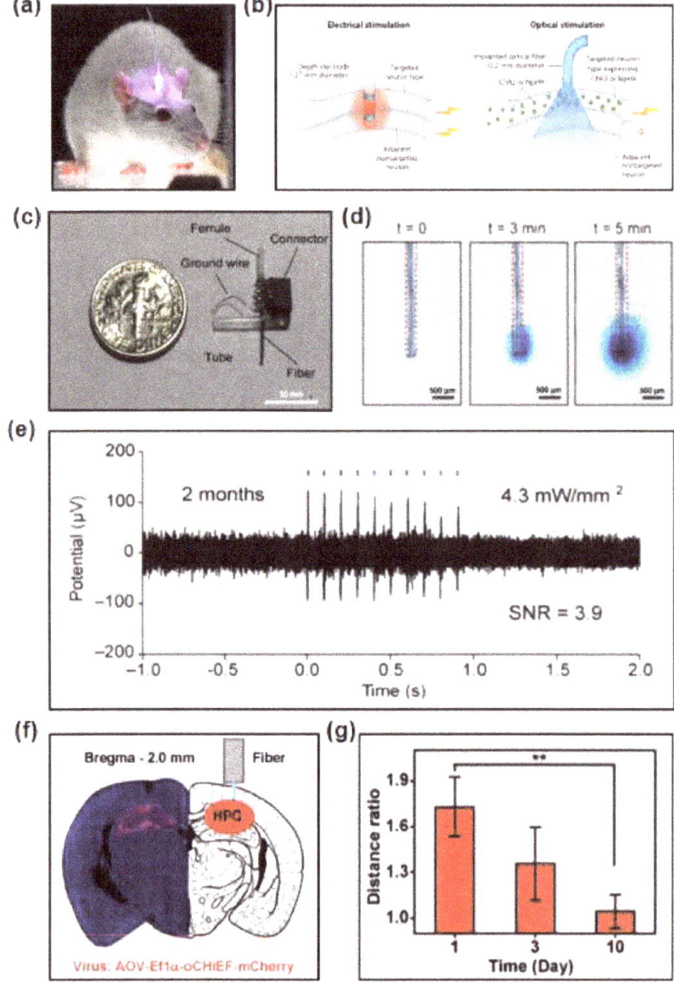

Figure 8. Application in optogenetics. (**a**) A fiber implanted into a freely behaving mouse. Reproduced with permission [76]. Copyright 2015. (**b**) Comparison of electrical simulation (**left**) and optical stimulation (**right**) for neural cells. Reproduced with permission [20]. Copyright 2014, Annual Reviews. (**c**) A fiber probe equipped by an optical ferrule, an electrical connector, and an injection tube. Reproduced with permission [72]. Copyright 2017, Nature Publishing Group. (**d**) A multifunctional fiber with integrated microfluidic channels for fluidic injection into the brain. Reproduced with permission [72]. Copyright 2017, Nature Publishing Group. (**e**) Electrophysiological traces under optical stimulation. Reproduced with permission [72]. Copyright 2017, Nature Publishing Group. (**f**) Confocal microscopic image of a coronal section after full expression (**left**) and schematic diagram of fiber implanted into coronal (**right**). Reproduced with permission [69]. Copyright 2018, Wiley-VCH. (**g**) Ratio of measured distance of mice travelling with and without optical stimulation. Reproduced with permission [69]. Copyright 2018, Wiley-VCH.

Fully biodegradable and bioresorbable photonic systems are also the pursuit of continued research in optogenetics. PLLA based optical fibers are explored as tools for light delivery and detection, sparing the secondary damage during the retraction [69]. The virus specifically targeting hyperactivating hippocampus (HPC) neurons is injected into the bilateral HPC and the encoded protein expresses after two weeks. Then PLLA fibers coupled with a 473 nm laser source are implanted into the section (Figure 8f). When the light is guided to HPC via PLLA fibers, a seizure is induced, which activates the mice and results in increased travelling distance. Figure 8g illustrates ratio of travelling distance of mice with and without optical stimulation. The significant decrease of distance ratio indicates performance degradation of PLLA fibers within 10–15 days. These results suggest promising tools for fundamental biomedical research and even clinical uses.

In addition, there are some other designs for specific applications. Pisanello et al. demonstrated that optical fibers with small taper angles can be used to illuminate focal or broad brain volumes [77]. They also use focus ion beam to open multiple light windows on a tapered optical fiber, which allows simultaneous, selective stimulation of different brain regions and reduces the invasiveness of optogenetic device [78]. Optogenetics also shows potential in in vivo optical-sensing and light-controlled therapy. Choi et al. reported hydrogel waveguides encapsulating optogenetic cells for sensing of cytotoxicity and therapy in diabetic animals [58]. These works offer promising complement for existing methods in certain scenarios.

3.2. Phototherapy and Laser Surgery

With numerous lasers and optical devices applied in clinical practice to assist operation and cure diseases, light has exerted increasingly significant influence on medicine. Light-activated therapies, laser surgery, optical diagnostics and other emerging technologies are widely used in treating tumors, dermatosis, and so on, and sophisticated technologies have been developed to ease pains in patients and treat diseased tissue with accuracy [79,80].

Phototherapies with selected light wavelengths have become commonplace in clinic and are used in numerable cases. Furthermore, more efficient capabilities can be achieved by exogenous photosensitizers than intrinsic phototherapy, such as oxidation in photodynamic therapy [14]. With appropriated design strategies, photodynamic therapy has the ability to target diseased cells without damage to healthy cells. Today, photodynamic therapy is an established treatment now clinically employed to treat various cancers (Figure 9a). In clinical practice, its efficiency depends on many factors such as the total light exposure dose and light fluency rate [4], rendering convenient light delivery system as one of the future research directions. Therefore, implantable and biodegradable waveguides may effectively extend the therapeutic depth and time frame in photodynamic therapies, enabling chronic deep-tissue treatment in the future [68].

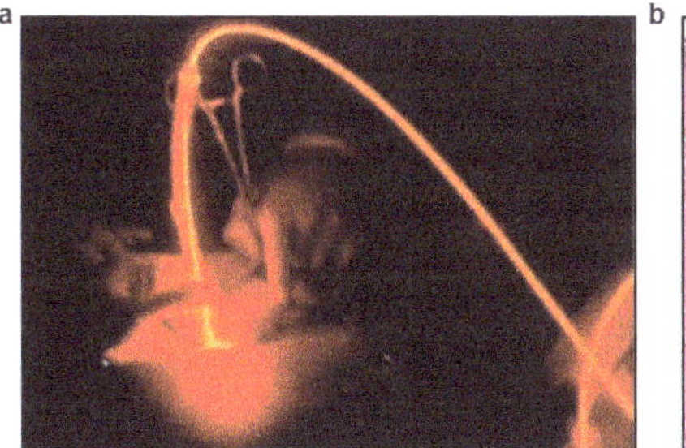
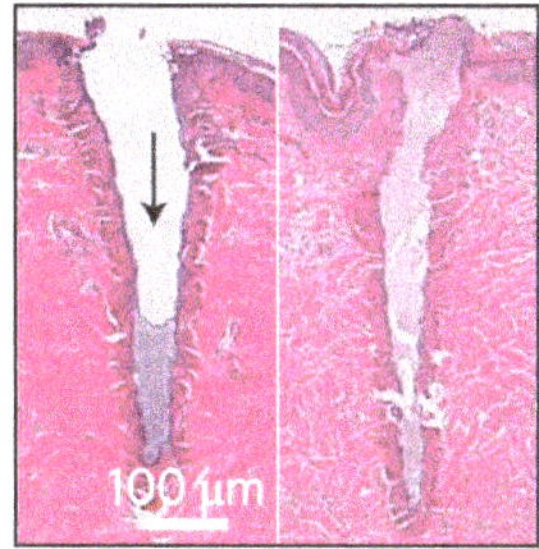

Figure 9. Applications in phototherapy and surgery. (**a**) An example of photodynamic therapy. Reproduced with permission [81]. National Cancer Institute. (**b**) An example of application of light in surgery. Reproduced with permission [1]. Copyright 2017, Nature Publishing Group.

Other light-based technologies, such as flexible waveguides for periscleral cross-linking [70] and photochemical tissue bonding for full-thickness skin incision [68], provide new opportunities for further innovation in photomedicine.

Lasers, with extremely high emission intensities, can be generated in short pulses and selective wavelength and cause hazards to body tissues, a characteristic appreciated in surgery. Laser surgery has been routinely used in various medical fields including ophthalmology, dermatology, and tissue ablation, which is accomplished through fiber-optic delivery [1]. Photothermal induced damage can be filled with plug or clot within minutes, where new tissue will replace unwanted tissue (Figure 9b). Implantable fibers and waveguides possess the advantage of delivering light into deep tissue, and their applications in photothermal or photodynamic therapies show potential for deep-tissue light-induced therapies [68].

3.3. Optical Sensing

Here, we provide an overview on implantable fibers and waveguides with a focus mainly on their applications in sensing. To research on metabolism of living cells and their activities, bio-sensing is a good way to get the information. However, deep-tissue and real-time sensing is still a challenge, blocking the rapid disease diagnosis and researches in physiology, pathology, etc. Light detecting, which uses light to interact with biological systems, has the advantages of minimal invasiveness and high spatiotemporal resolution. Optical sensing offers advantages over its electrochemical counterpart since they can be label-free, conduct real-time continuous monitoring for long periods of time, and cause minimal damage to the body.

Figure 10a illustrates a fiber photometry system used to record neural projection activity underlying certain behavior [23]. An implanted fiber simultaneously delivers 475 nm excitation light and collects fluorescence emission from excited neurons in targeted region. After a dichroic mirror and optical filters, green fluorescence reaches the photodetector. Fluorescence intensity positively correlates with sucrose licking epochs (Figure 10b), indicating the relation between certain activities and targeted brain region. Deep brain fluorescence sensing can also be coupled with optogenetic interrogation to facilitate in vivo neuro activity study [69].

Optical fibers are investigated for application in real-time continuous glucose sensing. Yetisen et al. demonstrated hydrogel fibers consisting of poly(acrylamide-co-poly(ethylene glycol) diacrylate) cores functionalized with phenylboronic acid and cladded with Ca alginate [62]. Phenylboronic acid, a glucose-sensitive chelating agent, is incorporated into the core for sensing glucose. At different glucose concentrations, the chelation of glucose enables reversible changes of fiber diameter, and in response changes of refractive index of the hydrogel fiber. The intensity of light transmitted across the hydrogel fiber is the function of binding time, and gradually reaches the equilibrium (Figure 10d).

The application of optical fibers for blood oxygenation sensing has also been explored. Figure 10e shows a schematic illustration of reflectance oximetry of tissues using fibers [36]. One fiber delivers excitation light at 560 nm and 640 nm into the tissue, while the other collects the light through tissue, which is received by spectrometer. The blood oxygenation is regulated by oxygen level of supplied gas, which is switched between nitrogen and oxygen. The changes of optical intensity at these wavelengths are measured and converted to relative oxy- and deoxy-hemoglobin concentrations. The inhalation of nitrogen results in sudden decrease of oxy-hemoglobin and increase in deoxy-hemoglobin, representing the drop of blood oxygen concentration (Figure 10f). Such findings provide potential insight into the use of photonic tools toward sensing and monitoring in vivo system, pushing forward the field of photomedicine.

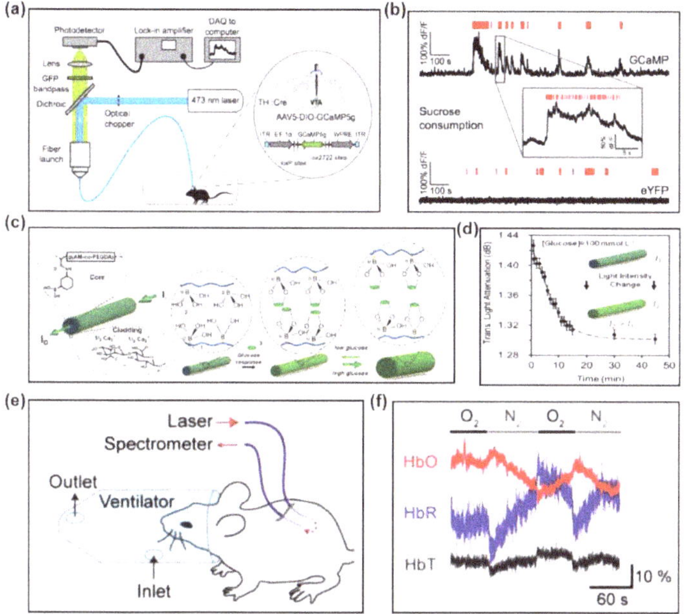

Figure 10. Fibers and waveguides in optical-sensing. (**a**) A scheme diagram of a fiber photometry system. Reproduced with permission [23]. Copyright 2014, Elsevier. (**b**) Recorded signals from the VTA of mice expressing GCaMP (**top**) and eYFP (**bottom**). Reproduced with permission [23]. Copyright 2014, Elsevier. (**c**) Scheme of the glucose-sensitive optical fiber. Reproduced with permission [62]. Copyright 2017, Wiley-VCH. (**d**) Attenuation of the transmitted light through the glucose-sensitive optical fiber. Reproduced with permission [62]. Copyright 2017, Wiley-VCH. (**e**) Reflectance oximetry of biological tissues. Reproduced with permission [36]. Copyright 2015, Wiley-VCH (**f**) Variation of typical time-lapse of calculated concentrations of oxy-hemoglobin, deoxy-hemoglobin and total hemoglobin. Reproduced with permission [36]. Copyright 2015, Wiley-VCH.

3.4. Optical Imaging

From biological to medical fields, bio-imaging is a valuable tool to analyze the characteristics and conditions of cells or specific regions, and even expression and distribution of molecules. Optical imaging, with the characteristics such as high sensitivity, high resolution and high speed, has great potential in exploring mechanism of diseases, obtaining physiological information and diagnosing disease. Over the past decades, the development of optical imaging has deepened our understanding of the structure and physiological activities of living organisms. Early practice is in vitro imaging on biological section or cell culture, which is accurate but fails to get information directly from living tissue. The second method is by opening an optical window in anesthetized animals. Such practice provides high-quality imaging, but cannot support long-time monitoring and in moving animal, presenting a significant limitation for optical imaging. Implantable optical fibers and waveguides in tissues for light delivery and collection become efficient ways to alleviate these problems. Figure 11 shows an example of flexible biodegradable fibers for deep-tissue optical imaging, indicating the potential of image delivery function [71].

To meet the diversified requirement on optical, mechanical and biological functions, researchers present citrate-based polymeric optical fibers with POC cladding and POMC core. Figure 11a depicts the experimental setup of optical imaging. Spatial patterns on a digital micromirror device (DMD) are projected onto the proximal end of the fiber by laser. At the input end of the fiber, a beam splitter is applied to confirm the pattern projection, which transmits across the fiber and is recorded by a charge-coupled device (CCD) camera. Due to the multi-modal propagation, the output projection does not resemble to the input pattern. Figure 11b shows the initial projected letters, the corresponding random speckle patterned output and the reconstructed image. The pre-recorded impulse responses of the multi-mode optical fiber used for image reconstruction are illustrated in Figure 11c. Based on the calibrated impulse responses, the retrieved image through multi-mode fiber still has good resolution. These findings provide a potential implantable imaging platform for advanced tissue imaging, monitoring and so on.

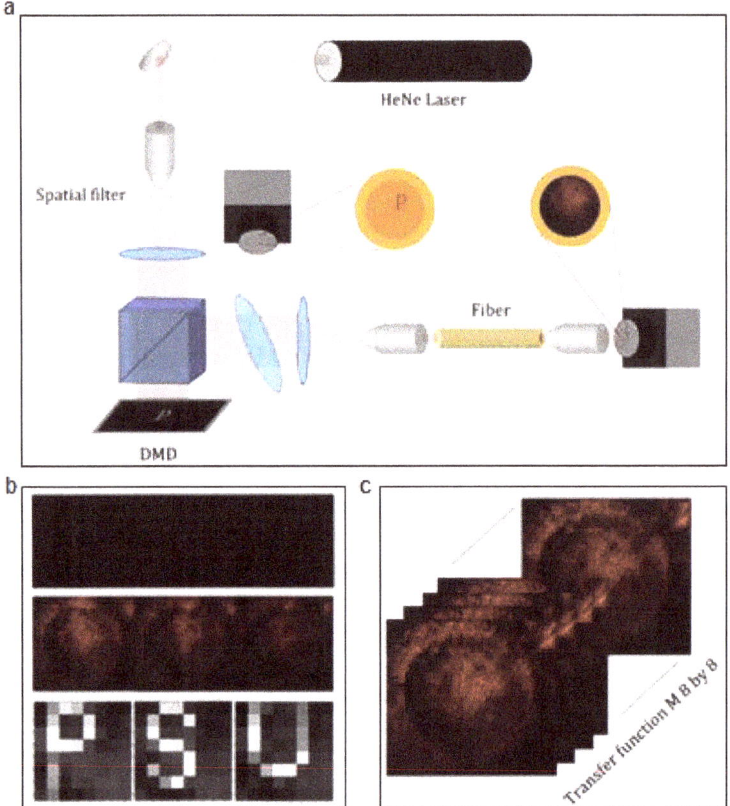

Figure 11. Bioimaging using a degradable fiber. (**a**) Experimental setup for the fiber imaging. Reproduced with permission [71]. Copyright 2017, Elsevier. (**b**) Initial projected letters (upper), the corresponding light output with speckle patterns (middle), and the reconstructed image (lower). Reproduced with permission [71]. Copyright 2017, Elsevier. (**c**) Collective images showing prerecorded impulse responses of the optical fiber. Reproduced with permission [71]. Copyright 2017, Elsevier.

4. Summary and Outlook

We have reviewed biocompatible and implantable optical fibers and waveguides made from inorganic materials, bio-derived natural materials, hydrogels, synthetic polymers, elastomers as well as hybrid materials. Their novel applications in biomedical fields, including optogenetics, laser surgery and phototherapy, biosensing and imaging, have also been summarized. The limited penetration depth of visible light to biological tissues has urged researchers to develop various tools to guide photons into targeted areas. Compared with implantable light sources and bioluminescence devices [2], optical fibers and waveguides offer a simple but effective approach to overcome the obstacle and also exhibit versatile functions by introducing novel materials, designing delicate structures and/or integrating other functional devices.

Although silica fiber optics have been serving as the most widely used light-guiding device due to their minimal optical loss and stability, it needs to be taken into account for more clinical practice that the intrinsic stiffness and brittleness as inorganic materials result in inferior biocompatibility. Therefore, several kinds of soft materials are exploited to fabricate optical fibers and waveguides with ideal flexibility, stretchability and favorable biocompatibility. Similar to silica fibers, the thermal drawing method is adopted to fabricate polymer fibers. Multifunctional fibers are drawn from the well-designed preform so that multiple materials and novel structures can be integrated, realizing simultaneous optical interrogation, electrical recording and chemical delivery [33,72]. Besides the conventional thermal drawing method, molding, printing and microfabricating process are all employed to form a fiber or create a novel waveguide structure. Molding provides a universal solution to implantable fibers and waveguides, for example, combined with dip coating, step-index fibers are made from hydrogels or elastomers and demonstrate diverse functions for phototherapy, biosensing and imaging [36,63,71]. Printing and microfabrication method are mainly used for planar waveguide devices, among which the utility still needs further development. Another significant trend is the emerging biodegradable optical fibers and waveguides which can gradually disappear in vivo and need no retraction from body. Based on natural materials (silk, cellulose, etc.) which possess superior biocompatibility, researchers have manufactured biodegradable fibers and waveguides [46,52]. Biodegradable synthetic polymers (PLGA, PLLA, etc.) demonstrate advantages including ideal optical transparency, tunable degradation time and ease of processing. The PLLA fibers are applied as optical neural interfaces and the optical performance has been systematically studied during biodegradation [69].

The rapidly expanding biomedical applications are calling for both general and specialized studies on implantable fibers and waveguides. First, further decreasing the propagation loss is still highly demanding. Compared to silica optical fibers with a loss coefficient of a few dB/km, most of the newly developed fibers and waveguides have much higher losses (~dB/cm). The propagation loss includes intrinsic optical loss and scattering loss. Although it is difficult to eliminate the intrinsic loss, improvement of materials synthesis and processing methods can lower the light scattering resulted from impurities, rough surfaces, or interfaces. For example, core-cladding fibers with a chemical bonding cladding layer demonstrate a better optical performance [61]. Second, specialized functionalization with materials or design is definitely worth to enable further exploration for biomedical applications. For example, different phototherapy treatments require customized materials and devices to achieve efficacy considering the mechanical properties, robustness, light profiles and so on. In particular applications, light extraction behaviors are of critical importance and should be taken into account [68,70]. With implantable optical fibers and waveguides, various in vitro and in vivo optical sensors can be realized with capabilities of measuring optical intensity, wavelength shift, fluorescence, etc. Materials with novel functions, like phenylboronic acid for glucose sensing [62] and carbon dots for toxicity tests [64], need more explorations in future. Integrating with other implantable devices (photodetectors, optical filters, etc.), the sensing strategies of fibers/waveguides can be further expanded. Third, optimizing biodegradable fibers and waveguides is of great significance for clinical usage. Although different biodegradable materials have been exploited for optical fibers and waveguides, there are still a few challenges for them. Because of biodegradation, the optical performance decreases with time after implantation [69,71]. In the future, triggered materials can be applied to fabricate core-cladding fibers with improved stability and quick degradation after use. By cooperating with transient electronics [82], multiple functions can be achieved as fully biodegradable photonic devices and systems.

It is envisioned that implantable fibers and waveguides will be built with better optical performance, flexibility and biocompatibility, and enhanced functionality. They will not only guide light to deep tissue, but also contribute to diagnosis, therapy and surgery in clinic. As a building block, they will also be an important part of complicated implantable photonic or optoelectronic devices, allowing diverse medical applications in future world.

Author Contributions: R.N. and Q.Z. contributed equally to this work.

Funding: This research was funded by National Natural Science Foundation of China (NSFC Project 51601103) and State Key Laboratory of Advanced Optical Communication Systems and Networks (Shanghai Jiao Tong University, project 2018GZKF03005).

Conflicts of Interest: The authors declare no conflict of interest.

References

1. Yun, S.H.; Kwok, S.J.J. Light in diagnosis, therapy and surgery. *Nat. Biomed. Eng.* **2017**, *1*, 0008. [CrossRef] [PubMed]
2. Humar, M.; Kwok, S.J.J.; Choi, M.; Yetisen, A.K.; Cho, S.; Yun, S.H. Toward biomaterial-based implantable photonic devices. *Nanophotonics* **2017**, *6*, 414–434. [CrossRef]
3. Sakimoto, T.; Rosenblatt, M.I.; Azar, D.T. Laser eye surgery for refractive errors. *Lancet* **2006**, *367*, 1432–1447. [CrossRef]
4. Dolmans, D.E.J.G.J.; Dai, F.; Jain, R.K. Photodynamic therapy for cancer. *Nat. Rev. Cancer* **2003**, *3*, 380–387. [CrossRef] [PubMed]
5. Dougherty, T.J.; Gomer, C.J.; Henderson, B.W.; Jori, G.; Kessel, D.; Korbelik, M.; Moan, J.; Peng, Q. Photodynamic therapy. *J. Natl. Cancer Inst.* **1998**, *90*, 889–905. [CrossRef] [PubMed]
6. Bansal, A.; Yang, F.; Xi, T.; Zhang, Y.; Ho, J.S. In vivo wireless photonic photodynamic therapy. *Proc. Natl. Acad. Sci. USA* **2018**, *115*, 201717552. [CrossRef] [PubMed]
7. Kim, J.; Gutruf, P.; Chiarelli, A.M.; Heo, S.Y.; Cho, K.; Xie, Z.; Banks, A.; Han, S.; Jang, K.I.; Lee, J.W. Miniaturized battery-free wireless systems for wearable pulse oximetry. *Adv. Funct. Mater.* **2017**, *27*, 1604373. [CrossRef] [PubMed]
8. Zhou, Z.; Shi, Z.; Cai, X.; Zhang, S.; Corder, S.G.; Li, X.; Zhang, Y.; Zhang, G.; Chen, L.; Liu, M.; et al. The Use of Functionalized Silk Fibroin Films as a Platform for Optical Diffraction-Based Sensing Applications. *Adv. Mater.* **2017**, *29*, 1605471. [CrossRef] [PubMed]
9. Dombeck, D.A.; Khabbaz, A.N.; Collman, F.; Adelman, T.L.; Tank, D.W. Imaging large-scale neural activity with cellular resolution in awake, mobile mice. *Neuron* **2007**, *56*, 43–57. [CrossRef] [PubMed]
10. Chen, T.-W.; Wardill, T.J.; Sun, Y.; Pulver, S.R.; Renninger, S.L.; Baohan, A.; Schreiter, E.R.; Kerr, R.A.; Orger, M.B.; Jayaraman, V. Ultrasensitive fluorescent proteins for imaging neuronal activity. *Nature* **2013**, *499*, 295–300. [CrossRef] [PubMed]
11. Jacques, S.L. Optical properties of biological tissues: A review. *Phys. Med. Biol.* **2013**, *58*, 37–61. [CrossRef] [PubMed]
12. Prasad, P.N.; Birge, R.R. *Introduction to Biophotonics*; John Wiley & Sons, Inc.: Hoboken, NJ, USA, 2004.
13. Kim, T.-I.; McCall, J.G.; Jung, Y.H.; Huang, X.; Siuda, E.R.; Li, Y.; Song, J.; Song, Y.M.; Pao, H.A.; Kim, R.-H. Injectable, cellular-scale optoelectronics with applications for wireless optogenetics. *Science* **2013**, *340*, 211–216. [CrossRef] [PubMed]
14. Nizamoglu, S.; Gather, M.C.; Yun, S.H. All-biomaterial laser using vitamin and biopolymers. *Adv. Mater.* **2013**, *25*, 5943–5947. [CrossRef] [PubMed]
15. Gather, M.C.; Yun, S.H. Single-cell biological lasers. *Nat. Photonics* **2011**, *5*, 406–410. [CrossRef]
16. Ding, H.; Lu, L.; Shi, Z.; Wang, D.; Li, L.; Li, X.; Ren, Y.; Liu, C.; Cheng, D.; Kim, H. Microscale optoelectronic infrared-to-visible upconversion devices and their use as injectable light sources. *Proc. Natl. Acad. Sci. USA* **2018**, *115*, 6632–6637. [CrossRef] [PubMed]
17. Chen, S.; Weitemier, A.Z.; Zeng, X.; He, L.; Wang, X.; Tao, Y.; Huang, A.J.; Hashimotodani, Y.; Kano, M.; Iwasaki, H. Near-infrared deep brain stimulation via upconversion nanoparticle-mediated optogenetics. *Science* **2018**, *359*, 679–684. [CrossRef] [PubMed]
18. Xu, H.; Yin, L.; Liu, C.; Sheng, X.; Zhao, N. Recent Advances in Biointegrated Optoelectronic Devices. *Adv. Mater.* **2018**, *30*, 1800156. [CrossRef] [PubMed]

19. Boyden, E.S.; Zhang, F.; Bamberg, E.; Nagel, G.; Deisseroth, K. Millisecond-timescale, genetically targeted optical control of neural activity. *Nat. Neurosci.* **2005**, *8*, 1263–1268. [CrossRef] [PubMed]

20. Deisseroth, K. Optogenetics. *Nat. Methods* **2011**, *8*, 26–29. [CrossRef] [PubMed]

21. Warden, M.R.; Cardin, J.A.; Deisseroth, K. Optical neural interfaces. *Annu. Rev. Biomed. Eng.* **2014**, *16*, 103–129. [CrossRef] [PubMed]

22. Murayama, M.; Pérez-Garci, E.; Lüscher, H.-R.; Larkum, M.E. Fiberoptic system for recording dendritic calcium signals in layer 5 neocortical pyramidal cells in freely moving rats. *J. Neurophysiol.* **2007**, *98*, 1791–1805. [CrossRef] [PubMed]

23. Gunaydin, L.A.; Grosenick, L.; Finkelstein, J.C.; Kauvar, I.V.; Fenno, L.E.; Adhikari, A.; Lammel, S.; Mirzabekov, J.J.; Airan, R.D.; Zalocusky, K.A.; et al. Natural neural projection dynamics underlying social behavior. *Cell* **2014**, *157*, 1535–1551. [CrossRef] [PubMed]

24. Scheggi, A.; Mignani, A. Optical fiber biosensing. *Opt. News* **1989**, *15*, 28–34. [CrossRef]

25. Mitsui, K.; Handa, Y.; Kajikawa, K. Optical fiber affinity biosensor based on localized surface plasmon resonance. *Appl. Phys. Lett.* **2004**, *85*, 4231–4233. [CrossRef]

26. Utzinger, U.; Richards-Kortum, R.R. Fiber optic probes for biomedical optical spectroscopy. *J. Biomed. Opt.* **2003**, *8*, 121–148. [CrossRef] [PubMed]

27. Sparta, D.R.; Stamatakis, A.M.; Phillips, J.L.; Hovelsø, N.; Zessen, R.V.; Stuber, G.D. Construction of implantable optical fibers for long-term optogenetic manipulation of neural circuits. *Nat. Protoc.* **2011**, *7*, 12–23. [CrossRef] [PubMed]

28. Sridharan, A.; Rajan, S.D.; Muthuswamy, J. Long-term changes in the material properties of brain tissue at the implant–tissue interface. *J. Neural Eng.* **2013**, *10*, 066001. [CrossRef] [PubMed]

29. Grosenick, L.; Marshel, J.H.; Deisseroth, K. Closed-Loop and Activity-Guided Optogenetic Control. *Neuron* **2015**, *86*, 106–139. [CrossRef] [PubMed]

30. Gilletti, A.; Muthuswamy, J. Brain micromotion around implants in the rodent somatosensory cortex. *J. Neural Eng.* **2006**, *3*, 189–195. [CrossRef] [PubMed]

31. Abouraddy, A.F.; Bayindir, M.; Benoit, G.; Hart, S.D.; Kuriki, K.; Orf, N.; Shapira, O.; Sorin, F.; Temelkuran, B.; Fink, Y. Towards multimaterial multifunctional fibres that see, hear, sense and communicate. *Nat. Mater.* **2007**, *6*, 336–347. [CrossRef] [PubMed]

32. Orf, N.D.; Shapira, O.; Sorin, F.; Danto, S.; Baldo, M.A.; Joannopoulos, J.D.; Fink, Y. Fiber draw synthesis. *Proc. Natl. Acad. Sci. USA* **2011**, *108*, 4743–4747. [CrossRef]

33. Canales, A.; Jia, X.; Froriep, U.P.; Koppes, R.A.; Tringides, C.M.; Selvidge, J.; Lu, C.; Hou, C.; Wei, L.; Fink, Y.; et al. Multifunctional fibers for simultaneous optical, electrical and chemical interrogation of neural circuits in vivo. *Nat. Biotechnol.* **2015**, *33*, 277–284. [CrossRef] [PubMed]

34. Parker, S.T.; Domachuk, P.; Amsden, J.; Bressner, J.; Lewis, J.A.; Kaplan, D.L.; Omenetto, F.G. Biocompatible silk printed optical waveguides. *Adv. Mater.* **2010**, *21*, 2411–2415. [CrossRef]

35. Jain, A.; Yang, A.H.J.; Erickson, D. Gel-based optical waveguides with live cell encapsulation and integrated microfluidics. *Opt. Lett.* **2012**, *37*, 1472–1474. [CrossRef] [PubMed]

36. Choi, M.; Humar, M.; Kim, S.; Yun, S.H. Step-index optical fiber made of biocompatible hydrogels. *Adv. Mater.* **2015**, *27*, 4081–4086. [CrossRef] [PubMed]

37. Keiser, G. *Optical Fiber Communications*; Academic Press: New York, NY, USA, 1985; Volume 9, pp. 1533–1534.

38. Ceci-Ginistrelli, E.; Pugliese, D.; Boetti, N.G.; Novajra, G.; Ambrosone, A.; Lousteau, J.; Vitale-Brovarone, C.; Abrate, S.; Milanese, D. Novel biocompatible and resorbable UV-transparent phosphate glass based optical fiber. *Opt. Mater. Express* **2016**, *6*, 2040–2051. [CrossRef]

39. Neel, E.A.; Pickup, D.M.; Valappil, S.P.; Newport, R.J.; Knowles, J.C. Bioactive functional materials: A perspective on phosphate-based glasses. *J. Mater. Chem.* **2009**, *19*, 690–701. [CrossRef]

40. T Tonnesen, H.H.; Karlsen, J. Alginate in Drug Delivery Systems. *Drug Dev. Ind. Pharm.* **2002**, *28*, 621–630. [CrossRef] [PubMed]

41. Hofmann, S.; Foo, C.T.; Rossetti, F.; Textor, M.; Vunjaknovakovic, G.; Kaplan, D.L.; Merkle, H.P.; Meinel, L. Silk Fibroin as an Organic Polymer for Controlled Drug Delivery. *J. Control. Release* **2006**, *111*, 219–227. [CrossRef] [PubMed]

42. Li, C.; Vepari, C.; Jin, H.; Kim, H.J.; Kaplan, D.L. Electrospun silk-BMP-2 scaffolds for bone tissue engineering. *Biomaterials* **2006**, *27*, 3115–3124. [CrossRef] [PubMed]

43. Li, Z.; Ramay, H.R.; Hauch, K.D.; Xiao, D.; Zhang, M. Chitosan-alginate hybrid scaffolds for bone tissue engineering. *Biomaterials* **2005**, *26*, 3919–3928. [CrossRef] [PubMed]

44. Tao, H.; Kainerstorfer, J.M.; Siebert, S.M.; Pritchard, E.M.; Sassaroli, A.; Panilaitis, B.; Brenckle, M.A.; Amsden, J.J.; Levitt, J.M.; Fantini, S.; et al. Implantable, multifunctional, bioresorbable optics. *Proc. Natl. Acad. Sci. USA* **2012**, *109*, 19584–19589. [CrossRef] [PubMed]

45. Tao, H.; Kaplan, D.L.; Omenetto, F.G. Silk materials-a road to sustainable high technology. *Adv. Mater.* **2012**, *24*, 2824–2837. [CrossRef] [PubMed]

46. Dupuis, A.; Guo, N.; Gao, Y.; Godbout, N.; Lacroix, S.; Dubois, C.; Skorobogatiy, M. Prospective for biodegradable microstructured optical fibers. *Opt. Lett.* **2007**, *32*, 109–111. [CrossRef] [PubMed]

47. Xin, H.; Li, Y.; Liu, X.; Li, B. Escherichia coli-based biophotonic waveguides. *Nano Lett.* **2013**, *13*, 3408–3413. [CrossRef] [PubMed]

48. Kim, U.; Park, J.; Li, C.; Jin, H.; Valluzzi, R.; Kaplan, D.L. Structure and Properties of Silk Hydrogels. *Biomacromolecules* **2004**, *5*, 786–792. [CrossRef] [PubMed]

49. Murphy, A.R.; Kaplan, D.L. Biomedical applications of chemically-modified silk fibroin. *J. Mater. Chem.* **2009**, *19*, 6443–6450. [CrossRef] [PubMed]

50. Marelli, B.; Kaplan, D.L.; Omenetto, F.G.; Hu, T.; Brenckle, M.A.; Applegate, M.B. Silk: A different kind of "fiber optics". *Opt. Photonics News* **2014**, *25*, 28–35.

51. Applegate, M.; Mitropoulos, A.; Perotto, G.; Kaplan, D.; Omenetto, F. Biocompatible silk fibroin optical fibers. *Adv. Photonics* **2015**, NT1B.4. [CrossRef]

52. Applegate, M.B.; Perotto, G.; Kaplan, D.L.; Omenetto, F.G. Biocompatible silk step-index optical waveguides. *Biomed. Opt. Express* **2015**, *6*, 4221–4227. [CrossRef] [PubMed]

53. Cenis, J.L.; Aznar-Cervantes, S.D.; Lozano-Pérez, A.A.; Rojo, M.; Muñoz, J.; Meseguer-Olmo, L.; Arenas, A. Silkworm gut fiber of *Bombyx morias* an implantable and biocompatible light-diffusing fiber. *Int. J. Mol. Sci.* **2016**, *17*, 1142. [CrossRef] [PubMed]

54. Huby, N.; Vie, V.; Renault, A.; Beaufils, S.; Lefevre, T.; Paquet-Mercier, F.; Pezolet, M.; Beche, B. Native spider silk as a biological optical fiber. *Appl. Phys. Lett.* **2013**, *102*, 4145–4149. [CrossRef]

55. Xin, Q.; Qian, Z.; Li, J.; Sun, H.; Yao, H.; Xia, X.; Jin, Z.; Wang, C.; Yan, W.; Wang, C. Synthetic engineering spider silk fiber as implantable optical waveguides for low-loss light guiding. *ACS Appl. Mater. Interfaces* **2017**, *9*, 14665–14676.

56. Yang, J.A.; Yeom, J.; Hwang, B.W.; Hoffman, A.S.; Hahn, S.K. In situ -forming injectable hydrogels for regenerative medicine. *Prog. Polym. Sci.* **2014**, *39*, 1973–1986. [CrossRef]

57. Hoffman, A.S. Hydrogels for biomedical applications. *Ann. N. Y. Acad. Sci.* **2001**, *944*, 62–73. [CrossRef] [PubMed]

58. Choi, M.; Choi, J.W.; Kim, S.; Nizamoglu, S.; Hahn, S.K.; Yun, S.H. Light-guiding hydrogels for cell-based sensing and optogenetic synthesis in vivo. *Nat. Photonics* **2013**, *7*, 987–994. [CrossRef] [PubMed]

59. Vashist, A.; Ahmad, S. Smart materials for drug delivery. *Orient. J. Chem.* **2013**, *29*, 861–870. [CrossRef]

60. Manocchi, A.K.; Domachuk, P.; Omenetto, F.G.; Yi, H. Facile fabrication of gelatin-based biopolymeric optical waveguides. *Biotechnol. Bioeng.* **2009**, *103*, 725–732. [CrossRef] [PubMed]

61. Guo, J.; Liu, X.; Jiang, N.; Yetisen, A.K.; Yuk, H.; Yang, C.; Khademhosseini, A.; Zhao, X.; Yun, S.H. Highly stretchable, strain sensing hydrogel optical fibers. *Adv. Mater.* **2016**, *28*, 10244–10249. [CrossRef] [PubMed]

62. Yetisen, A.K.; Jiang, N.; Fallahi, A.; Montelongo, Y.; Ruizesparza, G.U.; Tamayol, A.; Zhang, Y.S.; Mahmood, I.; Yang, S.A.; Kim, K.S.; et al. Glucose-sensitive hydrogel optical fibers functionalized with phenylboronic acid. *Adv. Mater.* **2017**, *29*, 1606380. [CrossRef] [PubMed]

63. Jiang, N.; Ahmed, R.; Rifat, A.A.; Guo, J.; Yin, Y.; Montelongo, Y.; Butt, H.; Yetisen, A.K. Functionalized flexible soft polymer optical fibers for laser photomedicine. *Adv. Opt. Mater.* **2017**, *5*, 1701118. [CrossRef]

64. Guo, J.; Zhou, M.; Yang, C. Fluorescent hydrogel waveguide for on-site detection of heavy metal ions. *Sci. Rep.* **2017**, *7*, 7902. [CrossRef] [PubMed]

65. Williams, D.F. *Biomaterials Science: An Introduction to Materials in Medicine*; Ratner, B.D., Hoffman, A.S., Lemons, J.E., Lemons, J., Eds.; Academic Press: New York, NY, USA, 2004; p. 864, ISBN 0-12-582463-7.

66. Gentile, P.; Chiono, V.; Carmagnola, I.; Hatton, P.V. An overview of poly(lactic-co-glycolic) acid (PLGA)-based biomaterials for bone tissue engineering. *Int. J. Mol. Sci.* **2014**, *15*, 3640–3659. [CrossRef] [PubMed]

67. Nizamoglu, S.; Gather, M.C.; Humar, M.; Choi, M.; Kim, S.; Kim, K.S.; Hahn, S.K.; Scarcelli, G.; Randolph, M.; Redmond, R.W. *Implantable Waveguides for Deep-Tissue Photoactivation*; Novel Optical Materials and Applications; OSA Technical Digest: Denver, CO, USA, 2017; p. NoTh1C.2.

68. Nizamoglu, S.; Gather, M.C.; Humar, M.; Choi, M.; Kim, S.; Kim, K.S.; Hahn, S.K.; Scarcelli, G.; Randolph, M.; Redmond, R.W.; et al. Bioabsorbable polymer optical waveguides for deep-tissue photomedicine. *Nat. Commun.* **2016**, *7*, 10374. [CrossRef] [PubMed]

69. Fu, R.; Luo, W.; Nazempour, R.; Tan, D.; Ding, H.; Zhang, K.; Yin, L.; Guan, J.; Sheng, X. Implantable and biodegradable poly(L-lactic acid) fibers for optical neural interfaces. *Adv. Opt. Mater.* **2017**, *5*, 1700941. [CrossRef]

70. Kwok, S.J.J.; Kim, M.; Lin, H.H.; Seiler, T.G.; Beck, E.; Shao, P.; Kochevar, I.E.; Seiler, T.; Yun, S.H. Flexible optical waveguides for uniform periscleral cross-linking. *Investig. Ophthalmol. Vis. Sci.* **2017**, *58*, v2596–v2602. [CrossRef] [PubMed]

71. Shan, D.; Zhang, C.; Kalaba, S.; Mehta, N.; Kim, G.B.; Liu, Z.; Yang, J. Flexible biodegradable citrate-based polymeric step-index optical fiber. *Biomaterials* **2017**, *143*, 142–148. [CrossRef] [PubMed]

72. Park, S.; Guo, Y.; Jia, X.; Choe, H.K.; Grena, B.; Kang, J.; Park, J.; Lu, C.; Canales, A.; Chen, R.; et al. One-step optogenetics with multifunctional flexible polymer fibers. *Nat. Neurosci.* **2017**, *20*, 612–619. [CrossRef] [PubMed]

73. Lu, C.; Froriep, U.P.; Koppes, R.A.; Canales, A.; Caggiano, V.; Selvidge, J.; Bizzi, E.; Anikeeva, P. Polymer fiber probes enable optical control of spinal cord and muscle function in vivo. *Adv. Funct. Mater.* **2015**, *24*, 6594–6600. [CrossRef]

74. Lu, C.; Park, S.; Richner, T.J.; Derry, A.; Brown, I.; Hou, C.; Rao, S.; Kang, J.; Mortiz, C.T.; Fink, Y.; et al. Flexible and stretchable nanowire-coated fibers for optoelectronic probing of spinal cord circuits. *Sci. Adv.* **2017**, *3*, e1600955. [CrossRef] [PubMed]

75. Deisseroth, K. Optogenetics: 10 years of microbial opsins in neuroscience. *Nat. Neurosci.* **2015**, *18*, 1213–1225. [CrossRef] [PubMed]

76. Deisseroth Lab Frontrat [EB/OL]. Available online: https://web.stanford.edu/group/dlab/optogenetics/ (accessed on 25 July 2018).

77. Pisanello, F.; Mandelbaum, G.; Pisanello, M.; Oldenburg, I.A.; Sileo, L.; Markowitz, J.E.; Peterson, R.E.; Della, P.A.; Haynes, T.M.; Emara, M.S.; et al. Dynamic illumination of spatially restricted or large brain volumes via a single tapered optical fiber. *Nat. Neurosci.* **2017**, *20*, 1180–1188. [CrossRef] [PubMed]

78. Pisanello, F.; Sileo, L.; Oldenburg, I.A.; Pisanello, M.; Martiradonna, L.; Assad, J.A.; Sabatini, B.L.; Vittorio, M.D. Multipoint emitting optical fibers for spatially addressable in vivo optogenetics. *Neuron* **2014**, *82*, 1245–1254. [CrossRef] [PubMed]

79. Leon, G.M.D. *The Biomedical Laser*; Springer: New York, NY, USA, 1981.

80. Chung, H.; Dai, T.; Sharma, S.K.; Huang, Y.Y.; Carroll, J.D.; Hamblin, M.R. The nuts and bolts of low-level laser (light) therapy. *Ann. Biomed. Eng.* **2012**, *40*, 516–533. [CrossRef] [PubMed]

81. PhotodynamicTherapy (Red). Available online: https://commons.wikimedia.org/wiki/File:Photodynamic_therapy_(red).jpg (accessed on 25 July 2018).

82. Hwang, S.W.; Tao, H.; Kim, D.H.; Cheng, H.; Song, J.K.; Rill, E.; Brenckle, M.A.; Panilaitis, B.; Won, S.M.; Kim, Y.S.; et al. A Physically transient form of silicon electronics. *Science* **2012**, *337*, 1640–1644. [CrossRef] [PubMed]

Review

Materials and Devices for Biodegradable and Soft Biomedical Electronics

Rongfeng Li †, Liu Wang † and Lan Yin *

School of Materials Science and Engineering, The Key Laboratory of Advanced Materials of Ministry of Education, State Key Laboratory of New Ceramics and Fine Processing, Tsinghua University, Beijing 100084, China; luckylrf@163.com (R.L.); liuwang@mail.tsinghua.edu.cn (L.W.)
* Correspondence: lanyin@tsinghua.edu.cn
† These authors contributed equally to this work.

Received: 3 October 2018; Accepted: 23 October 2018; Published: 26 October 2018

Abstract: Biodegradable and soft biomedical electronics that eliminate secondary surgery and ensure intimate contact with soft biological tissues of the human body are of growing interest, due to their emerging applications in high-quality healthcare monitoring and effective disease treatments. Recent systematic studies have significantly expanded the biodegradable electronic materials database, and various novel transient systems have been proposed. Biodegradable materials with soft properties and integration schemes of flexible or/and stretchable platforms will further advance electronic systems that match the properties of biological systems, providing an important step along the path towards clinical trials. This review focuses on recent progress and achievements in biodegradable and soft electronics for biomedical applications. The available biodegradable materials in their soft formats, the associated novel fabrication schemes, the device layouts, and the functionality of a variety of fully bioresorbable and soft devices, are reviewed. Finally, the key challenges and possible future directions of biodegradable and soft electronics are provided.

Keywords: biodegradable electronics; transient electronics; soft biomedical electronics; biodegradable materials

---

1. Introduction

With the growth of the global economy, and the development of science and technology, a massive assortment of electronics has been widely used in human society, which plays an important role in industrial processes, telecommunication, entertainment, healthcare, etc. [1–5]. Soft electronics that ensure conformal contact with nonplanar surfaces, such as soft biological tissues, are expected to play crucial roles in healthcare. The characteristic of these electronics is that they can significantly expand the capabilities of conventional rigid electronics in sensing, monitoring, diagnosing, and potentially intervening functions. The intimate contact between the soft device and the nonplanar object allows for high-quality data to be collected. Additionally, in the area of medical devices, soft electronics have similar mechanical properties to biological tissues and, thus, they cause minimal irritation to the human body.

On the other hand, biodegradable electronics possess unique characteristics and attract numerous research interests. The devices can dissolve, resorb, or physically disappear into physiological or environmental solutions, partially or completely, at controlled rates after the expecting working period [6–14]. Although long-lasting operation is one hallmark of traditional electronics, devices with biodegradability can potentially offer great benefits for temporary biomedical implants, green environmental electronics, and secured hardware. Biodegradable electronics can serve as temporary diagnostic and therapeutic platforms for important biological processes, e.g., wound healing and tissue regeneration, and they can be safely resorbed by the body after usage, eliminating a second surgery for device retrieval, and therefore avoiding associated infection risks and hospital costs [15]. Biodegradable electronics also provides an alternative way to alleviate issues that are associated with electronic waste (e-waste) [16], and they enable potential usage for security hardware, preventing unauthorized access of personal or security information [17,18].

Serving as medical implants, biodegradable electronics eliminate the potential retention of device materials, while soft electronics ensure conformal wrapping with the human body, as they are soft, curvilinear, and evolving [19]. Recent research on advanced materials [14], fabrication approaches [20], and design layouts [21] yield biodegradable and soft electronics that enable intimate integration into the body, with unique capabilities for diagnostic and therapeutic functions, which would otherwise be impossible when using conventional wafer-based electronics that are built upon non-degradable and rigid printed circuit boards. These emerging technologies provide critical tools that could have great potential to improve human health and enhance the understanding of biological systems.

This review specifically focuses on recent progress and achievements in electronics that combine both soft and biodegradable characteristics, targeting biomedical applications, while reviews on general transient electronics can be found elsewhere in references [6,13]. The available biodegradable materials in the soft formats, and their associated fabrication schemes are first reviewed, followed by the introduction of device layouts and the functionality of a variety of fully bioresorbable and soft devices, including solutions for power supply. Perspectives and the outlook of biodegradable soft electronics for clinical medicine are also provided at the end of the article.

2. Materials

A wide range of biodegradable materials have been explored to build biodegradable electronics. Traditional biodegradable materials are mostly based on polymers and magnesium alloys, and they serve mainly as structural components, e.g., cardiovascular stents and 3D scaffolds. As electronic properties are essential for constructing electronics, dissolvable inorganic materials with excellent operational characteristics are therefore also of great interest.

In addition to biodegradability, soft characteristics are another critical property to be considered for biomedical applications, in order to achieve minimal irritation to the body, and to obtain intimate contact with biological tissues. Soft materials should be able to survive mechanical deformations, and simultaneously, their functional properties should remain unaffected. The term "soft" can refer to flexible, foldable, stretchable, and twistable, and here, flexible materials are the focus.

The basic building blocks for electronic components are semiconductors, dielectrics, and conductors, and studies have developed strategies to ensure flexibility. The key method is to configure biodegradable inorganic semiconductors, dielectrics, and metals into thin film and open mesh formats, and to integrate them onto soft biodegradable polymeric or metal foil substrates. Through these techniques, biodegradable and flexible electronics that can adapt well to the soft nature of the human body.

In the following sections, inorganic biodegradable functional materials, substrate materials, and encapsulation materials, and organic functional materials will be reviewed respectively, in terms of their respective dissolution rates, mechanical properties, and biocompatibilities. Dissolution data from the literature of major inorganic materials are summarized in tables for better comparison. In all, dissolution rates of a wide range of biodegradable materials have been investigated in detail in simulated bio-fluids, such as phosphate-buffered saline (PBS), Hanks' solutions, artificial cerebrospinal fluid (ACSF), etc. A few studies have investigated the effects of proteins on the dissolution rates of Si. Strategies have been proposed to obtain soft biodegradable materials that combine both inorganic and organic components. Although detailed studies are needed to further reveal the pertinent biological influences, biodegradable materials of interest exhibit good biocompatibility through their evaluation in cell studies and animal trials.

2.1. Inorganic Functional Materials

Functional materials are key components for electronics, and they consist of semiconductors, conductive materials, and dielectric materials. Inorganic dissolvable thin-film materials that have been explored, include monocrystalline silicon (mono-Si), polycrystalline silicon (poly-Si), amorphous silicon (a-Si), germanium (Ge), silicon germanium alloy (SiGe), indium–gallium–zinc oxide (a-IGZO), and zinc oxide (ZnO) [22–27] for semiconductors; magnesium (Mg), molybdenum (Mo), tungsten (W), iron (Fe), and zinc (Zn) for conductive materials [23,27–29]; and magnesium oxide (MgO), silicon dioxide (SiO_2), and silicon nitride (SiN_x) for dielectric materials [14,20,30]. The acceptable levels of these elements can be informed from nutritional supplements. The recommended dietary allowance and tolerable upper intake levels of functional materials are summarized in Table 1. As is shown, Mg, Mo, Fe, and Zn are all necessary elements for the human body. It should be noticed that the mean intakes of Si in adult men and women are 40 and 19 mg day^{-1} respectively, and limited toxicity research on Si suggests that there is no risk of inducing adverse effects for the general population, based on the common intake level [31].The mean total Ge exposure for people is 4 µg day^{-1}, which can be absorbed from the intestinal tract and excreted largely through the kidneys [32–34]. In addition, W is usually found in rice, with concentrations of 7–283 µg kg^{-1} [35]. However, because of a lack of adequate and sufficient data for Si, Ge, and W, it is necessary to establish a recommended dietary allowance (RDA) and tolerable upper intake levels (UL).

Table 1. The recommended dietary allowance (RDA) and tolerable upper intake levels (UL) of biodegradable elements [31,36].

Life Stage	Category	Element			
		Mg (mg/day)	Mo (µg/day)	Fe (mg/day)	Zn (mg/day)
Infants	RDA	30–75	2–3	0.27–11	2–3
0–12 months	UL	–	–	40	4–5
Children	RDA	80–130	17–22	7–10	3–5
1–8 years	UL	65–110	300–600	40	7–12
Males	RDA	240–420	34–45	8–11	8–11
≥9 years	UL	350	1100–2000	40–45	23–40
Females	RDA	240–320	34–45	8–15	8–9
≥9 years	UL	350	1100–2000	40–45	23–40
Pregnancy	RDA	350–400	50	27	11–12
14–50 years	UL	350	1700–2000	45	34–40
Lactation	RDA	310–360	50	9–10	12–13
14–50 years	UL	350	1700–2000	45	34–40

Functional materials in the thin film format are adopted to assure flexibility, as well as reasonable degradation time frames. Studies have revealed that nanomembrane materials have high degrees of bendability, as the flexural rigidity and energy release rates scale down with thickness, enabling an intimate contact with non-planar and curvilinear surface [37]. Nanomembranes can be obtained by peeling the top membrane materials from commercially available wafers (e.g., silicon on insulator, SOI), which will be discussed later. Low-temperature deposition such as radiofrequency plasma-enhanced chemical vapor deposition (RF-PECVD) [38], electron cyclotron resonance (ECR) [39], and hot-wire chemical vapor deposition (HW-CVD) [40] can be used to fabricate high-quality functional thin films directly onto soft substrates. Additionally, solution-printing techniques have also been proposed as a low-cost alternative methods [41]. Another robust strategy for achieving soft materials is through structural design, which not only enhances the flexibility of intrinsic soft materials, but also gives flexibility to materials that are intrinsically rigid. Methods include separating rigid thin film materials into small islands, introducing serpentine, wavy, buckled interconnects, integrating rigid materials with soft substrates, etc. [42,43].

The research on silicon nanomembrane (Si NM) dissolution behavior greatly promotes the development of transient electronics, as it can leverage existing well-established Si semiconductor technology and realize high-performance biodegradable electronics. The dissolution rates of Si NMs in solutions with different ionic types and relevant concentrations [23,24], temperatures [24], pH values [23], concentrations of protein [44], as well as doping levels [22] have been investigated, all of which play an important role during the silicon dissolution process. The dissolution rates of Si NMs under different conditions are tabulated in Table 2. For example, dissolution rates have been observed for Si NMs (slightly p-doped 10^{-17} cm^{-3}, 100 orientation) in aqueous solutions containing various chloride and phosphate concentrations at different temperatures. Higher temperatures and concentrations of chlorides and phosphates can greatly promote Si dissolution, probably through a nucleophilic dissolution process [24]. The underlying mechanism regarding the influence of chlorides and phosphates has been evaluated through density functional theory (DFT) and molecular dynamics (MD) simulations [24]. Dissolution rates of Si are found to be sensitive to calcium and magnesium ions as well [44]; e.g., the addition of 1 mM of Ca^{2+} and Mg^{2+} can slightly increase the rates in phosphate buffered saline solutions. As shown in Table 2, the presence of albumin decelerates the dissolution rates, probably due to the absorption of the protein onto the Si surface [44]. In addition, the types and concentrations of dopants for Si NMs can affect the dissolution rate significantly, and a sharp decrease in dissolution rate can be found when dopant concentrations exceed a certain level, i.e., 10^{20} cm^{-3}.

Similarly, dissolution rates of poly-Si, a-Si, alloys of silicon, SiGe, and Ge show great dependence on the pH, temperatures, proteins, and type of ions [26]. For instance, the rates of these materials at physiological temperatures (37 °C) are higher than those at room temperature. At similar pHs, bovine serum leads to dissolution rates at 37 °C that are 30–40 times higher than those of a phosphate buffer solution for poly-Si, a-Si, and nano-Si. For SiGe, the dissolution rate exhibits an even more strongly accelerated rate (~185 times) in bovine serum. Besides Si and SiGe, dissolution rates of Ge and two-dimensional (2D) MoS_2 materials have also been evaluated in physiological solutions, and the dissolution rates are summarized in Table 3. The dissolution of monolayer MoS_2 crystals in PBS occurs as a defect-induced etching progress, in which the grain boundaries dissolve first, followed by the crystalline regions. Moreover, the increased concentrations of Na^+ and K^+ accelerate the degradation, because the existence of Na^+ or K^+ leads to lattice distortions of MoS_2 and the formation of Na_2S.

Table 2. The dissolution behavior of silicon nanomembranes (Si NMs) under different conditions [22–24,26,44].

Functional Materials	Temperature °C	Aqueous Solution	pH	Doping Type cm^{-3}	Dissolution Rate nm day^{-1}
Mono-Si NMs [24]	37	Phosphate 0.05 M	7.5	10^{17}(p)	2.512
		Phosphate 0.1 M			25.119
		Phosphate 0.5 M			50.119
		Phosphate 1 M			63.096
		Chloride 0.05 M			1.259
		Chloride 0.1 M			6.309
		Chloride 0.5 M			63.096
		Chloride 1 M			63.096
	50	Phosphate 0.05 M	7.5	10^{17}(p)	3.162
		Phosphate 0.1 M			31.623
		Phosphate 0.5 M			125.893
		Phosphate 1 M			251.189
		Chloride 0.05 M			5.012
		Chloride 0.1 M			15.849
		Chloride 0.5 M			199.526
		Chloride 1 M			398.107

Table 2. *Cont.*

Functional Materials	Temperature °C	Aqueous Solution	pH	Doping Type cm^{-3}	Dissolution Rate nm day^{-1}
Mono-Si NMs [22]	37	Buffer solution 0.1 M	7.4	10^{17}(p)	3.162
				10^{19}(p)	3.162
				10^{20}(p)	0.501
	37	Buffer solution 0.1 M	7.4	10^{17}(b)	3.162
				10^{19}(b)	3.162
				10^{20}(b)	0.251
	Room temperature (RT)	Coke	2.6	-	0.600
	RT	Milk	6.4	-	23.300
	RT	PBS 0.1 M	7.4	-	1.820
Mono-Si NMs [23]	37	Bovine serum	7.4	-	100.800
		Sea water	7.8	-	4.115
Mono-Si NMs [44]	37	Albumin phosphate buffered saline (PBS) Na$^+$	7.4	10^{15}(b)	42.100
		Albumin PBS Mg^{2+}			45.010
		Albumin PBS Ca^{2+}			51.000
	20	Purified water	5.5	10^{15}(b)	<0.010
				10^{20}(b)	<0.010
	20	Tap water	7.5	10^{15}(b)	0.710
				10^{20}(b)	0.420
	37	Serum	7.4	10^{15}(b)	21.020
				10^{20}(b)	0.500
	37	Hank's balanced salt solution (HBSS)	7.6	10^{15}(b)	58.010
				10^{20}(b)	7.010
	37	HBSS w/Ca, Mg	7.6	10^{15}(b)	66.020
				10^{20}(b)	8.020
	37	HBSS	8.2	10^{15}(b)	129.030
				10^{20}(b)	58.000
	37	HBSS w/Ca, Mg	8.2	10^{15}(b)	178.000
				10^{20}(b)	69.010
	20	Purified water	5.5	10^{15}(b)	0.010
		Tap water	7.5		0.720
		Serum	7.4		3.500
	37	Purified water	5.5	10^{15}(b)	0.210
		Tap water	7.5		2.620
		Sweat	4.5		0.530
		Serum	7.4		21.000
		HBSS	7.6		58.210
		HBSS w/Ca, Mg	7.6		66.010
		HBSS	8.2		129.020
		HBSS w/Ca, Mg	8.2		178.000
poly-Si NMs [26]	RT	Buffer solution	7	-	1.020
			7.4		1.505
			8		10.010
			10		398.107
	37	Buffer solution	7	-	1.259
			7.4		3.162
			8		19.953
			10		562.341
a-Si NMs [26]	RT	Buffer solution	7	-	1.778
			7.4		3.981
			8		15.849
			10		501.187
	37	Buffer solution	7	-	1.585
			7.4		5.011
			8		31.623
			10		630.957

Note: p stands for phosphate doping, and b stands for boron doping.

Table 3. The dissolution behavior of Ge and 2D MoS$_2$ under different conditions [26,45].

Functional Materials	Temperature °C	Aqueous Solution	pH	Dissolution Rate nm day^{-1}
Ge [26]	RT	Aqueous Buffer solution	7	0.794
			7.4	1.995
			8	15.849
			10	501.187
	37	Aqueous Buffer solution	7	1.259
			7.4	3.162
			8	19.953
			10	562.341
Poly-MoS$_2$ [45]	40	1.0 M PBS	7.4	0.010
			12	0.010
	60	1.0 M PBS	7.4	0.070
			12	0.160
	75	1.0 M PBS	7.4	0.200
			12	0.270
	85	1.0 M PBS	7.4	0.270
			12	0.400

Metals related to trace elements that are normally found in human body, such as Mg, Zn, W, Fe, Mo, and their oxides, are great candidates for interconnects and dielectric materials [23,29,46]. Strain-tolerant metallic thin films can be fabricated by electron-beam deposition, pulsed-laser deposition, or magnetron sputtering, following photolithography. Among them, Mg and Zn are utilized more frequently, owing to the easy processing property and better concentration tolerance for patients, which means lower costs and safer resorbable properties. However, the degradation rates of Mg and Zn are relative fast; metals with slower rates (W, Mo) are, therefore, more desirable if a longer life time is needed [29]. Fe thin films become rusted easily, and they are converted to iron oxides and hydroxides, which have extremely lower solubilities in neutral solutions, and slightly acidic environments are probably more desirable for inducing complete degradation [29].

Moreover, dielectric materials, including magnesium oxide (MgO), silicon dioxide (SiO$_2$), silicon nitride (Si$_3$N$_4$), and spin-on-glass (SOG), are also dissolvable in aqueous solutions. The dissolution rates of these materials depend not only on pHs, temperatures, and ion concentrations, but also on the physical and chemical properties of the films, which are affected by the deposition condition [10,14,47]. For example, the dissolution rates of oxides deposited by electron-beam (e-beam) evaporation are 100 times slower compared to those deposited by plasma-enhanced chemical vapor deposition (PECVD). For nitrides, the dissolution rate of a low-pressure chemical vapor deposition (LPCVD) nitride is slower than that of a PECVD nitride. The degradation rates of metals and dielectric materials are summarized in Tables 4 and 5, respectively.

Table 4. Dissolution behavior of metals [47–51].

Functional Materials	Test Conditions	Dissolution Rate nm day^{-1}	Dissolution Product
Mg [48]	Deionized water, RT	1680.000	Mg(OH)$_2$
Zn [49]	Deionized water, RT	168.000	Zn(OH)$_2$
Fe [47]	Simulated body fluids, 37 °C	5.000–80.000	Fe(OH)$_2$, Fe(OH)$_3$
Mo [50]	Deionized water, RT	7.200	H$_2$MoO$_4$
W [51]	Deionized water, RT	7.200–3.000–40.800	H$_2$WO$_4$

Table 5. Dissolution behaviors of dielectric materials in buffer solution at 37 °C [10,47].

Functional Materials	Fabrication Methods	Test Conditions	Dissolution Rate nm day^{-1}	Dissolution Product
SiO$_2$ [10]	E-beam	37 °C	10.000	Si(OH)$_4$
	PECVD	37 °C	0.100	
	Thermally grown	37 °C	0.003	
Si$_3$N$_4$ [10]	LPCVD	pH 7.4	0.158	Si(OH)$_4$ + NH$_3$
	LPCVD	pH 8	0.251	
	LPCVD	pH 10	0.316	
	LPCVD	pH 12	0.631	
	PECVD-LF	pH 7.4	0.794	
	PECVD-LF	pH 8	1.585	
	PECVD-LF	pH 10	1.995	
	PECVD-LF	pH 12	3.981	
	PECVD-HF	pH 7.4	0.794	
	PECVD-HF	pH 8	2.512	
	PECVD-HF	pH 10	6.310	
	PECVD-HF	pH 12	25.119	
Spin-on glass (SOG) [47]	cured at 300 °C	PBS, pH 7.4	50.000	Si(OH)$_4$
	cured at 800 °C	PBS, pH 7.4	6.000	

Figure 1a shows the representative flexible circuit based on dissolvable inorganic Si electronic materials on silk substrate, including transistors made by Si/MgO/Mg, diodes made by Si, and inductors and capacitors made by Mg/MgO, as well as resistor and connection wires made by Mg. The related transience properties in the operational characteristics of n-channel transistors are shown on the left side, which are comparable with transistors that are built with non-dissolvable materials. Logic circuits can be built based on the transistor unit cells, which provide a promising path to achieving soft and biodegradable multi-functional Si electronics [14]. Figure 1b illustrates a transient circuit composed of Ga$_2$O$_3$/In$_2$O$_3$/ZnO thin film transistors (TFTs), with transfer performance with widths (W)/lengths (L) (=30/10 μm). The output feature corresponds to a 0 to 10 V gate bias, with a step of 2 V, for 0 to 5 V drain bias (V$_{DS}$) [27], which represents an alternative inorganic semiconductor for soft transient electronics. In Figure 1c, W powders are integrated onto a flexible sodium carboxymethyl cellulose (Na–CMC) substrate to print a temperature sensor circuit with a sensing performance that is closed to the weather report, which indicates an alternative method for quickly achieving biodegradable circuits [28]. Two-dimensional materials, such as MoS$_2$, with attractive optical, electrical, and mechanical properties, have also been explored, to form biodegradable electronics. Figure 1d shows a transient pressure sensor that is integrated with Mo and MoS$_2$ as the functional materials, which can be utilized to prepare a temperature sensor in vivo. Meanwhile, the literature suggests that MoS$_2$ can gradually dissolve in PBS solution (pH = 7.4) at 75 °C, which may be adjusted by changing the grain size [45]. This investigation offers new insights incorporating ultrathin 2D materials for bioresorbable devices.

The biocompatibility of the materials and the products of their dissolution are important for applications in bioresorbable electronics. In vitro cytotoxicity studies on mono-Si, poly-Si, α-Si, SiGe, and Ge, with both neighboring stromal fibroblast cells and infiltrating immune cells, suggest that both of these materials and their dissolution products are biocompatible [26]. In addition, in vivo evaluations of Si NMs implanted into the subdermal regions of an albino, laboratory-bred strain of the house mouse (BALB/c mice), show that following five weeks of implantation, immunoprofiling of lymphocytes from the peripheral blood and draining lymph nodes revealed no significant differences in the percentages of CD4+ and CD8+ T cells for implanted animals and sham-operated controls, which suggests long-term immunological and tissue biocompatibility [23]. Furthermore, an in vitro assessment of cytotoxicity on a patterned array of Si NMs using cells from a metastatic breast cancer cell line (MAD-MB-231), and in vivo toxicity studies by implanting Si NMs on silk in the subdermal region of BALB/c mice, suggest this material is biocompatible, and that it has the potential to be used for long-term implantation [43]. Recently, in vitro cytotoxicity explored on 2D MoS$_2$ films with L-929 cells and mouse fibroblast cells showed that there is no adverse effect on cell adherence and proliferation in vitro for 24 days. In vivo long-term cytotoxicity and biocompatibility studies with MoS$_2$ layers implanting subcutaneously into BALB/c mice suggests that MoS$_2$ does not cause any serious immunological or inflammatory reactions, and that it is, therefore, suitable for long-term biomedical use.

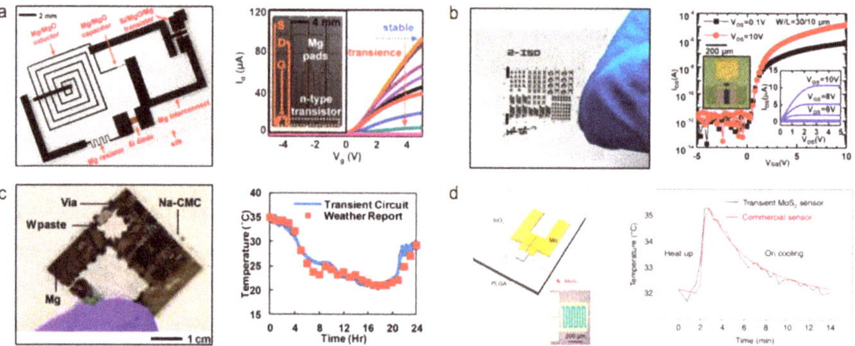

Figure 1. Different functional materials for soft and biodegradable devices. (**a**) Left: The circuit includes Si/MgO/Mg transistors, Si diodes, Mg/MgO inductors, and capacitors, as well as Mg resistors and interconnectors. Right: The transience of the operational characteristics of n-channel transistors. (**b**) Left: Schematic illustrations of transient Ga$_2$O$_3$/In$_2$O$_3$/ZnO thin film transistors (TFTs) and circuits. Right: Transfer characteristics for Ga$_2$O$_3$/In$_2$O$_3$/ZnO TFTs with widths (W)/lengths (L) (=30/10 μm). (**c**) Left: A transient printed circuit board (PCB) device with W and Mg for the temperature sensor; Right: Comparison for environmental temperatures measured by a transient circuit and the meteorological system. (**d**) Left: A multifunctional sensor using Mo and MoS$_2$; Right: Measurement of the intracranial temperature with transient MoS$_2$ and commercial sensors. Reproduced with permission from [14,27,28,45].

2.2. Organic Functional Materials

Conducting, semiconducting, and dielectric polymers are natural bridges between electronics and soft matter, because the vast chemical design space for polymers allows for the tunability of electronic, mechanical, and transient properties. A general strategy to create dielectric polymers is to incorporate high dielectric constant fills (e.g., SiO$_2$, aluminum oxide (Al$_2$O$_3$), hafnium oxide (HfO$_2$)) into a polymer matrix [12]. To circumvent the use of inorganic fillers, plant-based fibers (e.g., cotton, jute, bamboo, and banana fibers), sugars (e.g., glucose and lactose), and DNA and its precursors are promising natural polymers that intrinsically possess practical dielectric properties [52–58].

For conducting polymers, conjugated polymers that have been doped into a conducting state are used for device interconnectors and contacts. Common conducting polymers are polypyrrole (PPy), polyanniline (PANI), and poly(3,4-ethylenedioxythiophene) (PEDOT) [59]. One strategy for fabricating biodegradable conducting polymers is to blend conjugated polymers with biodegradable, insulating polymers. It has to be noted that these composites fabricated in this way are partially degradable, which means they are disintegrable, but the conductive polymers parts cannot be fully broken down to their monomers. Fully biodegradable conductive polymers might be obtained by conjugation breaking, but the conductivity of the materials is relatively lower than the partially degradable conductive polymers [12]. In addition, typical semiconducting polymers are polythiophenes (e.g., poly(3-hexylthiophene), P3HT), and diketopyrrolopyrroles (DPP) [60,61].

As with conducting polymers, blending has been utilized to generate partially degradable semiconducting polymers, e.g., by blending poly(3-thiophene methyl acetate) (P3TMA), a derivative of P3HT, with thermoplastic polyurethane (TPU) [62]. Fully degradable polymeric semiconductors have been achieved recently by introducing reversible imine bonds between DPP and p-phenylenediamine [63]. Conjugated molecules found in nature could also be utilized to build biodegradable electronics, including the natural dye indigo from the plants *Indigofera tinctorial* and *Isatis tinctorial* [64], natural pigment melanins [65,66], and β-carotene and anthraquinone derivatives [54,67].

2.3. Substrate Materials

Compared with functional materials, often with thicknesses of a few hundreds of nanometers, substrate materials with thicknesses at the micrometer scale contribute to the majority of the weight. As for soft electronics, substrate materials are a critically important consideration, because their mechanical properties can dominate that of the integrated system. Polymeric materials are often used as flexible substrate materials. Candidate materials need to be compatible with device fabrication processes, which usually involves high temperatures, water, and harsh solvents and, therefore, considerations of material properties centering around thermal stability and solvent compatibility are necessary. Degradation times, swelling rates, mechanical robustness, and the biocompatibility of the substrate materials are also critical to guiding the selection of substrates and, thus, achieving devices with controlled operational timeframes, as well as characteristics that match tissue environments. Although the properties of polymeric substrates vary greatly from material to material, these substrates tend to be flexible and biodegradable.

Based on these requirements, a series of substrate polymeric materials have been explored, which can be classified into natural materials and synthesized polymers. Materials that already exist in the natural environment have been applied as the substrates for biomedical electronics. These materials are biologically derived, which possess affinities and hypo-allergenicities to the human body, such as silk fibroin, cellulose, and chitosan, etc. Silk fibroin protein shows attractive properties that are related to biotechnology and biomedical fields [68–72], and possesses proper mechanical strength, a tunable life time, and minimum immune rejection. The n-channel metal-oxide-semiconductor field-effect transistors (MOSFETs) have been deposited onto thin silk film successfully, as shown in Figure 2a [20]. With similar characteristics, cellulose and chitosan are two other natural substrate materials, as shown in Figures 2b [63] and 2g [73], respectively. As a type of polysaccharide, cellulose is a richly-abundant substance in nature that composes of more than one-third of all plant components, and it even reaches an abundance of 90% in cotton [74]. Cellulose is a promising substrate for biomedical implants, as it takes advantage of attractive biocompatibility properties and degradation in a physiological environment [75–77]. Recently, a novel transistor system using Al_2O_3 as the dielectric layer and Fe as the electrodes, was prepared on ultrathin cellulose film to fabricate transient electronics [63]. In addition, as a derivative product of chitin by the deacetylation process, chitosan is another common substrate for temporary devices with proven biocompatibilities, as shown in Figure 2g [73]. Moreover, some other natural compounds such as potato starch, gelatin, or caramelized glucose have been used as substrates for designing biodegradable electronics [54]. Generally speaking, the dominant advantages of natural substrate materials are favorable, in terms of low-immunoreaction and great abundance, as well as low cost.

However, the intrinsic properties of natural materials limit their applications, as biodegradable electronics draw higher demands for substrates with specific properties relating to stability, mechanical strength, and degradation rate, etc., which promotes research for synthetic polymers [12]. Synthetic polymeric materials have been used in numerous biomedical applications for years, such as in bioresorbable stents and sutures [78–84]. Poly lactic-co-glycolic acid (PLGA) is a typical polymer that is used as a transient substrate, which is a copolymer of poly lactic acid (PLA) and poly glycolic acid (PGA). Compared to natural materials, the lifetime and mechanical strength of PLGA can be modified over a wide range by adjusting the ratio of PLA and PGA, which paves a promising path for biomedical implants with controllable working lives. In Figure 2d, transient complementary metal-oxide-semiconductors (CMOSs) are prepared on PLGA substrates with excellent operational characteristics [85]. Because of the different substrate requirements for biodegradable clinical devices, numerous polymers with different mechanical and disintegration performances are prepared and utilized as elastic substrate layers for soft and transient electronics, such as sodium carboxymethylcellulose (Na–CMC) [86], poly(caprolactone)–poly(glycerol sebacate)(PGS–PCL) [87], and poly(vinyl alcohol) (PVA) [27], etc., as shown in Figure 2c,e,f, respectively. Among them, PVA is a water-soluble polymer that forms flexible layers, which allows for the controlled transport rate of water through altering the crosslinking density of the chains and their subsequent swelling ratios. A recent report suggests that biodegradable elastomer poly (octamethylene maleate (anhydride) citrate) (POMaC) is an excellent candidate for applications in terms of its biocompatibility, mechanical properties, and degradation characteristics, which can be tuned by varying the polymerization conditions [88]. Besides, poly(1,8-octanediol-*co*-citrate) (POC) is a biodegradable elastomer that can take strains up to ~30% with linear elastic mechanical responses. Hydrogels as hydrophilic polymeric networks with 3D microstructures represent alternative options that are highly biocompatible and suitable for biomimetic applications, due to their water-rich natures and structural similarities to natural extracellular matrices. Moreover, they can be designed to degrade under controlled modes and rates by enzymatic hydrolysis, ester hydrolysis, photolytic cleavage, or a combination of these reactions.

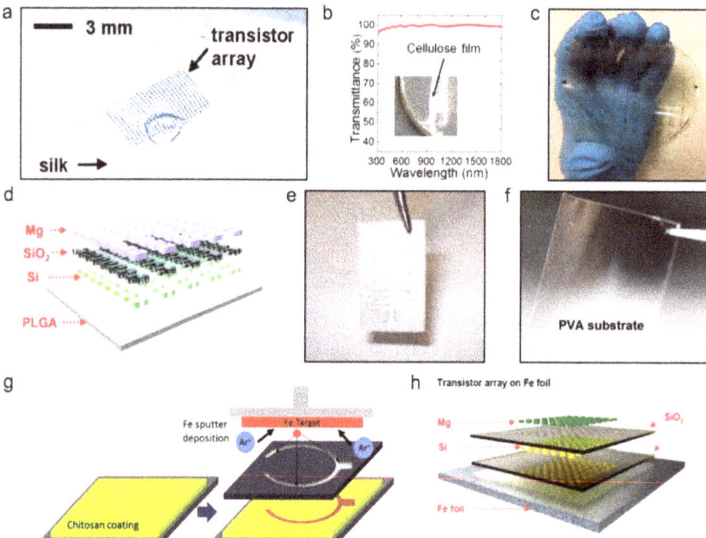

Figure 2. Various soft substrate materials for biodegradable electronics. (**a**) A silk substrate for transistor arrays. (**b**) A cellulose substrate for transistors. (**c**) A sodium carboxymethylcellulose (Na–CMC) bioresorbable substrate for Zn patterns. (**d**) A Poly lactic-co-glycolic acid (PLGA) substrate for transient electronic circuits. (**e**) An electrospun poly(caprolactone)–poly(glycerol sebacate) (PGS–PCL) sheet for a typical conductive pattern. (**f**) A PVA substrate for transient indium–gallium–zinc oxide (a-IGZO) TFTs. (**g**) A chitosan substrate for a biodegradable battery. (**h**) An Fe foil substrate for transistor arrays. Reproduced with permission from [20,27,47,63,85–87].

Although biodegradable polymeric substrates can often offer appropriate mechanical properties for soft and biodegradable electronics, direct device fabrication processed on polymer substrates is quite limited, as most types are sensitive to temperatures, solvents, or water. The swelling of most biodegradable polymers remains another challenge, as it can greatly shorten the functional lifetimes of electronic devices. Novel fabrication processes have been proposed to decouple polymeric substrates from the fabrication processes, which, however, introduce extra multiple fabrication steps; this will be discussed in the next section. Metal foils such as Mo, Fe, W, and Zn have also been proposed as alternatives to polymeric biodegradable substrates, to offer better compatibility with the fabrication process, because they are relatively temperature-resistant, and because they address swelling issues upon deployment in aqueous solutions, as illustrated in Figure 2h [47]. They can also show excellent electrical and thermal properties, favorable water and oxygen isolation performances, and they are relatively resistant to most solvents. However, the rigid properties of metal foils might limit their further applications.

2.4. Encapsulation Materials

Encapsulation materials, together with substrate materials, define the lifetimes of biodegradable electronics. Similarly, most polymeric substrate materials can be used to form strain-tolerant encapsulation layers, preventing rapid degradation of devices, such as PLGA, PCL, etc. [21,89]. Silk fibroin pockets have also been demonstrated to be useful for the controlled modulation of the device lifetime [90]. Similarly, POC can be used not only as the substrate layer, but also as the encapsulation layer, for transient biomedical devices, to protect them from the environment. However, water permeation resistance within biodegradable polymeric materials often cannot satisfy the requirements of devices when a longer lifetime is needed, e.g., an encapsulated Mg trace with silk fibroin could lose its conductivity within a few hours [14]. Encapsulation using a Si membrane (~1.5 μm) has been explored, and it can significantly extend the degradation times of dissolvable metals, e.g., Mg thin films with Si encapsulation result in a lifetime of 60 days in phosphate-buffered saline at 37 °C [44]. Bioresorbable electrocorticography electrodes based on Si encapsulation have demonstrated comparable recording results, compared to conventional standard electrodes, indicating the possibility of using a Si encapsulation layer for biodegradable electronics. In addition, alternating dielectric oxide layers of $SiO_2/Si_3N_4/SiO_2$ has also been proven to possess good water permeation resistance [30]. Further studies are needed to investigate biodegradable encapsulation materials with ultralow water permeation rates, as well as appropriate electrical properties and mechanical flexibilities, to achieve a wider range of operational time frames. Strategies include combining inorganic/organic multilayer materials to improve flexibility, and to introduce smart stimuli-responsive materials to precisely control the starting point of degradation.

3. Fabrication Schemes

Traditional device fabrication often involves photolithography, deposition, and etching processes. For the biomedical application, it is important to note that the implanted bioresorbable electronics require soft properties to realize their conformal contact with organs and tissues. In addition to that, the manufacturing process should not introduce any toxic materials or solvents, to guarantee favorable biocompatibilities. Consequently, novel fabrication techniques are needed to ensure material compatibility with the processing parameters.

For biodegradable devices, substrate materials with thicknesses at the micrometer scale contribute to the majority of the weight. Polymeric materials are often used as the substrates, because of their intrinsically soft and flexible properties, although metal foils have also been explored as an alternative [47]. Polymers are often dissolved into organic solvents and processed into a thin film format by drop casting, spin coating, or electrospinning methods. The thickness of the layer can be controlled by changing the relative solution concentrations, the speed of spin coating, or the electrospinning time [87,91]. The selections of proper solvents, spin-coating, or electrospinning parameters, and the surface treatments of handle substrates are critical to achieving free-standing polymer films with softness for the device fabrication.

The direct deposition of functional materials onto biodegradable and soft substrates, has been achieved through the use of shadow masks [14]. Decoupling fabrication processes from target biodegradable substrates through transfer-printing enables devices with a higher level of integration and more versatile material selections [85]. Figure 3a shows a transfer printing process utilizing foundry-based devices to achieve biodegradable 3D heterogeneous integrated circuits [21]. In order to obtain releasable micro-components from foundry-base wafers, a ~700 nm SiN$_x$ passive layer is deposited by plasma-enhanced chemical-vapor deposition (PECVD), followed by inductively-coupled plasma-reactive ion etching (ICP-RIE) process to form trenches. Poly(dimethylsiloxane) (PDMS) stamps are then utilized in a transfer printing process to remove and deliver the target collections (a total thickness of ~3 μm) onto target PLGA substrates. After that, a PLGA dielectric layer is prepared by the spin casting method, and a standard photolithography technology associated with RIE treatment, creates specific openings to make contact between the upper and lower layers. Another functional layer can then be prepared by applying a repeated procedure to finish the 3D integrated circuit. A combination of foundry-based Si wafers and transfer printing techniques provides a promising route towards high-performance and miniaturized biodegradable electronic systems.

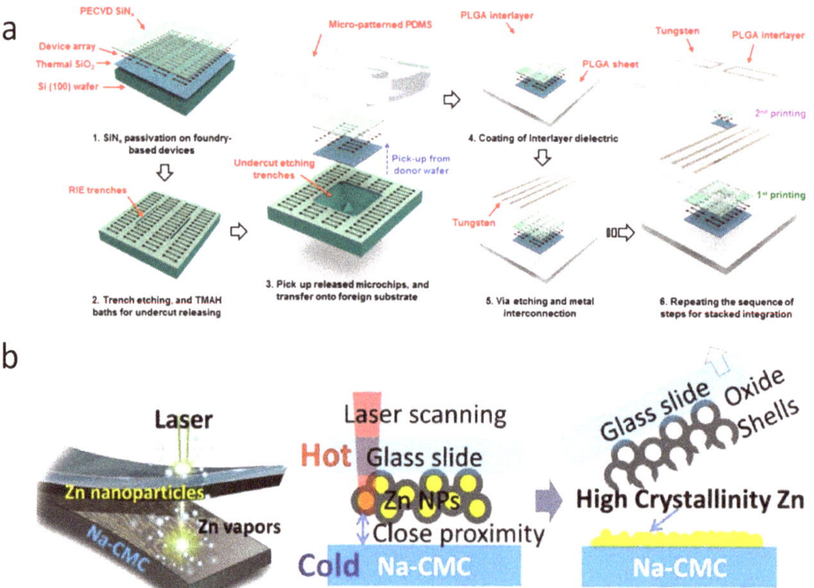

Figure 3. Fabrication schemes for soft and biodegradable electronics. (**a**) Preparation process of a 3D interconnected platform. Planarizing layers of PLGA serve as adhesives and interlayer dielectrics to facilitate 3D heterogeneous integration. (**b**) The schematic printing process for evaporation–condensation-mediated laser printing. Reproduced with permission from [21,86].

Printing techniques of biodegradable materials represent alternative methods to realize quick circuit patterns [46,86,89,92]. Figure 3b illustrates a fabrication approach to manufacture bioresorbable electronics. A continuous-wave (CW) fiber laser is utilized to supply sintering power, so as to crystalize Zn nanoparticles onto a soft Na–CMC substrate directly, with superior conductivity ($\sim$1.124 $\times$ 10^6 S m^{-1}) [86]. Such a fabrication technology offers a direct way for a roll-to-roll preparation platform to produce soft biodegradable electronics with high integration levels and low costs. In addition, highly conductive bioresorbable inks with an extended lifetime, consisting of polyanhydride and dispersed molybdenum microparticles, has also been reported. Such an ink can be applied to flexible wire and connection joints, as well as the antennas of bioresorbable devices [92]. These novel printing processes imply a fast and low-cost way to achieve biodegradable circuits.

4. Representative Soft and Biodegradable Devices for Biomedical Applications

Compared to conventional rigid implants, soft properties are favorable for biomedical implants to ensure conformally wrapping around biosystems that achieve intimate contact and minimize mechanical irritations, which are crucial and necessary for their applications, such as physiological signal detection and drug delivery. Combining biodegradable characteristics, devices can achieve fully bioresorption after usage, eliminating device retrieval. These devices could potentially serve as implantable diagnostic and therapeutic platforms, and they provide unprecedented physiological information and treatments, which are especially valuable for temporal biological processes, such as wound healing, neural network mapping, drug delivery, tissue regeneration, etc. Demonstrated soft and biodegradable electronic implants and pertinent power supply solutions will be reviewed in the following sections.

4.1. Diagnostic Platforms

For diagnostic purposes, because of their close connection with tissues and organs, soft transient electronics can detect abnormal physiological signals sensitively and precisely, even at the early stages of specific diseases, before they can be observed by conventional equipment, which is important for human healthcare and follow-up therapy. Figure 4 shows a soft and high-resolution recording system for electrocorticography (ECoG) based on Si devices [30], which offers a potential utility for treating neural disorders where biodegradation is required, to avoid tissue injury upon device removal. The flexible platform also enables the intimate contact of the device with the cerebral cortex, and allows for high-fidelity data recording. Figure 4a illustrates the structure of the device, consisting of a flexible PLGA substrate ($\sim$30 μm), a Si nanomembrane semiconductor, a Mo electrode ($\sim$300 nm), a SiO$_2$ gate dielectric, and SiO$_2$ ($\sim$300 nm)/Si$_3$N$_4$ ($\sim$400 nm)/SiO$_2$ ($\sim$300 nm) interlayer dielectrics. The device includes 128 metal oxide-semiconductor field-effect transistors (MOSFETs). Figure 4b exhibits the optical images of the unit cells of the device at different stages of fabrication, and a complete system. Figure 4c represents the linear (red) and log-scale (blue) transfer curves for representative n-channel MOSFET indication, with the mobility and on/off ratio of $\sim$400 cm^2 V^{-1} and $\sim$$10^8$, respectively. The device is implanted onto the whisker area, and Figure 4d reveals the schematic illustration of the whisker stimulation locations in a rat model. By stimulating the defined positions, the related evoked potential in a spatial distribution is recorded, as given in Figure 4e, indicating high resolution mapping of ECoG that matches or exceeds any existing devices. The sensing system can completely dissolve into an aqueous buffer solution gradually at pH = 12 and at 37 °C, as shown in Figure 4f. The sensitive sensing and biodegradable features of this transient device offer a promising application for internal physiological signal collection, which is significantly important for disease diagnosis.

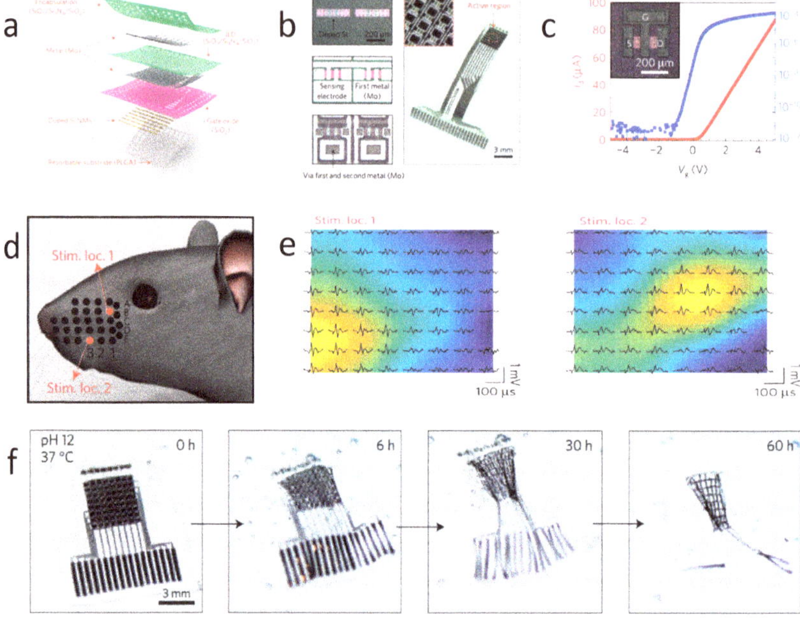

Figure 4. A soft and biodegradable neural electrode array sensor. (**a**) The schematic structure of an actively multiplexed sensing system for high-resolution electrocorticography. (**b**) Left: Optical micrograph images of a pair of subunits for the fabrication process. Right: The entire complete system. (**c**) Linear (red) and log scale (blue) transfer curves for a representative n-channel, MOSFET. (**d**) Schematic illustration of the whisker stimulation locations 1 and 2 (B1 and E3) in a rat model. (**e**) Left: Spatial distribution of the potentials evoked by the stimulation location B1. Right: Spatial distribution of the potentials evoked by stimulation location E3. (**f**) The degradation process at various stages of the sensor. Reproduced with permission from [30].

Stretchable and biodegradable pressure and strain sensors have also been reported for real-time monitoring for potential tendon recovery [88]. Figure 5a,b show the vertical structure and optical images of the sensor, respectively. The entire sensor is divided into four parts, including bottom and top encapsulation layers, as well as strain- and pressure-sensing areas, with Mg as the electrode, POMaC as the stretchable packaging layer, Poly(lactic acid) (PLLA) as the substrate, and PGS as the stretchable dielectric layer. Such a design allows for independent measurements of the strain and pressure. The sensor is implanted onto the back of a Sprague Dawley rat, as shown in Figure 5c. Figure 5d,e illustrate the collected pressure and strain signals after implantation for 2 and 3.5 weeks; the similar curves indicate a stable working performance. The biocompatibility of the sensor is demonstrated in Figure 5f, with CD68 positive cells decreasing gradually as the implantation period extends, suggesting that the inflammatory reaction is mitigated and, therefore, that excellent biocompatibility is reached. After more than two weeks of implantation, this sensor begins to degrade gradually. This research demonstrates the potential use of biodegradable devices for orthopedic applications.

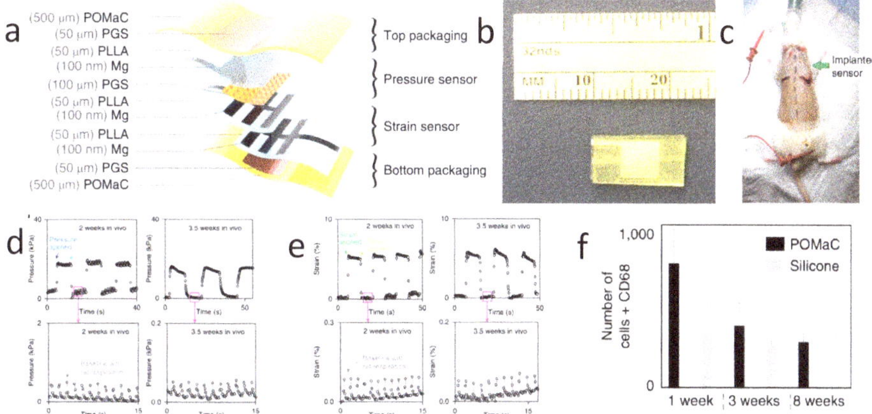

Figure 5. A stretchable and biodegradable strain and pressure sensor for orthopedic application. (**a**) Schematic diagram of the sensor space structure. (**b**) Optical image of the sensor. (**c**) The location of the implantable sensor. (**d**) Pressure signal detection two and 3.5 weeks after sensor implantation. (**e**) Strain signal detection two and 3.5 weeks after sensor implantation. (**f**) Results of immunohistochemistry. Reproduced with permission from [88].

4.2. Therapeutic Devices

In addition to diagnostic functions, soft and biodegradable electronics could also play an important role in therapeutic processes. For therapeutic platforms, flexible and biodegradable devices can offer personalized and precision treatments based on controlled performance and degradation rates. Bioresorbable devices can combine with drug delivery vehicles to achieve controlled drug release systems, which are critical for disease treatments.

The soft format of the sensor is favorable for controlled drug release with precise doses to specific areas, because of the conformal contact with tissues, which can improve treatment efficacy. Figure 6a displays the structure of a drug delivery system with a 2 × 2 array, consisting of inductive coupling coils and serpentine thermal heaters on a PLGA substrate [93]. The device allows for heating of the drug storage area through external wireless controls by inductive coupling. Drug release is thermally triggered by the phase transition of lipid layers imbedded with drugs. Figure 6b (left) shows the total cumulative percentage of doxorubicin released as a function of time from a device that is immersed in deionized water (12 mL) when it is activated by wireless external power between 0.1 to 1.3 W at 12.5 MHz and a distance of 2 mm. The release rate of the drug can be adjusted by adjusting the supplied power. Figure 6b (right) represents cumulative amounts of doxorubicin that are released wirelessly once a day, indicating good temperature-controlled drug release.

Wireless thermal therapy of bacterial management has also been demonstrated to assist with wound healing. Previous studies have revealed that some bacteria are highly sensitive to environmental changes, and that an increase in temperature lowers their survival [94–96]. Figure 6c (left) shows a transient radio frequency (RF) device for thermal therapy based on the Mg heater, embedded between silk fibroin layers [97]. Figure 6c (right) reveals related implantation processes for rats. The rats are infected by *Staphylococcus aureus* (*S. aureus*) with a subcutaneous injection (~5 μL) at the device implantation site, to mimic surgical site infections. The experimental rats are divided into untreated, low power (100 mW), and high power (500 mW) groups. Figure 6d (left) shows a thermal image of a rat with a high power supply after 10 min heat treatments, with the temperature reaching 49 °C. The infected tissues were collected after 24 h, and assessed by counting the normalized number of colony-forming units in the homogenates (n = 3) using standard plate-counting methods, and the related results are shown in Figure 6d (right). It is obvious that thermal treatment achieves a better bactericidal effect with higher temperatures. The concept of a biodegradable thermal therapy device can be widely applied to eliminate temperature-sensitive bacteria, which could be crucial to promoting the wound healing process.

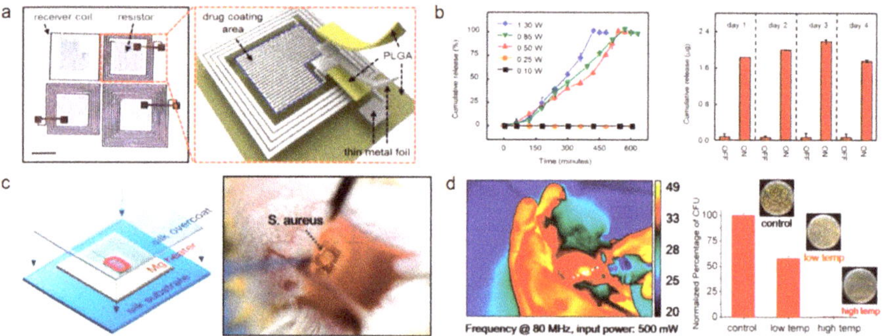

Figure 6. A thermally triggered drug delivery transient device and a radio frequency-controlled thermal therapy platform. (**a**) Images of the device structure. (**b**) Left: Cumulative release of doxorubicin from the device, operated with wireless power. Right: Controllable doxorubicin release from the device over 1 day on/off cycles. (**c**) Left: The schematic illustration of the radio frequency thermal therapy platform. Right: Photo of a device implanted in BALB/c mice. (**d**) Left: The thermal image of the device position while wirelessly powering the device. Right: The normalized number of colony-forming units after 24 h with different levels of input power. Reproduced with permission from [93,97].

4.3. Power Supply

Power supply is an essential component for biodegradable electronic systems, and great efforts have been made to explore biodegradable power sources, including batteries [98–102], supercapacitors [9], photovoltaic devices [103], radio frequency (RF) power scavengers [104], piezoelectric harvesters [25], etc. For biomedical applications, the power device should be flexible, stretchable, and miniaturized, to realize conformal contact and to minimize mechanical irritation.

A flexible and biodegradable RF wireless energy harvester appears in Figure 7a–d. Figure 7a,b illustrates the schematic structure of a wireless RF power transmitter, which consists of an RF antenna, an inductor, six capacitors, a resistor, and eight diodes [104]. By integrating with an Mg antenna, the RF system is capable of transmitting enough power to light up a red LED, as shown in Figure 7c. The entire system degrades rapidly in deionized water, as shown in Figure 7d. The demonstrated system provides a route for wireless energy harvesting for biodegradable implants; however, the current device volume limits its usage, and it needs further improvement. Figure 7e exhibits a flexible and biodegradable piezoelectric energy harvester based on ZnO on a silk substrate [25]. By bending the integrated film repeatedly, an output potential of 1.14 V, and curves of 0.55 nA are obtained, as shown in Figure 7f. Figure 7g illustrates the theoretical shape for the buckling of a device under compression. Although the piezoelectric energy harvester does not rely on an external device such as that for an RF energy scavenger, mechanical deformation inside the body is limited and, therefore, this limits the available power that can be obtained.

In summary, although various power solutions have been proposed, the performance of soft and biodegradable power supplies still remains an obstacle. The power density and the working life of the device should be further improved to satisfy the requirements of various biodegradable electronic systems. Materials and fabrication methods for power devices should also be advanced to achieve miniaturized, flexible, and stretchable platforms that are suitable as biomedical implants.

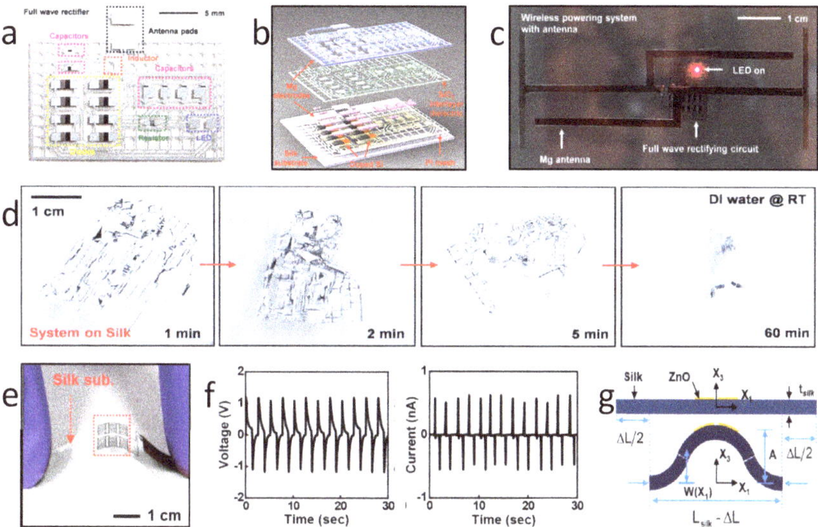

Figure 7. Flexible and biodegradable power harvesters. (**a**) A schematic illustration of transient RF power scavenging circuits. (**b**) A schematic illustration of an exploded view of the device. (**c**) An image of the device powered wirelessly with an RF transmitter and an Mg-receiving antenna. (**d**) Degradation process of the RF power harvester (**e**) Image of a soft and transient ZnO mechanical energy harvester on a silk substrate. (**f**) Output voltage and current ability during cycles of bending. (**g**) The theoretical shape for the buckling of a device under compression. Reproduced with permission from [25,104].

5. Summary and Outlook

As an emerging field, soft and biodegradable electronics have attracted more and more research interest because of their foreseeable applications for clinical implants, eco-friendly devices, and security hardware. For biodegradable biomedical devices, favorable biocompatibility, appropriate degradable rates, and robust mechanical properties, as well as superior performance, are desirable. Many studies have been made, to achieve remarkable progress towards biomedical applications.

However, there are still many issues that need to be addressed. More versatile materials with both biodegradable and soft properties need to be explored, to further broaden potential applications. For example, biodegradable functional materials that are highly stretchable and flexible could expand suitable implantation locations and significantly improve data recording sensitivities and accuracies, and minimize the irritation and inflammation that are associated with implantation. This could be achieved by developing composite structures that integrate hard and soft components, as well as through appropriate mechanical structure designs. Moreover, encapsulation materials with superior water resistance and soft properties are critical for waterproof sealing, to avoid the potential rupture of the encapsulation layer causing water leakage. Different from the fabrication methods for rigid biodegradable devices, novel fabrication technologies should be further explored to produce soft electronics with low costs and easy manufacturing procedures, as well as high levels of integration. In addition, the performance of soft devices, such as fast response and excellent sensitivity, as well as high accuracy, needs further improvement, and multifunctional electronics should be fabricated to meet the requirements of clinical standards. A comprehensive investigation of the device/tissue interface, and the metabolic processes of degradation products are also necessary to clarify the safety issues of biodegradable devices. These studies will improve the overall performance of soft and biodegradable devices, and they will promote the development of transient electronics, which could potentially make disease diagnosis and treatments more precise, effective, and intelligent.

Author Contributions: Literature search, R.L. & L.W.; Figures, R.L.; Tables, L.W.; Writing-Review & Editing, R.L. & L.W.; R.L. and L.W. contributed equally to the paper. Conceptualization & Writing, L.Y.

Funding: This research was funded by [National Natural Science Foundation of China] grant number [51601103], [1000 Youth Talents Program in China], and [China Postdoctoral Science Foundation] grant number [2017M620769].

Acknowledgments: This work was supported by the National Natural Science Foundation of China (NSFC) 51601103 (L.Y.), 1000 Youth Talents Program in China (L.Y.), and the China Postdoctoral Science Foundation 2017M620769 (R.L.).

Conflicts of Interest: The authors declare no conflict of interest. The funders had no role in the design of the study; in the collection, analyses, or interpretation of data; in the writing of the manuscript, and in the decision to publish the results.

References

1. Dubal, D.P.; Chodankar, N.R.; Kim, D.H.; Gomez-Romero, P. Towards flexible solid-state supercapacitors for smart and wearable electronics. *Chem. Soc. Rev.* **2018**, *47*, 2065–2129. [CrossRef] [PubMed]
2. Annapureddy, V.; Na, S.M.; Hwang, G.T.; Kang, M.G.; Sriramdas, R.; Palneedi, H.; Yoon, W.H.; Hahn, B.D.; Kim, J.W.; Ahn, C.W.; et al. Exceeding milli-watt powering magneto-mechano-electric generator for standalone-powered electronics. *Energy Environ. Sci.* **2018**, *11*, 818–829. [CrossRef]
3. Salauddin, M.; Toyabur, R.M.; Maharjan, P.; Park, J.Y. High performance human-induced vibration driven hybrid energy harvester for powering portable electronics. *Nano Energy* **2018**, *45*, 236–246. [CrossRef]
4. Wang, S.H.; Xu, J.; Wang, W.C.; Wang, G.J.N.; Rastak, R.; Molina-Lopez, F.; Chung, J.W.; Niu, S.M.; Feig, V.R.; Lopez, J.; et al. Skin electronics from scalable fabrication of an intrinsically stretchable transistor array. *Nature* **2018**, *555*, 83–99. [CrossRef] [PubMed]
5. Falco, A.; Rivadeneyra, A.; Loghin, F.C.; Salmeron, J.F.; Lugli, P.; Abdelhalim, A. Towards low-power electronics: Self-recovering and flexible gas sensors. *J. Mater. Chem. A* **2018**, *6*, 7107–7113. [CrossRef]
6. Li, R.F.; Wang, L.; Kong, D.Y.; Yin, L. Recent progress on biodegradable materials and transient electronics. *Bioact. Mater.* **2018**, *3*, 322–333. [CrossRef] [PubMed]
7. Gao, Y.; Zhang, Y.; Wang, X.; Sim, K.; Liu, J.S.; Chen, J.; Feng, X.; Xu, H.X.; Yu, C.J. Moisture-triggered physically transient electronics. *Sci. Adv.* **2017**, *3*, e1701222. [CrossRef] [PubMed]
8. Yoon, J.; Lee, J.; Choi, B.; Lee, D.; Kim, D.H.; Kim, D.M.; Moon, D.I.; Lim, M.; Kim, S.; Choi, S.J. Flammable carbon nanotube transistors on a nitrocellulose paper substrate for transient electronics. *Nano Res.* **2017**, *10*, 87–96. [CrossRef]
9. Lee, G.; Kang, S.K.; Won, S.M.; Gutruf, P.; Jeong, Y.R.; Koo, J.; Lee, S.S.; Rogers, J.A.; Ha, J.S. Fully Biodegradable Microsupercapacitor for Power Storage in Transient Electronics. *Adv. Energy Mater.* **2017**, *7*, 1700157. [CrossRef]
10. Kang, S.-K.; Hwang, S.-W.; Cheng, H.; Yu, S.; Kim, B.H.; Kim, J.-H.; Huang, Y.; Rogers, J.A. Dissolution Behaviors and Applications of Silicon Oxides and Nitrides in Transient Electronics. *Adv. Funct. Mater.* **2014**, *24*, 4427–4434. [CrossRef]
11. Khanra, S.; Cipriano, T.; Lam, T.; White, T.A.; Fileti, E.E.; Alves, W.A.; Guha, S. Self-Assembled Peptide-Polyfluorene Nanocomposites for Biodegradable Organic Electronics. *Adv. Mater. Interfaces* **2015**, *2*, 1500265. [CrossRef]
12. Feig, V.R.; Tran, H.; Bao, Z.N. Biodegradable Polymeric Materials in Degradable Electronic Devices. *ACS Cent. Sci.* **2018**, *4*, 337–348. [CrossRef] [PubMed]
13. Kang, S.K.; Koo, J.; Lee, Y.K.; Rogers, J.A. Advanced Materials and Devices for Bioresorbable Electronics. *Acc. Chem. Res.* **2018**, *51*, 988–998. [CrossRef] [PubMed]
14. Hwang, S.W.; Tao, H.; Kim, D.H.; Cheng, H.Y.; Song, J.K.; Rill, E.; Brenckle, M.A.; Panilaitis, B.; Won, S.M.; Kim, Y.S.; et al. A Physically Transient Form of Silicon Electronics. *Science* **2012**, *337*, 1640–1644. [CrossRef] [PubMed]
15. Kang, S.K.; Murphy, R.K.J.; Hwang, S.W.; Lee, S.M.; Harburg, D.V.; Krueger, N.A.; Shin, J.H.; Gamble, P.; Cheng, H.Y.; Yu, S.; et al. Bioresorbable silicon electronic sensors for the brain. *Nature* **2016**, *530*, 71. [CrossRef] [PubMed]
16. Irimia-Vladu, M.; Glowacki, E.D.; Voss, G.; Bauer, S.; Sariciftci, N.S. Green and biodegradable electronics. *Mater. Today* **2012**, *15*, 340–346. [CrossRef]

17. Park, C.W.; Kang, S.K.; Hernandez, H.L.; Kaitz, J.A.; Wie, D.S.; Shin, J.; Lee, O.P.; Sottos, N.R.; Moore, J.S.; Rogers, J.A.; et al. Thermally Triggered Degradation of Transient Electronic Devices. *Adv. Mater.* **2015**, *27*, 3783–3788. [CrossRef] [PubMed]

18. Hernandez, H.L.; Kang, S.K.; Lee, O.P.; Hwang, S.W.; Kaitz, J.A.; Inci, B.; Park, C.W.; Chung, S.J.; Sottos, N.R.; Moore, J.S.; et al. Triggered Transience of Metastable Poly(phthalaldehyde) for Transient Electronics. *Adv. Mater.* **2014**, *26*, 7637–7642. [CrossRef] [PubMed]

19. Kim, D.H.; Lu, N.S.; Ma, R.; Kim, Y.S.; Kim, R.H.; Wang, S.D.; Wu, J.; Won, S.M.; Tao, H.; Islam, A.; et al. Epidermal Electronics. *Science* **2011**, *333*, 838–843. [CrossRef] [PubMed]

20. Hwang, S.W.; Kim, D.H.; Tao, H.; Kim, T.I.; Kim, S.; Yu, K.J.; Panilaitis, B.; Jeong, J.W.; Song, J.K.; Omenetto, F.G.; et al. Materials and Fabrication Processes for Transient and Bioresorbable High-Performance Electronics. *Adv. Funct. Mater.* **2013**, *23*, 4087–4093. [CrossRef]

21. Chang, J.K.; Chang, H.P.; Guo, Q.; Koo, J.; Wu, C.I.; Rogers, J.A. Biodegradable Electronic Systems in 3D, Heterogeneously Integrated Formats. *Adv. Mater.* **2018**, *30*, 1704955. [CrossRef] [PubMed]

22. Hwang, S.-W.; Park, G.; Edwards, C.; Corbin, E.A.; Kang, S.-K.; Cheng, H.; Song, J.-K.; Kim, J.-H.; Yu, S.; Ng, J.; et al. Dissolution Chemistry and Biocompatibility of Single-Crystalline Silicon Nanomembranes and Associated Materials for Transient Electronics. *ACS Nano* **2014**, *8*, 5843–5851. [CrossRef] [PubMed]

23. Hwang, S.W.; Park, G.; Cheng, H.; Song, J.K.; Kang, S.K.; Yin, L.; Kim, J.H.; Omenetto, F.G.; Huang, Y.G.; Lee, K.M.; et al. 25th Anniversary Article: Materials for High-Performance Biodegradable Semiconductor Devices. *Adv. Mater.* **2014**, *26*, 1992–2000. [CrossRef] [PubMed]

24. Yin, L.; Farimani, A.B.; Min, K.; Vishal, N.; Lam, J.; Lee, Y.K.; Aluru, N.R.; Rogers, J.A. Mechanisms for hydrolysis of silicon nanomembranes as used in bioresorbable electronics. *Adv. Mater.* **2015**, *27*, 1857–1864. [CrossRef] [PubMed]

25. Dagdeviren, C.; Hwang, S.W.; Su, Y.W.; Kim, S.; Cheng, H.Y.; Gur, O.; Haney, R.; Omenetto, F.G.; Huang, Y.G.; Rogers, J.A. Transient, Biocompatible Electronics and Energy Harvesters Based on ZnO. *Small* **2013**, *9*, 3398–3404. [CrossRef] [PubMed]

26. Kang, S.K.; Park, G.; Kim, K.; Hwang, S.W.; Cheng, H.Y.; Shin, J.H.; Chung, S.J.; Kim, M.; Yin, L.; Lee, J.C.; et al. Dissolution Chemistry and Biocompatibility of Silicon- and Germanium-Based Semiconductors for Transient Electronics. *ACS Appl. Mater. Interfaces* **2015**, *7*, 9297–9305. [CrossRef] [PubMed]

27. Jin, S.H.; Kang, S.K.; Cho, I.T.; Han, S.Y.; Chung, H.U.; Lee, D.J.; Shin, J.; Baek, G.W.; Kim, T.I.; Lee, J.H.; et al. Water-Soluble Thin Film Transistors and Circuits Based on Amorphous Indium-Gallium-Zinc Oxide. *ACS Appl. Mater. Interfaces* **2015**, *7*, 8268–8274. [CrossRef] [PubMed]

28. Huang, X.; Liu, Y.H.; Hwang, S.W.; Kang, S.K.; Patnaik, D.; Cortes, J.F.; Rogers, J.A. Biodegradable Materials for Multilayer Transient Printed Circuit Boards. *Adv. Mater.* **2014**, *26*, 7371–7377. [CrossRef] [PubMed]

29. Yin, L.; Cheng, H.; Mao, S.; Haasch, R.; Liu, Y.; Xie, X.; Hwang, S.-W.; Jain, H.; Kang, S.-K.; Su, Y.; et al. Dissolvable Metals for Transient Electronics. *Adv. Funct. Mater.* **2014**, *24*, 645–658. [CrossRef]

30. Yu, K.J.; Kuzum, D.; Hwang, S.W.; Kim, B.H.; Juul, H.; Kim, N.H.; Won, S.M.; Chiang, K.; Trumpis, M.; Richardson, A.G.; et al. Bioresorbable silicon electronics for transient spatiotemporal mapping of electrical activity from the cerebral cortex. *Nat. Mater.* **2016**, *15*, 782. [CrossRef] [PubMed]

31. Institute of Medicine (US) Panel on Micronutrients. *Dietary Reference Intakes for Vitamin A, Vitamin K, Arsenic, Boron, Chromium, Copper, Iodine, Iron, Manganese, Molybdenum, Nickel, Silicon, Vanadium, and Zinc*; A Report of the Panel on Micronutrients, Subcommittees on Upper Reference Levels of Nutrients and of Interpretation and Uses of Dietary Reference Intakes, and the Standing Committee on the Scientific Evaluation of Dietary Reference Intakes; National Academy Press: Washington, DC, USA, 2001.

32. Ysart, G.; Miller, P.; Crews, H.; Robb, P.; Baxter, M.; De L'Argy, C.; Lofthouse, S.; Sargent, C.; Harrison, N. Dietary exposure estimates of 30 elements from the UK Total Diet Study. *Food Addit. Contam.* **1999**, *16*, 391–403. [CrossRef] [PubMed]

33. Millour, S.; Noel, L.; Chekri, R.; Vastel, C.; Kadar, A.; Sirot, V.; Leblanc, J.C.; Guerin, T. Strontium, silver, tin, iron, tellurium, gallium, germanium, barium and vanadium levels in foodstuffs from the Second French Total Diet Study. *J. Food Compos. Anal.* **2012**, *25*, 108–129. [CrossRef]

34. Tao, S.H.; Bolger, P.M. Hazard assessment of germanium supplements. *Regul. Toxicol. Pharm.* **1997**, *25*, 211–219. [CrossRef] [PubMed]

35. James, B.; Zhang, W.L.; Sun, P.; Wu, M.Y.; Li, H.H.; Khaliq, M.A.; Jayasuriya, P.; James, S.; Wang, G. Tungsten (W) bioavailability in paddy rice soils and its accumulation in rice (*Oryza sativa*). *Int. J. Environ. Health Res.* **2017**, *27*, 487–497. [CrossRef] [PubMed]

36. Institute of Medicine (US) Standing Committee on the Scientific Evaluation of Dietary Reference Intakes. *Dietary Reference Intakes for Calcium, Phosphorus, Magnesium, Vitamin D, and Fluoride*; National Academy Press: Washington, DC, USA, 1997.

37. Rogers, J.A.; Lagally, M.G.; Nuzzo, R.G. Synthesis, assembly and applications of semiconductor nanomembranes. *Nature* **2011**, *477*, 45–53. [CrossRef] [PubMed]

38. Hofmann, S.; Ducati, C.; Neill, R.J.; Piscanec, S.; Ferrari, A.C.; Geng, J.; Dunin-Borkowski, R.E.; Robertson, J. Gold catalyzed growth of silicon nanowires by plasma enhanced chemical vapor deposition. *J. Appl. Phys.* **2003**, *94*, 6005–6012. [CrossRef]

39. Guo, X.L.; Tabata, H.; Kawai, T. Pulsed laser reactive deposition of p-type ZnO film enhanced by an electron cyclotron resonance source. *J. Cryst. Growth* **2001**, *223*, 135–139. [CrossRef]

40. Schropp, R.E.I.; Feenstra, K.E.; Molenbroek, E.C.; Meiling, H.; Rath, J.K. Device-quality polycrystalline and amorphous silicon films by hot-wire chemical vapour deposition. *Philos. Mag. B* **1997**, *76*, 309–321. [CrossRef]

41. Lee, W.J.; Park, W.T.; Park, S.; Sung, S.; Noh, Y.Y.; Yoon, M.H. Large-Scale Precise Printing of Ultrathin Sol-Gel Oxide Dielectrics for Directly Patterned Solution-Processed Metal Oxide Transistor Arrays. *Adv. Mater.* **2015**, *27*, 5043–5048. [CrossRef] [PubMed]

42. Kim, D.H.; Ghaffari, R.; Lu, N.S.; Rogers, J.A. Flexible and Stretchable Electronics for Biointegrated Devices. *Annu. Rev. Biomed. Eng.* **2012**, *14*, 113–128. [CrossRef] [PubMed]

43. Ahn, J.H.; Kim, H.S.; Lee, K.J.; Jeon, S.; Kang, S.J.; Sun, Y.G.; Nuzzo, R.G.; Rogers, J.A. Heterogeneous three-dimensional electronics by use of printed semiconductor nanomaterials. *Science* **2006**, *314*, 1754–1757. [CrossRef] [PubMed]

44. Lee, Y.K.; Yu, K.J.; Song, E.M.; Farimani, A.B.; Vitale, F.; Xie, Z.Q.; Yoon, Y.; Kim, Y.; Richardson, A.; Luan, H.W.; et al. Dissolution of Monocrystalline Silicon Nanomembranes and Their Use as Encapsulation Layers and Electrical Interfaces in Water-Soluble Electronics. *ACS Nano* **2017**, *11*, 12562–12572. [CrossRef] [PubMed]

45. Chen, X.; Park, Y.J.; Kang, M.; Kang, S.K.; Koo, J.; Shinde, S.M.; Shin, J.; Jeon, S.; Park, G.; Yan, Y.; et al. CVD-grown monolayer MoS2 in bioabsorbable electronics and biosensors. *Nat. Commun.* **2018**, *9*, 1690. [CrossRef] [PubMed]

46. Mahajan, B.K.; Yu, X.W.; Shou, W.; Pan, H.; Huang, X. Mechanically Milled Irregular Zinc Nanoparticles for Printable Bioresorbable Electronics. *Small* **2017**, *13*, 1700065. [CrossRef] [PubMed]

47. Kang, S.-K.; Hwang, S.-W.; Yu, S.; Seo, J.-H.; Corbin, E.A.; Shin, J.; Wie, D.S.; Bashir, R.; Ma, Z.; Rogers, J.A. Biodegradable Thin Metal Foils and Spin-On Glass Materials for Transient Electronics. *Adv. Funct. Mater.* **2015**, *25*, 1789–1797. [CrossRef]

48. Kirkland, N.T.; Birbilis, N.; Staiger, M.P. Assessing the corrosion of biodegradable magnesium implants: A critical review of current methodologies and their limitations. *Acta Biomater.* **2012**, *8*, 925–936. [CrossRef] [PubMed]

49. Bowen, P.K.; Drelich, J.; Goldman, J. Zinc Exhibits Ideal Physiological Corrosion Behavior for Bioabsorbable Stents. *Adv. Mater.* **2013**, *25*, 2577–2582. [CrossRef] [PubMed]

50. Badawy, W.A.; Al-Kharafi, F.M. Corrosion and passivation behaviors of molybdenum in aqueous solutions of different pH. *Electrochim. Acta* **1998**, *44*, 693–702. [CrossRef]

51. Patrick, E.; Orazem, M.E.; Sanchez, J.C.; Nishida, T. Corrosion of tungsten microelectrodes used in neural recording applications. *J. Neurosci. Meth.* **2011**, *198*, 158–171. [CrossRef] [PubMed]

52. Hemstreet, J.M. Dielectric constant of cotton. *J. Electrost.* **1982**, *13*, 345–353. [CrossRef]

53. Jayamani, E.; Hamdan, S.; Rahman, M.R.; Bin Bakri, M.K. Comparative Study of Dielectric Properties of Hybrid Natural Fiber Composites. *Procedia Eng.* **2014**, *97*, 536–544. [CrossRef]

54. Irimia-Vladu, M.; Troshin, P.A.; Reisinger, M.; Shmygleva, L.; Kanbur, Y.; Schwabegger, G.; Bodea, M.; Schwodiauer, R.; Mumyatov, A.; Fergus, J.W.; et al. Biocompatible and Biodegradable Materials for Organic Field-Effect Transistors. *Adv. Funct. Mater.* **2010**, *20*, 4069–4076. [CrossRef]

55. Singh, T.B.; Sariciftci, N.S.; Grote, J.G. Bio-Organic Optoelectronic Devices Using DNA. *Org. Electron.* **2010**, *223*, 189–212.

56. Yumusak, C.; Singh, T.B.; Sariciftci, N.S.; Grote, J.G. Bioorganic field effect transistors based on crosslinked deoxyribonucleic acid (DNA) gate dielectric. *Appl. Phys. Lett.* **2009**, *95*, 263304. [CrossRef]

57. Singh, B.; Sariciftci, N.S.; Grote, J.G.; Hopkins, F.K. Bioorganic-semiconductor-field-effect-transistor based on deoxyribonucleic acid gate dielectric. *J. Appl. Phys.* **2006**, *100*, 024514. [CrossRef]

58. Wang, L.; Yoshida, J.; Ogata, N.; Sasaki, S.; Kajiyama, T. Self-Assembled Supramolecular Films Derived from Marine Deoxyribonucleic Acid (DNA)−Cationic Surfactant Complexes: Large-Scale Preparation and Optical and Thermal Properties. *Chem. Mater.* **2001**, *13*, 1273–1281. [CrossRef]

59. Worfolk, B.J.; Andrews, S.C.; Park, S.; Reinspach, J.; Liu, N.; Toney, M.F.; Mannsfeld, S.C.; Bao, Z.N. Ultrahigh electrical conductivity in solutionsheared polymeric transparent films. *Proc. Natl. Acad. Sci. USA* **2015**, *112*, 14138–14143. [CrossRef] [PubMed]

60. Qiao, Y.L.; Guo, Y.L.; Yu, C.M.; Zhang, F.J.; Xu, W.; Liu, Y.Q.; Zhu, D.B. Diketopyrrolopyrrole-Containing Quinoidal Small Molecules for High-Performance, Air-Stable, and Solution-Processable n-Channel Organic Field-Effect Transistors. *J. Am. Chem. Soc.* **2012**, *134*, 4084–4087. [CrossRef] [PubMed]

61. Chen, H.J.; Guo, Y.L.; Yu, G.; Zhao, Y.; Zhang, J.; Gao, D.; Liu, H.T.; Liu, Y.Q. Highly p-Extended Copolymers with Diketopyrrolopyrrole Moieties for High-Performance Field-Effect Transistors. *Adv. Mater.* **2012**, *24*, 4618–4622. [CrossRef] [PubMed]

62. Madrigal, M.M.P.; Giannotti, M.I.; Oncins, G.; Franco, L.; Armelin, E.; Puiggali, J.; Sanz, F.; del Valle, L.J.; Aleman, C. Bioactive nanomembranes of semiconductor polythiophene and thermoplastic polyurethane: Thermal, nanostructural and nanomechanical properties. *Polym. Chem.* **2013**, *4*, 568–583. [CrossRef]

63. Lei, T.; Guan, M.; Liu, J.; Lin, H.C.; Pfattner, R.; Shaw, L.; McGuire, A.F.; Huang, T.C.; Shao, L.L.; Cheng, K.T.; et al. Biocompatible and totally disintegrable semiconducting polymer for ultrathin and ultralightweight transient electronics. *Proc. Natl. Acad. Sci. USA* **2017**, *114*, 5107–5112. [CrossRef] [PubMed]

64. Irimia-Vladu, M.; Glowacki, E.D.; Troshin, P.A.; Schwabegger, G.; Leonat, L.; Susarova, D.K.; Krystal, O.; Ullah, M.; Kanbur, Y.; Bodea, M.A.; et al. Indigo—A Natural Pigment for High Performance Ambipolar Organic Field Effect Transistors and Circuits. *Adv. Mater.* **2012**, *24*, 375–380. [CrossRef] [PubMed]

65. Mostert, A.B.; Powell, B.J.; Pratt, F.L.; Hanson, G.R.; Sarna, T.; Gentle, I.R.; Meredith, P. Role of semiconductivity and ion transport in the electrical conduction of melanin. *Proc. Natl. Acad. Sci. USA* **2012**, *109*, 8943–8947. [CrossRef] [PubMed]

66. Bettinger, C.J.; Bruggeman, P.P.; Misra, A.; Borenstein, J.T.; Langer, R. Biocompatibility of biodegradable semiconducting melanin films for nerve tissue engineering. *Biomaterials* **2009**, *30*, 3050–3057. [CrossRef] [PubMed]

67. Ramachandran, G.K.; Tomfohr, J.K.; Li, J.; Sankey, O.F.; Zarate, X.; Primak, A.; Terazono, Y.; Moore, T.A.; Moore, A.L.; Gust, D.; et al. Electron transport properties of a carotene molecule in a metal-(single molecule)-metal junction. *J. Phys. Chem. B* **2003**, *107*, 6162–6169. [CrossRef]

68. Koh, L.D.; Yeo, J.; Lee, Y.Y.; Ong, Q.; Han, M.Y.; Tee, B.C.K. Advancing the frontiers of silk fibroin protein-based materials for futuristic electronics and clinical wound-healing. *Mater. Sci. Eng. C Mater.* **2018**, *86*, 151–172. [CrossRef] [PubMed]

69. Li, G.; Li, Y.; Chen, G.Q.; He, J.H.; Han, Y.F.; Wang, X.Q.; Kaplan, D.L. Silk-Based Biomaterials in Biomedical Textiles and Fiber-Based Implants. *Adv. Healthc. Mater.* **2015**, *4*, 1134–1151. [CrossRef] [PubMed]

70. Xie, M.B.; Li, Y.; Li, J.S.; Chen, A.Z.; Zhao, Z.; Li, G. Biomedical Applications of Silk Fibroin. In *Textile Bioengineering and Informatics Symposium Proceedings*; Textile Bioengineering & Informatics Society Ltd.: Hong Kong, China, 2014; Volumes 1 and 2, pp. 207–218.

71. Taddei, P.; Chiono, V.; Anghileri, A.; Vozzi, G.; Freddi, G.; Ciardelli, G. Silk Fibroin/Gelatin Blend Films Crosslinked with Enzymes for Biomedical Applications. *Macromol. Biosci.* **2013**, *13*, 1492–1510. [CrossRef] [PubMed]

72. Pal, R.K.; Farghaly, A.A.; Wang, C.Z.; Collinson, M.M.; Kundu, S.C.; Yadavalli, V.K. Conducting polymer-silk biocomposites for flexible and biodegradable electrochemical sensors. *Biosens. Bioelectron.* **2016**, *81*, 294–302. [CrossRef] [PubMed]

73. Edupuganti, V.; Solanki, R. Fabrication, characterization, and modeling of a biodegradable battery for transient electronics. *J. Power Sources* **2016**, *336*, 447–454. [CrossRef]

74. Jiang, L.; Zhang, J. Biodegradable and biobased polymers. In *Applied Plastics Engineering Handbook*, 2nd ed.; William Andrew Publishing: Oxford, UK, 2017; pp. 127–143.

75. Zhu, H.L.; Fang, Z.Q.; Preston, C.; Li, Y.Y.; Hu, L.B. Transparent paper: Fabrications, properties, and device applications. *Energy Environ. Sci.* **2014**, *7*, 269–287. [CrossRef]

76. Zhu, H.L.; Xiao, Z.G.; Liu, D.T.; Li, Y.Y.; Weadock, N.J.; Fang, Z.Q.; Huang, J.S.; Hu, L.B. Biodegradable transparent substrates for flexible organic-light-emitting diodes. *Energy Environ. Sci.* **2013**, *6*, 2105–2111. [CrossRef]

77. Jung, Y.H.; Chang, T.H.; Zhang, H.L.; Yao, C.H.; Zheng, Q.F.; Yang, V.W.; Mi, H.Y.; Kim, M.; Cho, S.J.; Park, D.W.; et al. High-performance green flexible electronics based on biodegradable cellulose nanofibril paper. *Nat. Commun.* **2015**, *6*, 7170. [CrossRef] [PubMed]

78. Li, Z.Q.; Wang, H.C.; Lv, S.Z.; Liu, L.; Guo, W.Y.; Yuan, M.; Yan, H.B.; Zhao, H.J.; Lang, S.P. Clinical Comparative Study on Efficacy and Safety for Treatment of Coronary Heart Disease with Cobalt-Base Alloy Bio Absorbable Polymer Sirolimus-Eluting Stent and Partner Stent. *Heart* **2011**, *97*, A149. [CrossRef]

79. Wu, Y.T.; Gao, Y.C. Five Years Follow Up Result after Application of Biodegradable Polymer Sirolimus-Eluting Stent in Patients with Coronary Heart Disease and Diabetes Mellitus. *Heart* **2013**, *99*, E175.

80. Inigo-Garcia, L.A.; Martinez-Garcia, F.J.; Milan-Pinilla, A.; Valle-Alberca, A.; Fernandez-Lopez, L.; Traverso-Castilla, V.V.; Delgado-Aguilar, A.; Bravo-Marques, R.; Ramirez-Moreno, A.; Siles-Rubio, J.R. Biodegradable Polymer Drug Eluting Stent: Efficacy and Safety with Short Regimen of Antiplatelet Therapy. *Cardiology* **2016**, *134*, 48.

81. Pilgrim, T.; Heg, D.; Roffi, M.; Tuller, D.; Muller, O.; Vuilliomenet, A.; Cook, S.; Weilenmann, D.; Kaiser, C.; Jamshidi, P.; et al. Ultrathin strut biodegradable polymer sirolimus-eluting stent versus durable polymer everolimus-eluting stent for percutaneous coronary revascularisation (BIOSCIENCE): A randomised, single-blind, non-inferiority trial. *Lancet* **2014**, *384*, 2111–2122. [CrossRef]

82. Waksman, R.; Pakala, R.; Baffour, R.; Seabron, R.; Hellinga, D.; Chan, R.; Su, S.H.; Kolodgie, F.; Virmani, R. In vivo comparison of a polymer-free Biolimus A9-eluting stent with a biodegradable polymer-based Biolimus A9 eluting stent and a bare metal stent in balloon denuded and radiated hypercholesterolemic rabbit iliac arteries. *Catheter. Cardiovasc. Interv.* **2012**, *80*, 429–436. [CrossRef] [PubMed]

83. Balch, O.K.; Collier, M.A.; DeBault, L.E.; Johnson, L.L. Bioabsorbable suture anchor (co-polymer 85/15 D,L lactide/glycolide) implanted in bone: Correlation of physical/mechanical properties, magnetic resonance imaging, and histological response. *Arthroscopy* **1999**, *15*, 691–708. [CrossRef]

84. Im, S.H.; Jung, Y.; Kim, S.H. Current status and future direction of biodegradable metallic and polymeric vascular scaffolds for next-generation stents. *Acta Biomater.* **2017**, *60*, 3–22. [CrossRef] [PubMed]

85. Hwang, S.W.; Song, J.K.; Huang, X.; Cheng, H.Y.; Kang, S.K.; Kim, B.H.; Kim, J.H.; Yu, S.; Huang, Y.G.; Rogers, J.A. High-Performance Biodegradable/Transient Electronics on Biodegradable Polymers. *Adv. Mater.* **2014**, *26*, 3905–3911. [CrossRef] [PubMed]

86. Shou, W.; Mahajan, B.K.; Ludwig, B.; Yu, X.W.; Staggs, J.; Huang, X.; Pan, H. Low-Cost Manufacturing of Bioresorbable Conductors by Evaporation-Condensation-Mediated Laser Printing and Sintering of Zn Nanoparticles. *Adv. Mater.* **2017**, *29*, 1700172. [CrossRef] [PubMed]

87. Najafabadi, A.H.; Tamayol, A.; Annabi, N.; Ochoa, M.; Mostafalu, P.; Akbari, M.; Nikkhah, M.; Rahimi, R.; Dokmeci, M.R.; Sonkusale, S.; et al. Biodegradable Nanofibrous Polymeric Substrates for Generating Elastic and Flexible Electronics. *Adv. Mater.* **2014**, *26*, 5823–5830. [CrossRef] [PubMed]

88. Boutry, C.M.; Kaizawa, Y.; Schroeder, B.C.; Chortos, A.; Legrand, A.; Wang, Z.; Chang, J.; Fox, P.; Bao, Z. A stretchable and biodegradable strain and pressure sensor for orthopaedic application. *Nat. Electron.* **2018**, *1*, 314–321. [CrossRef]

89. Lee, Y.K.; Kim, J.; Kim, Y.; Kwak, J.W.; Yoon, Y.; Rogers, J.A. Room Temperature Electrochemical Sintering of Zn Microparticles and Its Use in Printable Conducting Inks for Bioresorbable Electronics. *Adv. Mater.* **2017**, *29*, 1702665. [CrossRef] [PubMed]

90. Brenckle, M.A.; Cheng, H.Y.; Hwang, S.; Tao, H.; Paquette, M.; Kaplan, D.L.; Rogers, J.A.; Huang, Y.G.; Omenetto, F.G. Modulated Degradation of Transient Electronic Devices through Multilayer Silk Fibroin Pockets. *ACS Appl. Mater. Interfaces* **2015**, *7*, 19870–19875. [CrossRef] [PubMed]

91. Pan, R.Z.; Xuan, W.P.; Chen, J.K.; Dong, S.R.; Jin, H.; Wang, X.Z.; Li, H.L.; Luo, J.K. Fully biodegradable triboelectric nanogenerators based on electrospun polylactic acid and nanostructured gelatin films. *Nano Energy* **2018**, *45*, 193–202. [CrossRef]

92. Lee, S.; Koo, J.; Kang, S.K.; Park, G.; Lee, Y.J.; Chen, Y.Y.; Lim, S.A.; Lee, K.M.; Rogers, J.A. Metal microparticle—Polymer composites as printable, bio/ecoresorbable conductive inks. *Mater. Today* **2018**, *21*, 207–215. [CrossRef]

93. Lee, C.H.; Kim, H.; Harburg, D.V.; Park, G.; Ma, Y.J.; Pan, T.S.; Kim, J.S.; Lee, N.Y.; Kim, B.H.; Jang, K.I.; et al. Biological lipid membranes for on-demand, wireless drug delivery from thin, bioresorbable electronic implants. *Npg Asia Mater.* **2015**, *7*, e227. [CrossRef] [PubMed]

94. White, M.D.; Bosio, C.M.; Duplantis, B.N.; Nano, F.E. Human body temperature and new approaches to constructing temperature-sensitive bacterial vaccines. *Cell. Mol. Life Sci.* **2011**, *68*, 3019–3031. [CrossRef] [PubMed]

95. Duplantis, B.N.; Bosio, C.M.; Nano, F.E. Temperature-sensitive bacterial pathogens generated by the substitution of essential genes from cold-loving bacteria: Potential use as live vaccines. *J. Mol. Med.* **2011**, *89*, 437–444. [CrossRef] [PubMed]

96. Hooke, A.M. Temperature-Sensitive Mutants of Bacterial Pathogens—Isolation and Use to Determine Host Clearance and In-Vivo Replication Rates. *Method Enzymol.* **1994**, *235*, 448–457.

97. Tao, H.; Hwang, S.W.; Marelli, B.; An, B.; Moreau, J.E.; Yang, M.M.; Brenckle, M.A.; Kim, S.; Kaplan, D.L.; Rogers, J.A.; et al. Silk-based resorbable electronic devices for remotely controlled therapy and in vivo infection abatement. *Proc. Natl. Acad. Sci. USA* **2014**, *111*, 17385–17389. [CrossRef] [PubMed]

98. Yin, L.; Huang, X.; Xu, H.X.; Zhang, Y.F.; Lam, J.; Cheng, J.J.; Rogers, J.A. Materials, Designs, and Operational Characteristics for Fully Biodegradable Primary Batteries. *Adv. Mater.* **2014**, *26*, 3879–3884. [CrossRef] [PubMed]

99. Bouhlala, M.A.; Kameche, M.; Tadji, A.; Benouar, A. Chitosan hydrogel-based electrolyte for clean and biodegradable batteries: Energetic and conductometric studies. *Phys. Chem. Liq.* **2018**, *56*, 266–278. [CrossRef]

100. Jia, X.T.; Wang, C.Y.; Ranganathan, V.; Napier, B.; Yu, C.C.; Chao, Y.F.; Forsyth, M.; Omenetto, F.G.; MacFarlane, D.R.; Wallace, G.G. A Biodegradable Thin-Film Magnesium Primary Battery Using Silk Fibroin-Ionic Liquid Polymer Electrolyte. *ACS Energy Lett.* **2017**, *2*, 831–836. [CrossRef]

101. Jia, X.T.; Wang, C.Y.; Zhao, C.; Ge, Y.; Wallace, G.G. Toward Biodegradable Mg-Air Bioelectric Batteries Composed of Silk Fibroin-Polypyrrole Film. *Adv. Funct. Mater.* **2016**, *26*, 1454–1462. [CrossRef]

102. Huang, X.Y.; Wang, D.; Yuan, Z.Y.; Xie, W.S.; Wu, Y.X.; Li, R.F.; Zhao, Y.; Luo, D.; Cen, L.; Chen, B.B.; et al. A Fully Biodegradable Battery for Self-Powered Transient Implants. *Small* **2018**, *14*, e1800994. [CrossRef] [PubMed]

103. Lu, L.Y.; Yang, Z.J.; Meacham, K.; Cvetkovic, C.; Corbin, E.A.; Vazquez-Guardado, A.; Xue, M.T.; Yin, L.; Boroumand, J.; Pakeltis, G.; et al. Biodegradable Monocrystalline Silicon Photovoltaic Microcells as Power Supplies for Transient Biomedical Implants. *Adv. Energy Mater.* **2018**, *8*, 1703035. [CrossRef]

104. Hwang, S.W.; Huang, X.; Seo, J.H.; Song, J.K.; Kim, S.; Hage-Ali, S.; Chung, H.J.; Tao, H.; Omenetto, F.G.; Ma, Z.Q., et al. Materials for Bioresorbable Radio Frequency Electronics. *Adv. Mater.* **2013**, *25*, 3526–3531. [CrossRef] [PubMed]

Article

An Optical Biosensing Strategy Based on Selective Light Absorption and Wavelength Filtering from Chromogenic Reaction

Hyeong Jin Chun [1], Yong Duk Han [1], Yoo Min Park [1,2], Ka Ram Kim [1], Seok Jae Lee [2] and Hyun C. Yoon [1,*]

[1] Department of Molecular Science and Technology, Ajou University, Suwon 16499, Korea; moogoosla@ajou.ac.kr (H.J.C.); Han.Yong@mayo.edu (Y.D.H.); ympark@nnfc.re.kr (Y.M.P.); kkr4649@ajou.ac.kr (K.R.K.)

[2] Nano-bio Application Team, National NanoFab Center (NNFC), Daejeon 34141, Korea; sjlee@nnfc.re.kr

* Correspondence: hcyoon@ajou.ac.kr; Tel.: +82-31-219-2512

Received: 12 February 2018; Accepted: 6 March 2018; Published: 6 March 2018

Abstract: To overcome the time and space constraints in disease diagnosis via the biosensing approach, we developed a new signal-transducing strategy that can be applied to colorimetric optical biosensors. Our study is focused on implementation of a signal transduction technology that can directly translate the color intensity signals—that require complicated optical equipment for the analysis—into signals that can be easily counted with the naked eye. Based on the selective light absorption and wavelength-filtering principles, our new optical signaling transducer was built from a common computer monitor and a smartphone. In this signal transducer, the liquid crystal display (LCD) panel of the computer monitor served as a light source and a signal guide generator. In addition, the smartphone was used as an optical receiver and signal display. As a biorecognition layer, a transparent and soft material-based biosensing channel was employed generating blue output via a target-specific bienzymatic chromogenic reaction. Using graphics editor software, we displayed the optical signal guide patterns containing multiple polygons (a triangle, circle, pentagon, heptagon, and 3/4 circle, each associated with a specified color ratio) on the LCD monitor panel. During observation of signal guide patterns displayed on the LCD monitor panel using a smartphone camera via the target analyte-loaded biosensing channel as a color-filtering layer, the number of observed polygons changed according to the concentration of the target analyte via the spectral correlation between absorbance changes in a solution of the biosensing channel and color emission properties of each type of polygon. By simple counting of the changes in the number of polygons registered by the smartphone camera, we could efficiently measure the concentration of a target analyte in a sample without complicated and expensive optical instruments. In a demonstration test on glucose as a model analyte, we could easily measure the concentration of glucose in the range from 0 to 10 mM.

Keywords: point-of-care testing; soft material-based channel; PDMS optical filter; smartphone-based biosensor; chromogenic biochemical assay; naked-eye detection

1. Introduction

In the in vitro diagnostic (IVD) industry, development and commercialization of the biosensing technology enabling disease diagnosis wherever the patient is located has been regarded as the ideal goal. Therefore, the achievement of high portability, cost-effectiveness, and user-friendliness are considered major aims for robust analytical performance in biosensor technology research. Currently, due to good analytical performance including high accuracy and signal-to-noise ratio, optical biosensing technologies such as fluorescence analysis [1–3], surface plasmon resonance–based affinity biosensing [4–6], and ultraviolet/visible (UV/Vis) spectrophotometric biosensing are extensively investigated as promising approaches to the realization of point-of-care (POC) diagnostics. Because the robust analytical performance of these optical biosensing technologies can be attained only by means of a specific optical-signal-transducing technology (which has several drawbacks including low cost-effectiveness, high complexity, high power consumption, low portability, and poor user-friendliness), the applications of these conventional optical biosensing technologies to POC diagnostics are limited [7–10]. Thus, securing and developing an optical-signal-transducing system that can overcome those limitations on conventional transducers are regarded as the cornerstone for the development of a user-friendly POC optical biosensor. In this context, there is a growing research trend toward replacing conventional signal-transducing equipment with commercialized high-tech electronic devices such as scanners, optical storage devices, and smartphones, which provide low costs, low power consumption, and ease of use [11–17]. In this study, we focused on the development of a new optical-signal-transducing method that can replace the conventional spectrophotometry-based signal transduction of a colorimetric biosensor, by utilizing a common liquid crystal display (LCD) panel and a smartphone. The conventional spectrophotometric signal transducers employ a halogen lamp and monochromator as light source units at a specific wavelength. Because they require complicated configuration of optical components such as the prism and mirror, this system has poor portability, low cost-effectiveness, and substantial power consumption. In the case of an optical receiver of a conventional colorimetric transducer, the photodiode array is widely used. Unfortunately, it is also expensive and requires a complicated electric circuit to operate. Because those light sources and the optical receiver units are the most essential components of the optical signal transducer, the realization of a POC optical biosensor would depend on the introduction of a novel light source and optical receiver components (having high cost-effectiveness, user-friendliness, portability, and good optical properties at low power consumption) into the optical transducing system. To this end, we focused on the LCD panel of a computer monitor and smartphone, which are commercialized electronic products that humans always face in everyday life, as a light source unit and optical receiver unit, respectively. The LCD panel provides well-defined visible light whose color can be easily manipulated by controlling their color pixels (red, green, and blue ones). Compared to a monochromator, the LCD panel is already distributed to the public, and its control of brightness and color levels is easy. As an optical receiver, the smartphone provides a digital camera that allows for high-definition imaging. Besides, by means of its inherent display panel and the embedded image software, in situ analysis of a registered image can be accomplished on the smartphone. According to these features, we constructed a novel optical transducing system by reassembling a smartphone and LCD monitor (Figure 1).

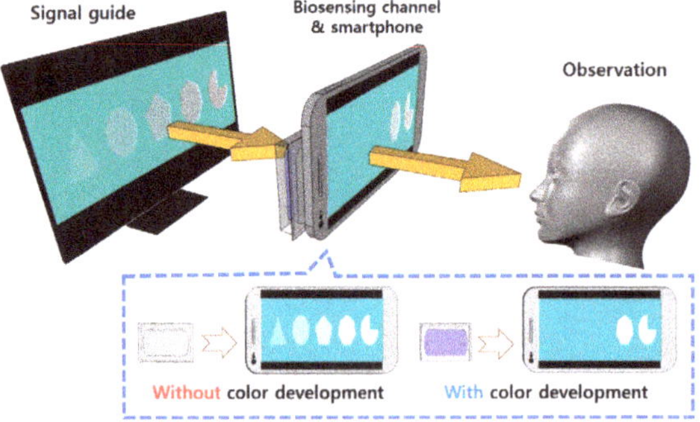

Figure 1. The scheme of the smartphone-based optical biosensing system. The numbers of signal guides on the screen were changed by superimposing a chromogen-containing biosensing channel. In the absence of a chromogen in the biosensing channel, intact clear-cut signal guides were observed, but the signal guides disappeared in the biosensing channel containing a blue product.

Into the newly developed transducing system, we introduced a horseradish peroxidase (HRP)-mediated chromogenic assay as a biosensing strategy. Conventionally, in the newly designed assay, the biorecognition reaction between the bioreceptor and target analyte induces color development under the action of HRP, and intensity of the resulting color development is affected by the concentration of the target analyte. The conventional colorimetric biosensors analyze changes in the color intensity according to the biochemical reaction to quantify the concentration of a target analyte. Nevertheless, measuring the color intensity or absorbance requires external analysis algorithms and specific software that is not user-friendly. To overcome this limitation in the conventional colorimetric biosensors, more intuitive and user-friendly signal quantification method—that can be operated without specific software—should be devised. To this end, we hypothesized that if the changes in color intensity can be converted into a change in a certain numerical parameter, a more convenient and intuitive quantitative analysis for the target analyte would be created. To realize this idea, we introduce the principle of a "secret message card". The latter consists of a card containing message words and is covered by randomly painted patterns with various colors and a colored semitransparent cellophane film to read the message words without the interference from randomly painted color patterns on it. When the message words on the secret card are observed through the colored cellophane film, the patterns of the words with the same color as that of the cellophane film will be filtered out and disappear. This is because the spectrum of color light from the pattern overlaps with the absorption spectrum of the colored cellophane film. In contrast, the message words in a color different from that of the cellophane film pass through the cellophane film as is, and reach the observer, because there is no spectral overlap between the light of words and cellophane film. In this phenomenon, the colored cellophane film acts as an optical filter interfering with and filtering colored light. We attempted to apply this color-filtering principle of the secret message card to our optical transducing system as a signal quantification strategy. In this study, the colored cellophane film was replaced by a transparent and soft material-based biosensing channel which can induce development of the color (blue) via the aforementioned HRP-mediated chromogenic biochemical assay. To apply a biosensing channel for developed sensing system, the biosensing channel should be made of a transparent and soft material that can transmit the light. The silicone elastomer, such as Sylgard, Ecoflex, and Silbione, satisfies above condition, and it has high flexibility and stretchability so that they can be used as optical filters in a variety situation. In addition to this, the message words were replaced by the signal guide pattern containing polygonal figures with specific colors (a triangle, circle, pentagon, heptagon, and 3/4 circle). The signal guide was designed and produced in drawing software and was displayed on the LCD monitor, which utilizes (as a light source) equipment in our transducing system. On the LCD monitor panel, five polygon images, containing lights of red, green, and blue, were displayed as signal guide patterns. The area surrounding pattern images contains only green and blue light at the same ratio, consistent with the figure inside. In the biosensing channel, the HRP-mediated chromogenic biochemical assays were conducted to obtain blue reactants. Then, the resulting biosensing channel was mounted on the smartphone camera. When we observe polygons on the LCD monitor though this camera, the color of signal guide patterns is harmonized with the surrounding color because the red color in the pattern is filtered by the blue reactants in the biosensing channel as in the secret message card. Consequently, the number of visible polygons on the monitor would be changed by the extent of the chromogenic reaction in a biosensing channel. After counting of the polygons, the concentration of target analytes could be easily measured. In this study, a glucose assay was devised by means of HRP and chromogenic substrates. The details of the biosensing principle with analytical approaches are reported herein.

2. Materials and Methods

2.1. Materials and Instruments

The Sylgard 184 silicone elastomer kit was acquired from Dow Corning (Midland, MI, USA). HRP was purchased from Toyobo (Osaka, Japan). Dopamine hydrochloride, glucose oxidase (GOx) from *Aspergillus niger*, 4-aminoantipyrine (4-AAP), amine-terminated G4 polyamidoamine (PAMAM) dendrimer, and Tris(hydroxymethyl)aminomethane (Tris) were acquired from Sigma-Aldrich (St. Louis, MO, USA), while *N*-Ethyl-*N*-(2-hydroxy-3-sulfopropyl)-3,5-dimethylaniline (MAOS) was from Dojindo (Kumamoto, Japan). Zoom Imaging Lens (PMAG 0.7X-4.5X) and CMOS Color USB Camera (resolution of 5.0 MP (megapixels)) were purchased from Edmund optics (Barrington, NJ, USA). To show the optical signal guide, LG computer monitor was used. A camera from the LG G2 smartphone was employed to register the obtained images. A phosphate-buffered saline containing 0.1 M phosphate and 0.15 M NaCl (PBS, pH 7.2) was prepared in doubly distilled and deionized water (DDW).

2.2. Surface Modification of the Biosensing Channel

The biosensing channel was fabricated by means of polydimethylsiloxane (PDMS) upper layer and polyethylene (PET) film bottom layer. For the casting of PDMS upper layer having dimensions of 20 mm in length, 5 mm in width, and 2 mm in depth, an acrylic mold was prepared by computer numerical machining process. A 10:1 (*v*/*v*) mixture of Sylgard 184 monomer and initiator was poured onto the acrylic mold and placed in a vacuum chamber for 30 min to remove bubbles from the PDMS mixture. Then PDMS-filled acrylic mold was incubated in an oven at 80 °C for 1 h to cure PDMS [18]. Next, the cured PDMS substrate was detached from the acrylic mold and washed with distilled water. The washed substrates were dried in an oven at 80 °C for 20 min. The prepared PDMS channel layer has in/out transport channels at both sides and circular reaction region at the center of channel. For the solution manipulation including injection and draining, two holes were made on both sides of the PDMS channel using a puncher. To make a complete fluidic channel as a bottom layer, a transparent PET film having an adhesive layer on a single side was attached to the PDMS upper channel layer. The volume of the newly prepared biosensing channel was circa 200 µL. To construct the biorecognition layer on the biosensing channel surface, a polydopamine layer was coated with PDMS and PET surface as an intermediate layer for subsequent chemical conjugation processes [19]. Because the polydopamine coatings have an ability to immobilize primary-amine- and/or thiol-containing molecules by covalent interactions such as imine bond and Michael addition, the polydopamine layer could serve as an initial intermediate layer at the following bioconjugation steps [20,21]. Briefly, 2 mg/mL dopamine hydrochloride was prepared in a 10 mM Tris-HCl buffer solution, and its pH was adjusted to 8.5. The resultant dopamine solution was immediately injected into the biosensing channels, and the channel was incubated for 16 h at room temperature in a darkroom. After coating with polydopamine, the biosensing channel was washed five times with an excess of DDW. The resultant polydopamine-coated biosensing channels were filled with DDW and stored at 4 °C until further biomolecular modification.

2.3. Fabrication of the Biorecognition Layer for Glucose Biosensing

In this study, a bienzymatic colorimetric glucose assay using glucose oxidase (GOx) and HRP was chosen as a model biochemical assay to assess the biosensing applicability of the newly developed optical transducing strategy. To construct the bioreceptor layer enabling bienzymatic glucose assay in the biosensing channel, the multiple enzyme layers consisting of glucose oxidase and HRP were immobilized inside the polydopamine-coated biosensing channel by the layer-by-layer (LBL) technique using the PAMAM dendrimer as a building block molecule [22]. First, a 0.5% (*w*/*v*) G4 PAMAM dendrimer aqueous solution was prepared in 0.1 M phosphate buffer. Then, the dendrimer solution was injected into the polydopamine-coated channels and incubated in the dark for 16 h at room temperature. During this procedure, PAMAM dendrimers were covalently immobilized on the polydopamine layer via imine bond formation between primary amine groups of the dendrimer and catechol moieties of the polydopamine layer. The dendrimer-decorated biosensing channel was rinsed with PBS three times and filled with a 20 mM ethanolamine solution for 15 min to block the unreacted functional moieties in the polydopamine layer. Because 64 primary amine groups were expressed on the surface of the PAMAM G4 dendrimer in a globular shape, many amine groups should be exposed on the dendrimer-treated polydopamine layer. To conjugate enzymes to the amine-exposed biosensing channel surface via a covalent bond, the carbohydrate moieties on GOx and HRP were oxidized and converted into aldehyde groups by means of 40 µM sodium periodate. To build an HRP layer, a 1 mg/mL periodate-oxidized HRP solution (PBS, pH 6.8) was injected into the dendrimer-decorated biosensing channel and allowed to react for 1 h. In this procedure, HRP is immobilized on the amine-terminated sensing surface via Schiff's base formation between aldehyde groups of HRP and primary amine groups of the dendrimer on the channel surface. Next, the intermediate dendrimer layer was built by applying an aqueous PAMAM dendrimer solution (0.5%, PBS) to the HRP-coated channel for 1 h. To form the GOx layer, a 1 mg/mL periodate-oxidized GOx solution (PBS, pH 6.8) was applied to the dendrimer-treated biosensing channel and allowed to react for 1 h. As in the HRP conjugation, GOx was immobilized on the dendrimer layer via Schiff's base formation. For the strong color development in the bienzymatic colorimetric glucose sensing method, processes of construction of this bienzyme layer consisting of the dendrimer, HRP, dendrimer, and GOx was repeated three times (Figure S1, Supplementary Materials). The resulting modified biosensing channel coated with triple bienzyme layers was filled with PBS and stored at 4 °C until use.

2.4. Preparation of the Optical Signal Guide

In this study, the optical signal guide (to be used as a reference light signal and light source for wavelength filtering–based colorimetric biosensing) was prepared in a graphics editor (software, Microsoft PowerPoint 2016) and displayed on the LCD monitor panel. First, five different polygonal figures including a triangle, circle, pentagon, heptagon, and 3/4 circle were drawn in the graphics editor. Then, using the color editing option, which allows for adjustment of the intensity of red (R), green (G), and blue (B) color elements step by step from 0 to 255, all the inner and outer regions of the polygon were painted with a color which contains only G and B color element. After initial color adjustment, the intensity of the R color element inside each polygon was manipulated, meanwhile the intensity values of G and B elements were maintained (255 and 255, respectively). The intensities of the R color element inside the triangle, circle, pentagon, heptagon, and 3/4 circle were adjusted to 135, 165, 195, 225, and 255, respectively. Compared to R color element varied from 135 to 255 in each polygon's interior, the intensities of G and B color elements was maintained at 255 everywhere on a signal guide pattern. Then, the color-adjusted signal guide pattern image was documented as an image file in PNG or TIFF format to minimize the loss of its information resulting from file compression. In the following experiments, the signal guide image files were loaded onto the LCD display panel of a computer or smartphone by means of image viewer software (Gallery application).

2.5. Verification of the Optical Biosensing Principle

To test whether the selective light absorption phenomenon–based optical signal transduction works, a model study involving a smartphone LCD display and a microscope was carried out. First, the signal guide pattern was displayed on the smartphone LCD panel and the glucose biosensing channel was mounted onto the signal guide–displaying smartphone LCD panel. Then, a portable microscope having charge-coupled device (CCD) imaging sensor was vertically installed above the biosensing channel and was connected to the computer for real-time acquisition of magnified images of RGB pixels in signal guide patterns on the smartphone display. A 10 mM glucose solution was prepared in PBS as a glucose sample meanwhile PBS alone (0 mM glucose) was prepared and served as a control sample. As a chromogenic substrate for the enzymatic colorimetric glucose assay, Trinder's reagent solution containing 20 mM 4-AAP and 2 mM MAOS was prepared in PBS. A 1:1 mixture of the glucose sample and chromogenic substrate solution was injected into the smartphone-mounted glucose biosensing channel. According to the progress of the glucose-mediated cascade enzyme reaction, the colorless solution inside biosensing channel gradually turned into blue. In this setup, the image of RGB pixels on the smartphone LCD panel passes through the biosensing channel on top of it and reaches the microscope imaging sensor. In this process, the light of each RGB pixel is partially absorbed by the blue end product inside the glucose biosensing channel, and thus the light intensity of each RGB pixel observed by the microscope changes. After 5 min of the reaction, images of signal guide patterns and each RGB pixel were documented as image files. Finally, the color intensity of signal guide pattern and pixel images was quantitatively analyzed in NIH ImageJ software (Version: 1.50i)

2.6. Glucose Analysis by Enzymatic Colorimetric Assay

To demonstrate the applicability of the newly developed signal transducing technology to point-of-care biosensing, quantitative glucose detection (using a smartphone and a computer LCD monitor panel) was accomplished. First, glucose samples at various concentrations (0, 1.25, 2.5, 5 and 10 mM) were prepared. As a chromogenic substrate, Trinder's reagent solution containing 20 mM 4-AAP and 2 mM MAOS was prepared. Right before the glucose assays, each glucose sample and Trinder's reagent solution was mixed in a 1:1 volume ratio. The glucose biosensing channel, which is modified with triple bienzyme (GOx and HRP) layers was mounted on the smartphone camera lens using a clip. The signal guide image was displayed on the computer LCD monitor panel. Then, into the glucose-biosensing channel, the prepared glucose-chromogen mixture (solution) was injected. Under this setup, the signal guide patterns displayed on the LCD monitor panel can be observed and documented through the biosensing channel-mounted digital camera of a smartphone. As the glucose-mediated biocatalytic reaction proceeded, concentration of the blue assay end product inside the channel was increased and thereby its red-light absorbance was also increased. Under these circumstances, the number of remaining polygons in the digitally recorded signal guide image on smartphone camera may be changed by correlation between changes in red-light absorbance of glucose assay end product and the varied red intensities of each polygon. The image of the signal guide pattern was registered every 1 min during 5 min of total assay time. To minimize the signal interference from external light, all glucose detection tests were carried out under darkroom condition. After the image acquisition, the polygons in registered pattern images were counted with the naked eye and in the image analysis software (Pattern-recognition option in the Microsoft PowerPoint 2016).

3. Results

3.1. The Basic Signal-Transducing Principle: Spectral Correlation between the Chromogenic Solution and Color Pixels on the LCD Panel

In this study, we focused on the development of a new signal-transducing technology that can efficiently translate the colorimetric signal of a biochemical assay into signals that can be easily quantified by the naked eye. To convert the color intensity signal, which linearly depends on target analyte concentration, into visually countable signals without complicated optical instruments, here we employed the correlation of the absorption spectrum of the chromogenic compound and variations in RGB color pixel intensities on the computer LCD monitor panel as a major signal-transducing principle.

As for the LCD monitor panel, it is composed of numerous tiny red (R), green (G), and blue (B) pixels and those R, G, and B pixels provide light with maximum intensity near 600, 530, and 400 nm, respectively. Because the light intensity of those RGB pixels can be quantitatively controlled step by step from 0 to 255 in a graphics editor, we can generate a light source of a specific wavelength by controlling RGB pixel intensities. Based on this feature, we introduced the LCD panel as a light source for a conventional chromogenic biochemical assay that requires light of a specific wavelength in the quantitative analysis of assay-derived colored end product. In this study, as a biochemical assay principle, Trinder's reaction–based bienzymatic chromogenic glucose assay was evaluated [21,23,24]. In the presence of glucose, GOx generates H_2O_2, and the latter is catalytically degraded by HRP. Under the influence of H_2O_2 decomposition by HRP, colorless Trinder's chromogenic substrate compounds (4-AAP and MAOS) are conjugated with each other and form a blue end product. This blue end product exhibits maximum absorbance at 630 nm (Figure 2, middle panel). Additionally, the amount of the blue reactant in the glucose assay and the intensity of its absorption spectrum are proportional to the concentration of glucose (Figure S2, Supplementary Materials). Considering that the light in the 630 nm region absorbed by the blue end product of the glucose assay is red light, a signal guide pattern (i.e., a set of polygons capable of presenting the change of red light according to the chromogenic reaction as a visualized information and providing it to the user) was designed as shown in Figure 2. Fundamentally, the signal guide pattern is a digital drawing of a polygon which is designed to have different color pixel characteristics inside and outside. The color intensity of each G and B color pixel inside and outside a polygon was adjusted to the same value of 255 in the graphics editor. For the R color pixel, the interior and surroundings of the polygon were set to 255 and 0, respectively, as shown in Figure 2 (upper panel). Because only green and blue colors were activated, the outside of the polygon had a cyan color and emission spectrum peaks near 400 and 530 nm. In contrast, the interior of the polygon showed white color because red, green, and blue lights are simultaneously emitted and mixed. Although the emission spectrum of the polygon's inside area was similar to that outside the polygon, emission intensity near 600 nm corresponding to red light was increased as intended. When the light emitted from the signal guide pattern of a computer LCD panel was passed through the blue end product of the glucose assay (λ_{max} = 630 nm), the red-light portion was absorbed into the blue chromogenic compound while green and blue light were barely affected. Due to the red color filter-like effect of the chromogenic compound, it was observed that the white polygon changed to a cyan polygon during examination of the signal guide pattern on the LCD panel through the biosensing channel that turned blue (Figure 2, bottom panel). As shown in the emission spectrum in Figure 2 (bottom right panel), when light passed through the blue chromogenic compound, the intensity of the emission spectrum near 600 nm, acquired from the polygon inside, decreased and became similar to the emission spectrum outside the polygon. Conversely, when we see surroundings of a polygon in a signal guide pattern through the biosensing channel that developed a color, its cyan color and initial emission spectral property were retained because the light from the cyan region outside the polygon does not contain red light. When the concentration of glucose is high enough to generate the colored assay reactant in sufficient amounts that can absorb all red light, the red-color-filtering effect of the chromogenic compound induces synchronization of the colors inside and outside of a polygon observed through the biosensing channel, and thus makes the polygon disappear (Figure 2, bottom panel). Via this principle, changes in the colorimetric signal induced by a biochemical assay can be easily visualized as disappearance of a polygonal pattern, without complicated optical instruments. Nonetheless, in the present state, it can be used only for qualitative detection of glucose at a specific concentration. Therefore, to apply this basic signal transducing principle to quantitative glucose biosensing, the signal guide pattern was engineered and improved as follows.

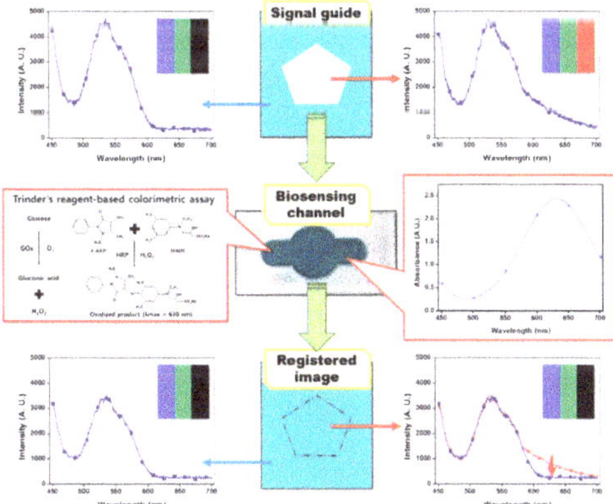

Figure 2. Illustration of the principles used in the newly developed optical biosensing system. The signal guide consists of red, green, and blue light at a certain ratio, and the area surrounding the polygons contains only green and blue light. After locating the biosensing channel that has the blue chromogen on the signal guide, we converted the color of signal guide into blue owing to absorption of the red light by chromogen.

3.2. The Principle of Quantitative Glucose Analysis by the New Signal-Transducing Method

For application of our new optical-signal-transducing principle to quantitative glucose biosensing, a new signal guide pattern composed of five polygons, including a triangle, circle, pentagon, heptagon, and 3/4 circle was prepared (Figure 3A) and displayed on the computer LCD monitor panel. For G and B pixels, the inside and outside regions of each polygon were fixed at 255 as in the initial design. In contrast, the R pixel intensities within each polygon were gradually varied from 135 to 255 (135 for triangles, 165 for circles, 195 for pentagons, 225 for heptagons, and 255 for 3/4 circles). Because each polygon has different R pixel intensity, the glucose concentration for disappearance of each pattern should vary. In the case of a low glucose concentration, the amount of the blue reactant produced in the glucose assay is small. This low concentration of the colored product yields low red-light absorbance; therefore, polygons whose R color pixel intensity is low (e.g., triangle and/or circle) would disappear preferentially. By contrast, when the blue product is excessively produced at a high concentration of glucose in the sample being analyzed, most polygons would disappear because the almost red light from R pixels will be fully absorbed by the blue reactant of the assay. To assess the correlation between glucose concentration and disappearance of polygons having various R pixel intensities, we monitored the prepared polygons on the LCD monitor panel using a smartphone camera through a glucose-sensing channel loaded with a glucose sample (0, 1.3, 5, or 10 mM) for 5 min.

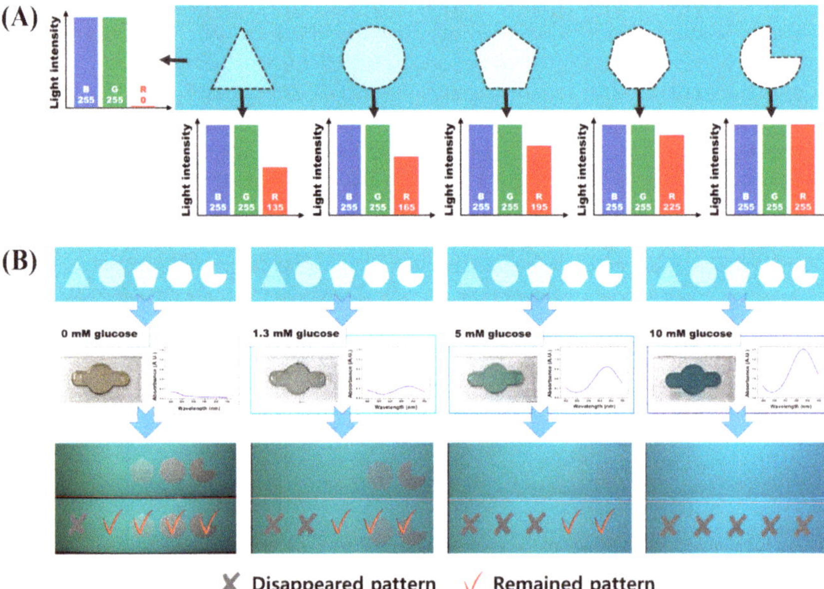

Figure 3. Fabrication of the signal guides on the LED display including the various intensity of red light: (**A**) The signal guides have a fixed intensity of blue and green light with varying intensity of red light, and the surrounding area contains blue and green of fixed intensity; and (**B**) the images of a signal guide during superposition of a biosensing channel. The numbers of signal guides change according to the chromogen concentration.

As shown in the middle panel of Figure 3B, numbers of polygons that disappeared increased according to the increase in glucose concentration. As the glucose concentration increases, intensities of color and absorption spectra of the glucose assay reactant increased (Figure 3B, middle panel). As shown in the bottom panel of Figure 3B, in an assay with 1.3 mM glucose, polygons with low R pixel intensity (triangle and circle) preferentially disappeared because the R pixel intensity was low enough to be fully absorbed by a small amount of the blue reactant of the assay. In contrast, polygons with relatively high R pixel intensity (pentagon, heptagon, and 3/4 circle) did not disappear because their red-light intensity exceeds the red-light absorption capacity of the blue reactant that was produced at a low concentration of glucose. (Note that the polygons that disappeared were marked with X and the remaining polygons were marked with V.). In the glucose assay applied to a 5 mM glucose sample, the pentagon disappeared, meanwhile heptagon and 3/4 circle remained. In a glucose assay with a 10 mM glucose sample, all polygons disappeared because the red-light absorption capacity of the reactant generated in the glucose assay was high enough to completely absorb all red light of R pixels in polygons. In this experiment, we found that the color intensity linearly changing with alterations in the target analyte concentration could be translated into visual signals that can be counted with the naked eye in this newly developed optical-signal-transducing technology. Using this principle, we can easily evaluate the concentration of glucose by counting the polygons that disappeared (or remained).

3.3. Validation of the Optical-Signal-Transducing Principle at a Pixel Level

To test whether the conversion of color intensity into a visually assessable signal was accomplished by the spectral correlation between the absorption spectrum of the chromogenic compound and light intensity of R pixels on the LCD panel displaying the signal guide pattern, a microscopic analysis was conducted. For this experiment, the signal guide pattern image (polygons) was loaded and displayed on the smartphone LCD panel instead of the computer LCD monitor panel. In the polygon-displaying region of the smartphone LCD panel, the glucose biosensing channels, containing a mixture of glucose samples (0 and 10 mM) and Trinder's chromogenic reagents, were mounted. After the compounds were allowed to react for 5 min, the RGB color pixels in the signal guide pattern of the LCD panel were monitored using a portable microscope with a CCD digital camera (Figure 4A).

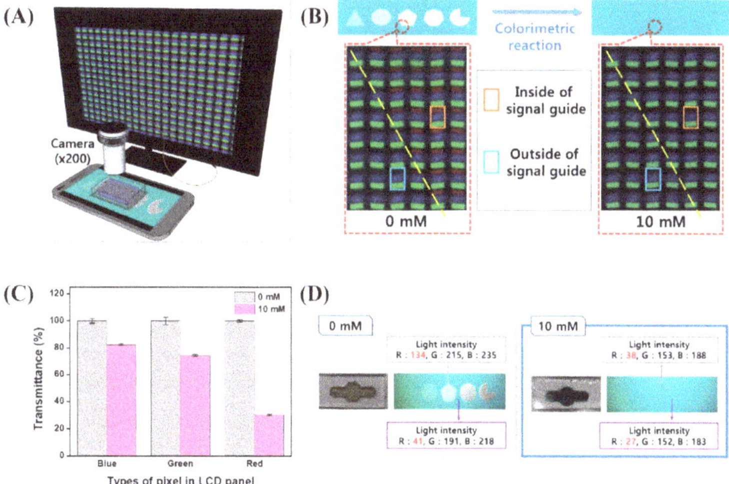

Figure 4. Alterations of RGB intensity on the mobile screen depending on the analysis of 0 or 10 mM glucose in the glucose assay: (**A**) The confirmation of changes in RGB pixels on the screen by microscopy. (**B**) The resulting images of the RGB pattern in the signal guide. The red light in the signal guide was changed by superimposing the blue chromogen-containing biosensing channel. (**C**) The transmittance ratio of blue, green, and red light in the signal guides. The blue and green lights slightly changed, while the red light is significantly altered. (**D**) The resulting images of a whole signal guide. The numbers of signal guides changed in the presence of a chromogen, and the red light highly decreased in the glucose assay involving a 10 mM glucose sample.

Figure 4B shows the magnified pixel images of the smartphone LCD panel where the polygons were displayed. The yellow dashed line indicates the boundary between the inside and outside of a pentagon. Regarding an assay result for 0 mM glucose (left panel), only G and B color pixels were observed in the region outside the pentagon as initially adjusted by means of the graphics software (Microsoft PowerPoint 2016). Inside the pentagon, all RGB color pixels were observed, just as in the initial design because the blue end product (of the glucose assay) absorbing red light was not generated in the absence of glucose. Because the differences in R pixel intensities between inside and outside regions of the pentagon persisted after the glucose assay of the 0 mM glucose sample, the pentagon shape did not disappear and was visible as initially designed. Regarding the assay result for a 10 mM glucose sample (Figure 4B, right panel), G and R pixels in the regions inside and outside the pentagon retained their light intensity even after the intense color development. Nevertheless, the intensity of the R pixels inside the pentagon sharply decreased and was barely distinguishable visually. As a result, the image of the R pixel inside the pentagon at the reduced light intensity became similar to the image of the R pixel outside the pentagon. In the obtained findings, readers can see that the disappearance of the pentagon pattern in the course of the chromogenic glucose assay reaction is closely related to the spectral correlation of R pixel intensity and the intensity of the absorption spectrum of the glucose assay reactant. To confirm this result quantitatively, the transmittance of light from RGB pixels of the signal guide pattern (pentagon) that passed through the glucose assay reactant (10 mM glucose) was calculated by analyzing color intensity of RGB pixels in the registered image by means of the NIH ImageJ software. The relative transmittance of light from RGB pixels of the pentagon pattern that passed through 10 mM glucose was calculated based on the intensity of light (from RGB pixels inside the pentagon) that passed through the 0 mM glucose reactant, which was set to 100%. As shown in Figure 4C, the transmittance for R, G, and B pixels inside the pentagon (toward the reactant corresponding to the 10 mM glucose sample) decreased by 70%, 25%, and 17%, respectively, relative to reference transmittance (0 mM glucose).

Additionally, the color synchronization of the signal guide pattern according to the progress of the chromogenic reaction in the glucose assay was quantitatively analyzed from a macroscopic view. The image of the signal guide pattern was registered using a smartphone camera through glucose-sensing channel loaded with the glucose sample (0 or 10 mM). By means of the registered images, the RGB color intensities inside and outside the pentagon were extracted and quantitatively analyzed in the NIH ImageJ software, as shown in Figure 4D. In this analysis, the image of the signal guide pattern that was photographed by the smartphone camera served as a model of how color synchronization of patterns looks to the naked eye. As shown in the left panel of Figure 4D, in the glucose assay applied to the 0 mM glucose sample, the intensity of RGB color elements inside the pentagon was measured and found to be 134 (red), 215 (green), and 235 (blue). Besides, the intensity of RGB color elements outside the pentagon was 41 (red), 191 (green), and 218 (blue), respectively. Because of the differences in intensity of red color elements between inside and outside regions of the pentagon, the shape and color of the pentagon were easily distinguishable from the cyan surrounding area when 0 mM glucose was tested in the assay. In contrast, as shown in the right panel of Figure 4D, the intensities of RGB color elements inside the pentagon were registered at 38 (red), 153 (green), and 188 (blue). These data are similar to RGB color intensity values outside the pattern (27 for red, 152 for green, and 183 for blue). Because the RGB color profiles inside and outside the pentagon were similar, the shape and color of pentagon were indistinguishable from the cyan surrounding area to the naked eye when 10 mM glucose was tested in the glucose assay. From the obtained results, we concluded that the newly designed optical-signal-transducing technique that translates the color intensity into the visually assessable signal is based on the spectral correlation of the absorption spectra of chromogenic compounds and the light intensity of R pixels on the signal guide patterns of the LCD panel.

3.4. Optimization of Conditions for Glucose Biosensing Based on the New Transducing Technology

In the newly developed optical-signal-transducing technique, the increase in color intensity and absorbance of the reactant in the glucose assay is converted into an increased number of polygons that disappeared. Therefore, for the accurate glucose biosensing using this transducing principle, the disappearance of a polygon should sensitively reflect a change in the concentration of glucose. Given that the disappearance of the polygon is directly related to the absorbance of the reactant in the glucose assay, the yield of the bienzyme-mediated chromogenic reaction should be improved for the sensitive glucose sensing. In this context, the amount of immobilized enzyme and the reaction time of a glucose assay (which are closely related to the production rate of the blue chromogenic reactant in the glucose assay) were optimized.

As a bioreceptor layer for the glucose biosensing, in this study, GOx and HRP were covalently immobilized on the surface of a biosensing channel via the LBL technique. The advantage of this technique for the construction of the enzyme-based biosensing interface is that the amount of enzymes to be immobilized on the biosensing surface can be adjusted by controlling the number of enzyme layers. Based on this feature, to determine the optimal amount of enzyme immobilization that would allow for a sensitive chromogenic reaction in the glucose assay, three enzyme surfaces with different numbers of enzyme layers were constructed and compared. As shown in Figure S1, HRP and GOx were alternately immobilized on the surface of a biosensing channel by means of the polydopamine technique and dendrimer. This bienzyme layer (BEL), composed of the dendrimer, HRP, and GOx, was regarded as a surface modification unit. By repeating the BEL construction procedures (Figure S1), single, double, and triple BEL layers (BEL1, BEL2, and BEL3, respectively) were built on the surface of the glucose-biosensing channel. To the prepared three types of glucose biosensing channel, a mixture of glucose (2.5 mM) and Trinder's reagent was applied. Then, through the glucose-sensing channel loaded with a glucose sample (0 and 10 mM), the image of signal guide pattern on computer LCD monitor was observed and registered using a smartphone camera every 1 min up to 10 min.

As shown in Figure 5A, in the assay involving BEL1, the distinguishable disappearance of polygons in the signal guide pattern was not observed for assay duration of 10 min. In the assay using BEL2, a polygon disappearance was observed at 9 min after the observation. These results indicate that the amount of enzymes in channels BEL1 and BEL2 was insufficient to generate blue end product absorbing red light of the signal guide pattern. In contrast, in the assay using BEL3, obvious polygon disappearance was observed 4 min earlier than for BEL2 (at 5 min after the start of observation). In addition, during the monitoring of the signal guide pattern through the BEL3 channel, the number of polygons that disappeared gradually increased with time, and finally only one polygon (3/4 circle) remained at 10 min after the start of observation. This result indicates that the amount of enzymes that is immobilized on the channel surface in the triple BEL format by the LBL method is sufficient to induce an intense color-development reaction that can effectively absorb red light of the signal guide patterns on the LCD panel. Therefore, BEL3 was chosen as a condition for the enzyme immobilization on the glucose-biosensing channel.

Next, the reaction duration for the glucose assay was optimized as follows, so that the assay can show the most distinguishable differences in disappearance of polygons between glucose concentrations. Into the BEL3-containing glucose biosensing channel, glucose samples at different glucose concentrations (0, 5, and 10 mM) was injected with Trinder's reagent. Then, through the prepared glucose sensing channel, the image of the signal guide pattern on the LCD panel was monitored and registered using the smartphone camera every minute up to 10 min.

(A)

Enzyme layer	Time										
	0 min.	1 min.	2 min.	3 min.	4 min.	5 min.	6 min.	7 min.	8 min.	9 min.	10 min.
BEL 1											
BEL 2											
BEL 3											

(B)

Glucose concentration	Time										
	0 min.	1 min.	2 min.	3 min.	4 min.	5 min.	6 min.	7 min.	8 min.	9 min.	10 min.
0 mM											
5 mM											
10 mM											

Figure 5. The resulting images of the optimization test for biosensing of surface modifications: (**A**) A 2.5 mM sample of glucose is assayed by means of the GOx/HRP catalysis. The one, two, and three enzyme layers were compared, and the images with color development were registered every 1 min up to 10 min. As the immobilized enzyme layer increases, the signal guides significantly changed. (**B**) The result of testing the reaction time by means of the three-enzyme layer. With reaction time at 5 min, the signal guides are clearly distinguished in accordance with the applied glucose concentration.

As shown in Figure 5B, when the reaction time was less than 5 min, notable differences in polygon disappearance among glucose concentrations were not detected. This finding indicates that the given reaction time (less than 5 min) is not enough to produce sufficient amount of the blue reactant in the glucose assay. In contrast, at 5 min of glucose assay time, the numbers of invisible polygons proportionally increased with the glucose concentration and those changes could be distinctively detected by the naked eye. When the reaction time exceeded 5 min, the number of invisible polygons in the glucose assay with the 5 mM glucose sample and that with 10 mM glucose was similar, and it was hard to distinguish their differences with the naked eye. This is because the blue reactant in the glucose assay was excessively produced at reaction duration over 5 min, and the excessively produced chromogenic compounds absorb most of red light emitted by the signal guide pattern. Based on this finding, reaction time of 5 min, which yielded clear-cut differences in the number of invisible polygons in accordance with the changes in glucose concentration, was selected as the optimal reaction time condition for quantitative glucose biosensing. The optimal reaction conditions derived from this experiment (BEL3-containing glucose biosensing channel and reaction time of 5 min) were applied to the following glucose-biosensing procedure.

3.5. LCD Panel-Based Glucose Detection by the Naked Eye

To demonstrate the applicability of the newly developed optical transducing technology to practical biosensing, a quantitative glucose assay was carried out. On the LCD panel of the computer monitor, the signal guide pattern (Figure 3) was displayed. To the BEL3-modified glucose biosensing channel, a mixture of a glucose sample at various concentrations (0, 1.3, 2.5, 5, and 10 mM) and Trinder's reagent was applied. The glucose-loaded biosensing channel was mounted on the smartphone camera lens and was incubated for 5 min for color development. After that, as depicted in Figure 6A, the image of signal guide patterns on the LCD panel was observed and documented by the smartphone camera, with the glucose-biosensing channel that developed blue color serving as a color filter. Finally, the remaining polygons in the registered signal guide image were counted with the naked eye. Simultaneously, the remaining polygons in the signal guide pattern were also counted by the pattern-recognition option of the graphics editor (software, Microsoft PowerPoint 2016) to compare the counting results between the naked-eye analysis and algorithmic analysis. The resulting images and the polygon counting results are shown in Figure 6B.

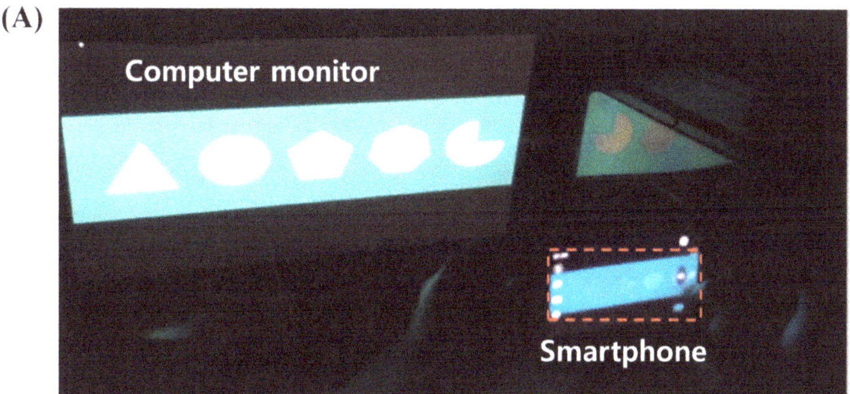

Figure 6. The glucose analysis based on the newly developed biosensing principle: (**A**) an image of application of the new biosensing system to a computer monitor; and (**B**) the resulting images of biochemical reactions at various glucose concentrations in the sample (0, 1.3, 2.5, 5, or 10 mM) after GOx/HRP catalysis. The smartphone-based quantitative analysis was conducted by naked-eye observation, and computational calculation was implemented to provide accurate counting.

As expected, depending on the increase in glucose concentration, the color intensity of the glucose assay reactant in the biosensing channel increased, and the number of remaining polygons in the observed signal guide pattern decreased. In this experiment, when the concentration of the blue glucose analyte in the biosensing channel increased, the absorption of red light, emitted inside the polygon in the signal guide pattern, was enhanced. This red-light absorption induced synchronization of the colors inside and outside the polygon, so that the polygon disappears in a given observation condition. In the glucose assay with low concentrations of the glucose sample (0, 1.3, and 2.5 mM), the triangle, circle, and pentagon disappeared in this order because the intensity of their R pixel was weak enough to be fully absorbed by the reactant in the glucose assay at low concentrations (135 for the triangle; 165 for circle; 195 for pentagon). When those polygons showing low R pixel intensity disappeared at low glucose concentrations, the heptagon and 3/4 circle (yielding high R pixel intensity) were still visible. The polygons that have high R pixel intensity such as the heptagon (R: 225) and 3/4 circle (R: 255) disappeared only at the high glucose concentrations (5 and 10 mM). The obtained results revealed that the disappearance of the polygons was dependent on the spectral correlation of absorbance of the glucose assay reactant and emission intensity of R pixels in polygons as we have initially intended.

To demonstrate that the naked-eye counting results were not affected by the subjective intervention of individual observers, the obtained naked-eye results were compared with the algorithmic analysis results. In the naked-eye analysis, the number of residual polygons observed at 0, 1.3, 2.5, 5, and 10 mM glucose in the samples was 4, 3, 3, 2, and 0, respectively. As shown in Figure 6B, the residual polygons which were recognized and detected by the pattern recognition algorithm were outlined in red color. As a result, the number of red polygons was the same as the number of remaining polygons that were counted by the naked eye. This result indicates that the newly designed optical transducing technique is a promising approach to implementing a user-friendly biosensing system based on convenient naked-eye analysis.

In our glucose-biosensing results, the numbers of remaining polygons at 1.3 mM glucose in the samples were the same as those at 2.5 mM glucose. Strictly speaking, this result means that the newly developed glucose-biosensing method is not suitable for detection of low concentrations of glucose. Nevertheless, considering that the diagnostic criterion for a glycemic profile to indicate diabetes mellitus is 5 mM glucose, we expect that our glucose-biosensing technique based on the new optical transducing technology can be used for the screening of diabetic patients and healthy people.

From the obtained findings, we concluded that the newly developed biosensing method based on the new optical-signal-transducing principle utilizing spectral correlation between the absorption spectrum of a chromogenic reactant and color profiles of pixels in the LCD panel for glucose will help to realize a POC diagnostic system that is highly user-friendly and cost-effective.

4. Conclusions

By combining the LCD display panel, smartphone, transparent and soft material-based optical filter, and enzymatic colorimetric assay principle, we developed a new optical-signal-transducing technology. The LCD panel that contains numerous color pixels whose intensity can be easily regulated by software served as a light source in the new signal transducing system. The colored end product of the enzymatic chromogenic assay was used to interfere with the RGB lights inside and outside a polygon on the LCD panel based on the color filter-like effect. By means of the spectral correlation between the absorption spectrum of the colorimetric reactant under study and color profiles of pixels on the LCD panel, the color intensity signals that change linearly with the alterations in analyte concentration were successfully converted to the visual signal (e.g., disappearance of a polygon) that can be counted as an integer number by the naked eye. Using the new signal-transducing system, we could achieve biosensing of glucose in the range of 0–10 mM, covering the clinical criteria for diabetes mellitus screening, without complicated optical instruments. Because a common LCD monitor panel and a smartphone were used as optical instruments, the newly developed signal-transducing system provides high user-friendliness and operational convenience. Besides, because these devices are widely distributed in the population nowadays, the newly developed system should be helpful for implementation of a POC biosensing system overcoming the constraints in time and space on diagnostics. Additionally, given that the HRP-mediated chromogenic reaction served as a principle of colorimetric signal generation, the proposed optical-signal-transducing system can be applied to the HRP-mediated affinity biosensing such as an enzyme-linked immunosorbent assay involving HRP as a signaling reporter. Judging by the obtained results and considerations, we expect that this optical transducing approach will provide insights into materialization of a POC biosensing system that is user-friendly and has a wide application range.

Supplementary Materials: The following are available online at www.mdpi.com/1996-1944/11/3/388/s1. Figure S1: The workflow of the manufacture of a biorecognition layer on the biosensing channel surface. Construction of the enzyme layer for detection of glucose by means of GOx, HRP, and a dendrimer; Figure S2: Changes in the absorbance at visible wavelengths caused by various concentrations of glucose in the sample.

Acknowledgments: This work was supported by the National Research Foundation (NRF-2016R1A2B4006564) of Korea and the Priority Research Centers Program (2009-0093826).

Author Contributions: H.J.C., Y.D.H. and H.C.Y. conceived and designed the experiments; H.J.C., Y.D.H. and K.R.K. performed the experiments; H.J.C., Y.M.P., S.J.L. and H.D.Y. analyzed the data; and H.J.C., Y.D.H., Y.M.P., S.J.L. and H.C.Y. contributed to prepare the manuscript.

Conflicts of Interest: The authors declare no conflict of interest.

References

1. Viveros, L.; Paliwal, S.; McCrae, D.; Wild, J.; Simonian, A. A fluorescence-based biosensor for the detection of organophosphate pesticides and chemical warfare agents. *Sens. Actuator B-Chem.* **2006**, *115*, 150–157. [CrossRef]
2. Newman, R.H.; Fosbrink, M.D.; Zhang, J. Genetically Encodable Fluorescent Biosensors for Tracking Signaling Dynamics in Living Cells. *Chem. Rev.* **2011**, *111*, 3614–3666. [CrossRef] [PubMed]
3. Jiao, X.; Fei, X.; Li, S.; Lin, D.; Ma, H.; Zhang, B. Design Mechanism and Property of the Novel Fluorescent Probes for the Identification of *Microthrix parvicella* In Situ. *Materials* **2017**, *10*, 804. [CrossRef] [PubMed]
4. Homola, J. Surface Plasmon Resonance Sensors for Detection of Chemical and Biological Species. *Chem. Rev.* **2008**, *108*, 462–493. [CrossRef] [PubMed]
5. Laurent, N.; Voglmeir, J.; Flitsch, S.L. Glycoarrays—Tools for determining protein-carbohydrate interactions and glycoenzyme specificity. *Chem. Commun.* **2008**, *37*, 4400–4412. [CrossRef] [PubMed]
6. Abbas, A.; Linman, M.J.; Cheng, Q. New trends in instrumental design for surface plasmon resonance-based biosensors. *Biosens. Bioelectron.* **2011**, *26*, 1815–1824. [CrossRef] [PubMed]
7. Song, Y.; Wei, W.; Qu, X. Colorimetric Biosensing Using Smart Materials. *Adv. Mater.* **2011**, *23*, 4215–4236. [CrossRef] [PubMed]
8. Schlücker, S. Surface-Enhanced Raman Spectroscopy: Concepts and Chemical Applications. *Angew. Chem.-Int. Ed.* **2014**, *53*, 4756–4795. [CrossRef] [PubMed]
9. Anker, J.N.; Hall, W.P.; Lyandres, O.; Shah, N.C.; Zhao, J.; Van Duyne, R.P. Biosensing with plasmonic nanosensors. *Nat. Mater.* **2008**, *7*, 442–453. [CrossRef] [PubMed]
10. Klostranec, J.M.; Chan, W.C.W. Quantum Dots in Biological and Biomedical Research: Recent Progress and Present Challenges. *Adv. Mater.* **2006**, *18*, 1953–1964. [CrossRef]
11. Breslauer, D.N.; Maamari, R.N.; Switz, N.A.; Lam, W.A.; Fletcher, D.A. Mobile Phone Based Clinical Microscopy for Global Health Applications. *PLoS ONE* **2009**, *4*, e6320. [CrossRef] [PubMed]
12. Bandodkar, A.J.; Wang, J. Non-invasive wearable electrochemical sensors: A review. *Trends Biotechnol.* **2014**, *32*, 363–371. [CrossRef] [PubMed]
13. Roda, A.; Michelini, E.; Cevenini, L.; Calabria, D.; Calabretta, M.M.; Simoni, P. Integrating biochemiluminescence detection on smartphones: Mobile chemistry platform for point-of-need analysis. *Anal. Chem.* **2014**, *86*, 7299–7304. [CrossRef] [PubMed]
14. Park, Y.M.; Han, Y.D.; Kim, K.R.; Zhang, C.; Yoon, H.C. An immunoblot-based optical biosensor for screening of osteoarthritis using a smartphone-embedded illuminometer. *Anal. Methods* **2015**, *7*, 6437–6442. [CrossRef]
15. Chun, H.J.; Park, Y.M.; Han, Y.D.; Jang, Y.H.; Yoon, H.C. Paper-based glucose biosensing system utilizing a smartphone as a signal reader. *BioChip J.* **2014**, *8*, 218–226. [CrossRef]
16. Wei, Q.; Qi, H.; Luo, W.; Tseng, D.; Ki, S.J.; Wan, Z.; Göröcs, Z.; Bentolila, L.A.; Wu, T.T.; Sun, R.; et al. Fluorescent imaging of single nanoparticles and viruses on a smart phone. *ACS Nano* **2013**, *7*, 9147–9155. [CrossRef] [PubMed]
17. Wang, X.; Chang, T.; Lin, G.; Gartia, M.R.; Liu, G.L. Self-Referenced Smartphone-Based Nanoplasmonic Imaging Platform for Colorimetric Biochemical Sensing. *Anal. Chem.* **2016**, *89*, 611–615. [CrossRef] [PubMed]
18. Ko, J.S.; Yoon, H.C.; Yang, H.; Pyo, H.-B.; Chung, K.H.; Kim, S.J.; Yoon, T.K. A polymer-based microfluidic device for immunosensing biochips. *Lab Chip* **2003**, *3*, 106–113. [CrossRef]
19. Park, Y.M.; Han, Y.D.; Chun, H.J.; Yoon, H.C. Ambient light-based optical biosensing platform with smartphone-embedded illumination sensor. *Biosens. Bioelectron.* **2017**, *93*, 205–211. [CrossRef] [PubMed]
20. Lee, H.; Dellatore, S.M.; Miller, W.M.; Messersmith, P.B. Mussel-Inspired Surface Chemistry for Multifunctional Coatings. *Science* **2007**, *318*, 426–430. [CrossRef] [PubMed]
21. Han, Y.D.; Chun, H.J.; Yoon, H.C. The transformation of common office supplies into a low-cost optical biosensing platform. *Biosens. Bioelectron.* **2014**, *29*, 259–268. [CrossRef]

22. Yoon, H.C.; Kim, H.S. Multilayered Assembly of Dendrimers with Enzymes on Gold: Thickness-Controlled Biosensing Interface. *Anal. Chem.* **2000**, *72*, 922–926. [CrossRef] [PubMed]
23. Tamaoku, K.; Murao, Y.; Akiura, K. New water-soluble hydrogen donors for the enzymatic spectrophotometric determination of hydrogen peroxide. *Anal. Chim. Acta* **1982**, *136*, 121–127. [CrossRef]
24. Nakamura, H.; Mogi, Y.; Akimoto, T.; Naemura, K.; Kato, T.; Yano, K.; Karube, I. An enzyme-chromogenic surface plasmon resonance biosensor probe for hydrogen peroxide determination using a modified Trinder's reagent. *Biosens. Bioeletron.* **2008**, *24*, 455–460. [CrossRef] [PubMed]

Article

Processing Techniques for Bioresorbable Nanoparticles in Fabricating Flexible Conductive Interconnects

Jiameng Li, Shiyu Luo, Jiaxuan Liu, Hang Xu * and Xian Huang *

Department of Biomedical Engineering, Tianjin University, 92 Weijin Road, Tianjin 300072, China; MrLeepursuit@163.com (J.L.); luoshiyu@tju.edu.cn (S.L.); jiaxuanliu@tju.edu.cn (J.L.)
* Correspondence: xuhang618@tju.edu.cn (H.X.); huangxian@tju.edu.cn (X.H.)

Received: 5 June 2018; Accepted: 25 June 2018; Published: 28 June 2018

Abstract: Bioresorbable electronics (or transient electronics) devices can be potentially used to replace build-to-last devices in consumer electronics, implantable devices, and data security, leading to reduced electronic waste and surgical processes through controllable dissolution. Recent development of printing bioresorbable electronics leads to bioresorbable conductive pastes or inks that can be used to make interconnects, circuit traces, and sensors, offering alternative solutions for the predominant complementary metal oxide semiconductor (CMOS) processes in fabrication of bioresorbable electronics. However, the conductivities offered by current bioresorbable pastes and processing techniques are still much lower than those of the bulk metals, demanding further improvement in both paste composition and process optimization. This paper aims at exploring several influential factors such as paste compositions and processing techniques in determining conductivities of bioresorbable patterns. Experimental results reveal that an optimized paste constituent with a ratio of Zn:PVP:glycerol:methanol = 7:0.007:2:1 by weight can generate stable conductive pastes suitable for a screen printing process. In addition, a high conductivity of 60,213.6 S/m can be obtained by combining hot rolling and photonic sintering. The results demonstrate that large-scale transient electronics can be obtained by combining screen printing, hot rolling and photonic sintering approaches with optimized paste compositions, offering important experimental proofs and approaches for further improving the conductivity of bioresorbable pastes or inks that can accommodate the demands for mass fabrication and practical use in electronic industry.

Keywords: bioresorbable electronics; printing electronics techniques; conductive inks

---

1. Introduction

The majority of more than 50 million tons of electronic-waste generated each year globally end up in the landfill, or are just simply incinerated, causing enormous environmental issues, such as soil compaction, acid rain, and water pollution [1,2]. Efforts in recycling electronic-waste have been focused on reducing the cost and time consumption of recycling processes, which features a few iconic techniques such as automatic sorting, mechanical disassembly, and magnetic separation [3–5]. However, these techniques have stringent requirements for large facility, expensive equipment, and hazardous chemicals. Recent development of transient electronic devices that can degrade under environmentally friendly approaches triggered by water, humidity, light, and air flow leads to a safe and effective solution to the rampant pollution caused by electronic-waste while facilitating recycling [6–11].

Bioresorbable electronic devices have been presented in various formats with applications ranging from data storage to internal medicine and health care [12–17]. The fabrication approaches of such devices involve modified complementary metal oxide semiconductor (CMOS) processes and printing electronics techniques [18]. An emerging trend combines CMOS and printing electronics technology to yield bioresorbable circuits that offer both improved complexity and high time/cost efficiency, demonstrating promising use in replacing build-to-last electronic devices for specific applications that only require short working periods [19]. Fabrication of bioresorbable electronic devices can be achieved by anhydrous and low temperature processes at the cost of device performance [20]. Thus, improvement of CMOS and printing electronics technology for bioresorbable electronics is critically needed.

One fundamental improvement involves developing bioresorbable pastes or inks for making interconnectors, circuit traces, and other electronic components such as sensors, resistors, and electrodes [12–17]. Using zinc nanoparticles (Zn NPs) together with various printing and sintering approaches, bioresorbable conductive patterns has been demonstrated in our previous studies as well as other literature [19–23]. However, existing bioresorbable inks offer conductivity ranging from 2.2×10^4 to 3.0×10^5 S/m [19–22], which is at least 20 times lower than that of bulk metal. Improved conductivity may be obtained by adjusting compositions of bioresorbable inks as well as other handling techniques. In this paper, we investigate the influence of weight ratio of Zn NPs as well as a new process flow that involves combination of screen printing, hot rolling and photonic sintering techniques. The influence of individual technique in determining the conductivities of Zn patterns was also investigated. A high conductivity at 60,213.6 S/m can be achieved with excellent flexibility to withstand repeated bending. The results suggest that large-scale transient electronics can be obtained by combining screen printing, hot rolling and photonic sintering approaches with optimized ink compositions, offering important experimental proof and approaches for further improving the conductivity of bioresorbable pastes and inks that can adapt to the demands for mass fabrication and practical use in electronic industry.

2. Materials and Methods

2.1. Preparation of Conductive Inks and Bioresorbable Substrates

Preparation of Zn nanoparticle inks started with mixing glycerol and methanol at a mass ratio of 2:1 to yield a bi-solvent system. Polyvinylpyrrolidone (PVP, 0.1 wt% for Zn NPs) was then dissolved in the mixed solvent as surfactant to increase ink viscosity and prevent aggregation of Zn NPs. Zn NPs (50 nm in diameter, Beijing Dk Nano technology Co., Ltd., Beijing, China) were then added to the solution at a 7:3 weight ratio, followed by mechanical stirring and sonication for 30 min, resulting in a conductive paste with a proper viscosity (~10 Pa·s) and particle sizes (~500 nm) to satisfy the requirements of screen printing.

Polyvinyl alcohol (PVA) has been used for making bioresorbable substrates. A solution of PVA can be obtained by adding 10 wt% PVA (Energy Chemical Inc., Shanghai, China) to DI water at 80 °C and stirring at 500 rpm until the solution became bubble free. The resulting solution was subjected to filtration (membrane size is 500 nm) to remove any insoluble impurity, and was then dispersed onto glass slides, allowing the formation of flexible and transparent films after drying in air at room temperature for 24 h.

2.2. Fabrication of Conductive Patterns

Figure 1 demonstrates a fabrication process for printing bioresorbable patterns made of Zn NPs on a PVA substrate. The entire process involves screen printing, hot rolling and photonic sintering, all of which are anhydrous to prevent Zn NPs from reacting with water. Curing and sintering of as-printed patterns were conducted under the protection of argon to avoid surface oxidation of Zn NPs. A screen printer (PHP-2020A, Shanghai Xuanting Co., Ltd., Shanghai, China) was used to print various patterns ranging from straight interconnect to curved electrodes, followed by a curing process at 100 °C to remove solvent in the patterns. A hot roller (MRX-JS200L, Shenahen Mingruixiang Automation Equipment Co., Ltd., Shenzhen, Guangdong, China) with a rolling speed of 40 mm/s was used to compress the printed patterns to remove additional void space generated because of solvent evaporation. The patterns were then sintered by a photonic sintering system (LH840, Xenon Co., Ltd., Dover, DE, USA) using an energy at 4.9 J/cm^2 per pulse. The morphology of the conductive patterns was measured by a scanning electron microscopy (SEM, SUPRA55VP, Zeiss, Oberkochen, Baden-Württemberg, Germany) with a 15 KV accelerating voltage and a working distance of 10 mm, and the conductivity of patterns was measured through a four-point probe measurement system (FPPM2015A, Suzhou Jingge Electronic Co., Ltd., Suzhou, Jiangsu, China) following the ASTM F1711-96(2008) standard.

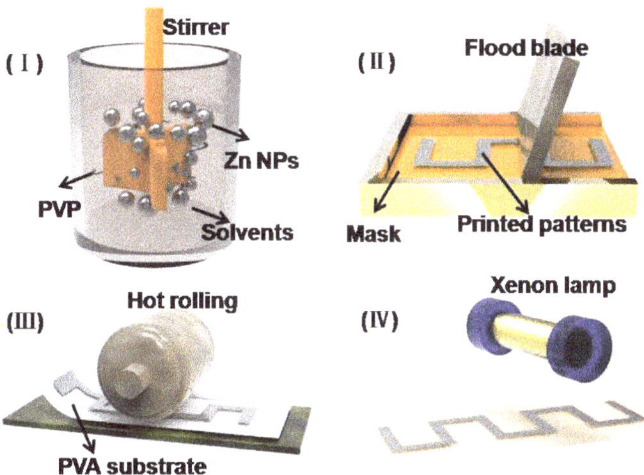

Figure 1. Schematics of fabricating transient electronic patterns through ink preparation (**I**), screen printing (**II**), hot rolling; (**III**) and photonic sintering; (**IV**) approaches.

3. Results and Discussion

An optimized weight ratio between Zn NPs and the mixed solution was first investigated. The weight ratio of Zn NPs is considered as a determining factor that significantly influences the conductivity of bioresorbable patterns. The conductive patterns obtained by pastes with varied weight ratios of Zn NPs from 10% to 70% were fabricated by screen printing and were treated with hot-rolling and photonic sintering. The conductivities of sintered patterns improve with increased ratios of Zn NPs (Figure 2b). Patterns made by the conductive paste with 10 wt% Zn NPs are only 132.8 S/m, while patterns with 50 wt% and 70 wt% Zn NPs show good conductivities of 43,144.2 and 56,849.5 S/m, respectively. SEM images of the above patterns further show increased density and reduced porosity in bioresorbable patterns with higher weight ratios (Figure 2(aI–aIV)). The sample made by the conductive paste with 70 wt% Zn NPs shows the highest conductivity, and, thus, this weight ratio is used throughout our following investigation.

It is also observed that subsequent treatments of the printed patterns after screen printing perform important roles in determining the conductivity of the patterns. Conductivity of printed patterns that have been subjected to pure hot rolling is determined to be 42,806.9 S/m as compared with 594.9 S/m for an as-printed sample. The result suggests that increasing the compactness of printed samples by hot rolling is an effective approach to achieve high conductivity. When no further sintering processes is involved, the major improvement of the conductivity can be attributed to the volume shrinkage that can increase spontaneous bonding of nanoparticles and enhance tunneling effect among nanoparticles. SEM images of patterns with (Figure 2(aVI–aIII)) and without (Figure 2(aV–aVII)) hot rolling further demonstrate that significant coalescence of nanoparticles exists in the hot-rolled sample.

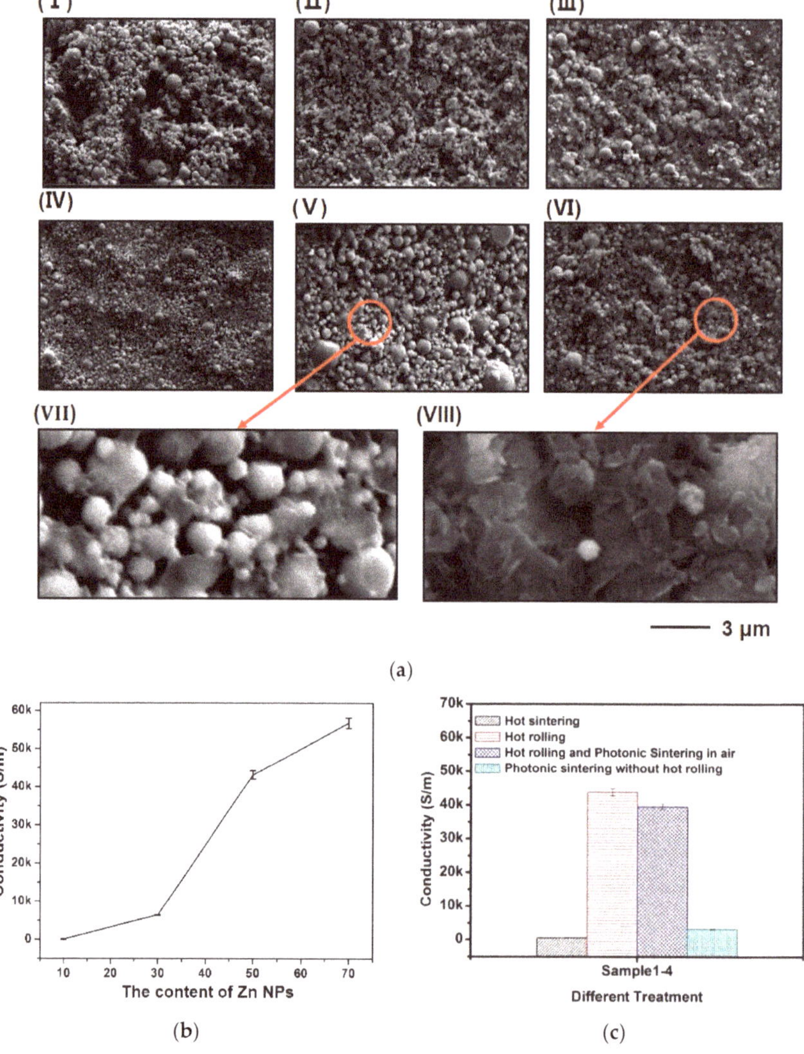

Figure 2. (a) The surface morphology of printed patterns using different content of Zn NPs at (I) 10 wt%, (II) 30 wt%, (III) 50 wt%, and (IV) 70 wt%. (V) and (VI) represent that patterns treated without and with hot rolling, respectively. (both of them without photonic sintering). (VIII) and (VII) represent amplified images of (VI) and (V), respectively; (b) The conductivity of patterns using different content of Zn NPs; (c) The conductivity of printed pattern under different treatments.

The effect of sintering approaches was also investigated by comparing prolonged thermal sintering and ultra-fast pulsed light sintering. Due to the low decomposition temperature (~120 °C) of PVA substrates, the temperature during the thermal sintering process was selected to be 100 °C to avoid substrate damage. The highest conductivity of the resulting pattern is only 621.7 S/m after sintering on a hot plate for an extended period for over 3 h. The conductivity is only slightly higher than the as-printed sample, and is insufficient for practical applications. Photonic sintering of the patterns was conducted in a custom-made stainless-steel enclosure filled with argon using a photonic sintering system that generates a high intensity (4.9 J/cm^2) wide spectrum light pulse (540 µs in duration). The photonic sintering process alone can lead to increased conductivity from 594.9 to 3237.5 S/m in one pulse without introducing the hot rolling process. When the hot rolling process is involved, the conductivity achieves 45,126.6 S/m after one pulse (Figure 2c) and a high conductivity of 60,213.6 S/m after twenty pulses. As the energy of the photonic sintering system is much lower than the one (25.88 J/cm^2) in our previous publication [20], it is demonstrated that the effect of reduced sintering energy can be compensated by increasing the number of pulses (Figure 3(aIV)). Only slight changes in conductivity can be observed if the number of pulses keeps increasing. However, more pulses may lead to substrate damage, thus 20 pulses may be the optimum value for this particular sintering system. SEM images further confirm that no obvious change occur in the morphology of the patterns after over 20 pulses (Figure 3(aI–aII)), and the compactness of printed patterns is enhanced after multiple light pulses (Figure 2(aIV) and Figure 3(aI)). The results reveal that the surface of the printed pattern gradually becomes more compacted with increased number of pulses. It is also worth mentioning that dendrite formation is observed within the sintered sample due to rapid heating and cooling processes as analyzed in our previous publication [20]. However, the special flake shape of dendrite may indicate different surface energy and environmental conditions as compared with our previous work. The formation of these flakes may further reduce the distance between nanoparticles and increase the conductivity by offering increase contact areas.

The effect of surface oxidation is considered as one major obstacle for further improving the conductivity of bioresorbable patterns. Surface oxidation appears both as pre-existing surface layers that grow spontaneously during the synthesis of Zn NPs and as subsequent layers formed during the preparation and handling processes of pastes or inks. The surface composition has been studied using element EDS mapping after a photonic sintering process that were conducted under ambient environment without argon protection. It can be observed that oxygen has broad distribution in the tested sample and can correspond to 10 mol% Zn NPs (determined by EDS without any treatment, Figure 3b). The conductivity of such sample is only 39,581.3 S/m (Figure 2c) which is lower than that for the sample sintered in argon. Evidently, creating an oxygen-free environment is necessary in the photonic sintering process. In addition, previous studies have demonstrated a pre-existing thin layer of surface oxide on the as-synthesized Zn NPs [19,20,23]. This layer may be removed by acid treatment or reaction with reduction agents prior to use in an effort to obtain better conductivity.

The flexibility and conductivity of the bioresorbable patterns were also explored. When bending a sintered pattern into different curvatures, no obvious damage and delamination can be observed (Figure 4(aI)). The conductivity of the pattern reduces continuously from 54,034.2 to 51,197.6 S/m with increased curvature from 0 to 0.8 cm^{-1}. It is noticed that the conductivity is stable with less than 0.8% changes at curvatures ranging from 0 to 0.6 cm^{-1}. Significant changes can be observed when the curvature becomes larger than 0.7 cm^{-1} (Figure 4b). This may be related to the formation of internal cracks due to applied bending strain. The reversibility of the printed patterns in response to repeated bending was conducted by bending the pattern from a planar state to a curved state with a curvature of 0.6 cm^{-1} repeatedly (Figure 4c). The results show reversible patterns in the conductivity with deviation between individual bending cycles for less than 1.5%. When connecting a light-emitting diode (LED) to the bioresorbable patterns (Figure 4(aII)), the LED can be lit after application of a power supply. The LED light was stable during the entire test process. When connecting the bioresorbable interconnect to a circuit that contained both a power supply and an LED, the LED could maintain its light intensity even when the interconnection was in an extremely curved state (Figure 4(aIII)), suggesting that the bioresorbable patterns may be able to adapt to complicate surrounding environments that demand large deformation.

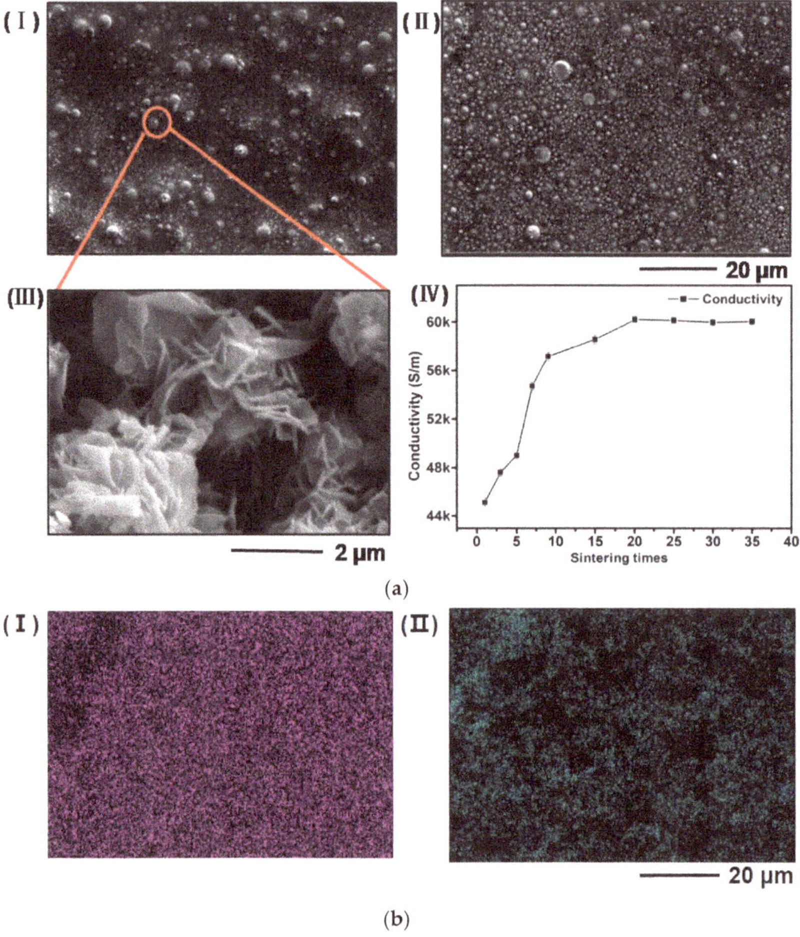

Figure 3. (**a**) (**I**) and (**II**) represent that patterns treated with 20 and 25 light pulses, respectively; (**III**) a higher resolution of SEM image of surface morphology of Zn NPs after treating with 20 pulses of photonic sintering; (**IV**) The changes of conductivity with different number of light pulses in the photonic sintering process; (**b**) The mapping of a printed pattern shows the distribution of Zn (**I**) and O (**II**).

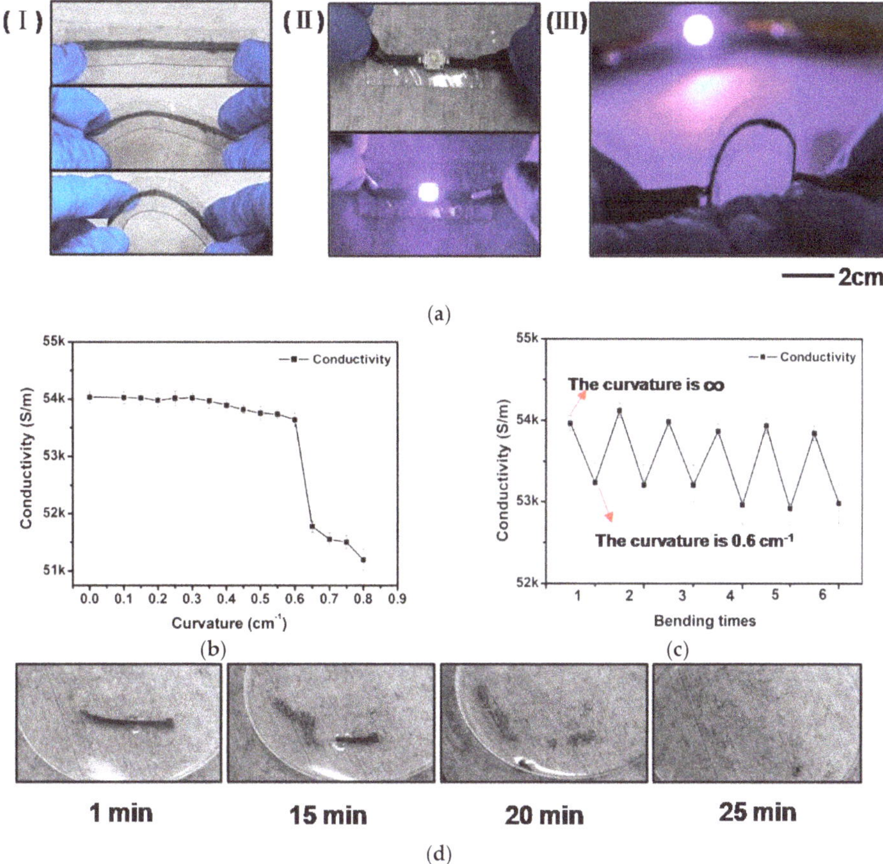

Figure 4. (**a**) (**I**) A printed pattern were bent into different curvatures; (**II**) The printed pattern as interconnects to connect a LED with a power supply; (**III**) An LED was lit by a printed interconnect in an extremely curved state; (**b**) The conductivity of a bioresorbable pattern in different curvatures; (**c**) Changes in conductivity when a bioresorbable pattern was bent repeatedly from 0 to 0.6 cm^{-1} in curvature; (**d**) A dissolution process of a printed pattern.

Finally, the dissolution capability of the bioresorbable patterns was tested by immersing an unpackaged sample directly into DI water. The printed pattern started to dissolve rapidly after immersing in water. The entire dissolution process can be completed within 25 min (Figure 4d). The dissolution time can be fine-tuned by adding surface coating layers with various dissolution times.

4. Conclusions

This paper investigates several influential factors in determining the conductivity of bioresorbable patterns. Both an optimized composition and processing techniques have been achieved, leading to highly conductive bioresorbable patterns with a maximum conductivity of 60,213.6 S/m. The resulting patterns also possess excellent flexibility and dissolution capability. The experimental proof and approaches presented in this paper offer important guidelines for further improving the conductivity of bioresorbable pastes or inks that can adapt to the demands for mass fabrication and practical use in electronic industry.

Further study is required to minimize surface oxidation on nanoparticles. This may be tackled by introducing additives that can either absorb water to turn into weak acid that can gently remove surface oxidation layers on nanoparticles or work as reduction agents that may react with surface oxidation when exposed to pulsed sintering light. In addition, complex circuit traces with more electronic components may be achieved using screen printing and ink dispersion, resulting in integrated bioresorbable PCB circuits that may be able to replace current build-to-last circuits in many applications.

Author Contributions: Conceptualization: H.X. and X.H.; Experiments: J.L. (Jiameng Li), S.L. and J.L. (Jiaxuan Liu); Data Curation: J.L. (Jiameng Li) and H.X.; Writing-Original Draft Preparation: J.L. (Jiameng Li) and H.X.; Writing-Review & Editing: X.H.; Supervision, X.H.

Funding: This research received no external funding.

Acknowledgments: This work is supported by the National Natural Science Foundation of China under Grant No. 61604108, the Natural Science Foundation of Tianjin under Grant No. 16JCYBJC40600, and the Independent Innovation Fund in Tianjin University, Tianjin, China.

Conflicts of Interest: The authors declare no conflict of interest.

References

1. Pramila, S.; Fulekar, M.H.; Bhawana, P. E-waste—A challenge for tomorrow. *Res. J. Recent. Sci.* **2012**, *1*, 86–93.
2. Kumar, U.; Singh, D.N. Electronic waste: Concerns & hazardous threats. *Int. J. Curr. Eng. Technol.* **2014**, *4*, 802–811.
3. Robinson, B.H. E-waste: An assessment of global production and environmental impacts. *Sci. Total Environ.* **2009**, *408*, 183–191. [CrossRef] [PubMed]
4. Aizawa, H.; Yoshida, H.; Sakai, S.-I. Current results and future perspectives for japanese recycling of home electrical appliances. *Resour. Conserv. Recycl.* **2008**, *52*, 1399–1410. [CrossRef]
5. Cui, J.; Zhang, L. Metallurgical recovery of metals from electronic waste: A review. *J. Hazard. Mater.* **2008**, *158*, 228–256. [CrossRef] [PubMed]
6. Yu, X.; Shou, W.; Mahajan, B.K.; Huang, X.; Pan, H. Materials, processes, and facile manufacturing for bioresorbable electronics: A review. *Adv. Mater.* **2018**. [CrossRef] [PubMed]
7. Kim, B.H.; Kim, J.-H.; Persano, L.; Hwang, S.-W.; Lee, S.; Lee, J.; Yu, Y.; Kang, Y.; Won, S.M.; Koo, J.; et al. Dry transient electronic systems by use of materials that sublime. *Adv. Funct. Mater.* **2017**, *27*, 1606008. [CrossRef]
8. Yoon, J.; Lee, J.; Choi, B.; Lee, D.; Kim, D.H.; Kim, D.M.; Moon, D.; Lim, M.; Kim, S.; Choi, S.-J. Flammable carbon nanotube transistors on a nitrocellulose paper substrate for transient electronics. *Nano Res.* **2016**, *10*, 87–96. [CrossRef]
9. Hernandez, H.L.; Kang, S.-K.; Lee, O.P.; Hwang, S.-W.; Kaitz, J.A.; Inci, B.; Park, C.W.; Chung, S.; Sottos, N.R.; Moore, J.S.; et al. Triggered transience of metastable poly(phthalaldehyde) for transient electronics. *Adv. Mater.* **2014**, *26*, 7637–7642. [CrossRef] [PubMed]
10. Gao, Y.; Zhang, Y.; Wang, X.; Sim, K.; Liu, J.; Chen, J.; Feng, X.; Xu, H.; Yu, C. Moisture-triggered physically transient electronics. *Sci. Adv.* **2017**, *3*, e1701222. [CrossRef] [PubMed]
11. Bauer, S.; Kaltenbrunner, M. Built to disappear. *ACS Nano* **2014**, *8*, 5380–5382. [CrossRef] [PubMed]
12. Jia, X.; Wang, C.; Ranganathan, V.; Napier, B.; Yu, C.; Chao, Y.; Forsyth, M.; Omenetto, F.G.; MacFarlane, D.R.; Wallace, G.G. A biodegradable thin-film magnesium primary battery using silk fibroin–ionic liquid polymer electrolyte. *ACS Energy Lett.* **2017**, *2*, 831–836. [CrossRef]
13. Wang, Z.; Fu, K.; Liu, Z.; Yao, Y.; Dai, J.; Wang, Y.; Liu, B.; Hu, L. Design of high capacity dissoluble electrodes for all transient batteries. *Adv. Funct. Mater.* **2017**, *27*, 1605724. [CrossRef]
14. Xi, H.; Chen, D.; Lv, L.; Zhong, P.; Lin, Z.; Chang, J.; Wang, H.; Wang, B.; Ma, X.; Zhang, C. High performance transient organic solar cells on biodegradable polyvinyl alcohol composite substrates. *RSC Adv.* **2017**, *7*, 52930–52937. [CrossRef]
15. Salvatore, G.A.; Sülzle, J.; Valle, F.D.; Cantarella, G.; Robotti, F.; Jokic, P.; Knobelspies, S.; Daus, A.; Büthe, L.; Petti, L.; et al. Biodegradable and highly deformable temperature sensors for the internet of things. *Adv. Funct. Mater.* **2017**, *27*, 1702390. [CrossRef]

16. Kang, S.K.; Murphy, R.K.J.; Hwang, S.-W.; Lee, S.M.; Harburg, D.V.; Krueger, N.A.; Shin, J.; Gamble, P.; Cheng, H.; Yu, S.; et al. Bioresorbable silicon electronic sensors for the brain. *Nature* **2016**, *530*, 71–76. [CrossRef] [PubMed]

17. Lee, G.; Kang, S.-K.; Won, S.M.; Gutruf, P.; Jeong, Y.R.; Koo, J.; Lee, S.-S.; Rogers, J.A.; Ha, J.S. Fully biodegradable microsupercapacitor for power storage in transient electronics. *Adv. Energy Mater.* **2017**, *7*, 1700157. [CrossRef]

18. Mahajan, B.K.; Yu, X.; Shou, W.; Pan, H.; Huang, X. Mechanically milled irregular zinc nanoparticles for printable bioresorbable electronics. *Small* **2017**, *13*, 1700065. [CrossRef] [PubMed]

19. Lee, C.H.; Kang, S.K.; Salvatore, G.A.; Ma, Y.; Kim, B.H.; Jiang, Y.; Kim, J.S.; Yan, L.; Wie, D.S.; Banks, A.; et al. Wireless microfluidic systems for programmed, functional transformation of transient electronic devices. *Adv. Funct. Mater.* **2015**, *25*, 5100–5106. [CrossRef]

20. Mahajan, B.K.; Ludwig, B.; Shou, W.; Yu, X.; Fregene, E.; Xu, H.; Pan, H.; Huang, X. Aerosol printing and photonic sintering of bioresorbable zinc nanoparticle ink for transient electronics manufacturing. *Sci. China Inf. Sci.* **2018**, *61*, 060412. [CrossRef]

21. Lee, Y.K.; Kim, J.; Kim, Y.; Kwak, J.W.; Yoon, Y.; Rogers, J.A. Room temperature electrochemical sintering of zn microparticles and its use in printable conducting inks for bioresorbable electronics. *Adv. Mater.* **2017**, *29*, 1702665. [CrossRef] [PubMed]

22. Huang, X.; Liu, Y.; Hwang, S.-W.; Kang, S.-K.; Patnaik, D.; Cortes, J.F.; Rogers, J.A. Biodegradable materials for multilayer transient printed circuit boards. *Adv. Mater.* **2014**, *26*, 7371–7377. [CrossRef] [PubMed]

23. Shou, W.; Mahajan, B.K.; Ludwig, B.; Yu, X.; Staggs, J.; Huang, X.; Pan, H. Low-cost manufacturing of bioresorbable conductors by evaporation-condensation-mediated laser printing and sintering of Zn nanoparticles. *Adv. Mater.* **2017**, *29*, 1700172. [CrossRef] [PubMed]

 *materials*

Article

Material Characterization of Hardening Soft Sponge Featuring MR Fluid and Application of 6-DOF MR Haptic Master for Robot-Assisted Surgery

Jong-Seok Oh [1], Jung Woo Sohn [2] and Seung-Bok Choi [3,*]

[1] Division of Mechanical & Automotive Engineering, Kongju National University, Cheonan-Si 31080, Korea; jongseok@kongju.ac.kr
[2] Department of Mechanical Design Engineering, Kumoh National Institute of Technology, Gumi-Si 39177, Korea; jwsohn@kumoh.ac.kr
[3] Smart Structures and Systems Laboratory, Department of Mechanical Engineering, Inha University, Incheon 22181, Korea
* Correspondence: seungbok@inha.ac.kr; Tel.: +82-32-860-7319

Received: 20 June 2018; Accepted: 20 July 2018; Published: 24 July 2018

Abstract: In this work, the material characterization of hardening magneto-rheological (MR) sponge is analyzed and a robot-assisted surgery system integrated with a 6-degrees-of-freedom (DOF) haptic master and slave root is constructed. As a first step, the viscoelastic property of MR sponge is experimentally analyzed. Based on the viscoelastic property and controllability, a MR sponge which can mimic the several reaction force characteristics of human-like organs is devised and manufactured. Secondly, a slave robot corresponding to the degree of the haptic master is manufactured and integrated with the master. In order to manipulate the robot motion by the master, the kinematic analysis of the master and slave robots is performed. Subsequently, a simple robot cutting surgery system which is manipulated by the haptic master and MR sponge is established. It is then demonstrated from this system that using both MR devices can provide more accurate cutting surgery than the case using the haptic master only.

Keywords: hardening sponge; MR sponge; 6 degrees-of-freedom (6-DOF) MR haptic master; RMIS (robot-assisted minimally invasive surgery)

---

1. Introduction

In recent years, haptic technology has led to advances in various fields of study including robotics, space exploration, manufacturing, and transportation. In particular, haptic technology has the potential to significantly influence medical industry such as robot-assisted minimally invasive surgery (RMIS) by providing surgeons with a sense of touch. Currently, a commonly used surgery robot system, such as da Vinci™, does not provide haptic sensation as feedback to a surgeon. So, the surgeon cannot perceive how strongly he or she is handling a patient's organ. This may be quite dangerous for the patient. Hence, the restoration of reaction force feedback and tactile information is needed for safer and more accurate robotic surgery. In order to realize the haptic feedback, many types of haptic master and tactile device have been proposed [1–21]. However, RMIS systems with reaction force or tactile feedback showed limited performance. Thus, this study proposes a novel method using both force feedback and tactile sensation for application in RMIS.

Haptic master devices utilizing servomotors had been proposed to solve absence in haptic feedback [1]. However, many disadvantages of the proposed master are reported. For example, it is difficult to obtain continuous and smooth haptic feedback owing to the cogging phenomenon, brush friction, and so forth. Also, an active system using a servomotor is inherently unstable due to power failure and malfunction of the controller. By contrast, the configuration of a semi-active control system is similar with that of a passive system. So, the semi-active system shows passive performance when the power failure occurs. Accordingly, new approaches using controllable magneto-rheological (MR) fluids have been proposed to devise a controllable haptic master. It is generally known that the actuator using MR fluid can obtain smooth and continuous actuating motion owing to the phase change from the liquid to the solid state in a considerably short time by controlling a magnetic field. Several types of MR actuator have been commercially applied to obtain high performance, such as clutch, brake, and several types of damper [2–8]. Among them, the rotary damper and MR sponge damper is actively applied for vibration control application [7,8]. Using the salient property of the MR fluid, several types of haptic master systems have been proposed and investigated in multiple studies. Li et al. developed a haptic master using 2-degrees-of-freedom (DOF) MR brakes [9]. Senkal et al. developed a joystick-type 3-DOF haptic master featuring spherical brake mechanism [10]. Furthermore, a joystick with a 2-DOF haptic system was proposed by using an MR damper mechanism [11]. Oh et al. devised a haptic feedback system by integrating a 4-DOF haptic master using a bidirectional MR clutch and gimbal mechanism to enhance surgical accuracy [12,13]. In clinical practice, a higher-DOF haptic master system is requisite for accomplishing the motions required for surgery in any direction. In other words, the slave robot for RMIS is required to have 3 DOFs for human wrist motion (pitch, yaw, and roll) and 3 DOFs for robot manipulation motion (X, Y, and Z) to stretch any point or position of an incised organ. A few studies proposed a 6-DOF haptic master using servomotors for RMIS [14,15].

In addition, a few tactile devices for sensing the hardness of living tissue were reported for application to RMIS. Several studies on tactile devices that can be integrated with haptic masters have been performed for robot surgery [16,17]. However, these studies have also identified difficulty in realizing the various sensations of human organs by touching sensors. In order to resolve this limitation, in recent years, several types of tactile device with controllable MR fluid have been proposed. The viscoelastic property of the MR fluid can be controlled using the magnitude of the magnetic field; thus, a tactile device can generate various levels of stiffness (or softness) that can represent most human-like organs/tissues. Han et al. and Oh et al. developed a tactile device composed of a diaphragm and MR fluid and evaluated its effectiveness via a psychological test [18,19]. Even though an appropriate reaction force was generated, the viscoelastic features of organs could not be represented because it is not enough to express the elastic property of human tissues using MR fluid and a diaphragm. Given this, Kim et al. presented a hardening sponge device composed of MR fluid and an elastic sponge [20]. The desired reaction force and viscoelastic sensation could be easily realized by changing the current intensity of the MR sponge. The performance evaluation was undertaken using porcine specimens, whose viscoelastic features are similar to those of humans [21–23]. From test results, it was demonstrated that the MR sponge can mimic the tactile feeling of actual human-like organs through multiple experimental tests. Accordingly, it is expected that the MR sponge tactile device can help surgeons feel the stiffness of the organ/tissues during RMIS.

However, so far there has been no research in terms of a robot surgery experiment that utilizes both the 6-DOF MR haptic master and the MR tactile device simultaneously. Consequently, the technical novelty and contributions of this work are summarized as follows. As a first step, the inherent properties of MR fluid and sponge used in this work are characterized as a function of the magnetic field (or current) and deformation length to design an appropriate size of MR sponge device. Then, a cell-type of hardening sponge device is developed using a smart MR fluid and a sponge to generate various magnitudes of the force of human-like tissues. Its effectiveness is demonstrated via experimental tests performed on two specimens, i.e., the porcine heart, lung, and liver. Secondly, a new type of 6-DOF haptic master applicable in RMIS is developed using the following controllable fluid actuators: MR brakes for 3-DOF rotational motion and MR clutches for 3-DOF translational motion. This is the first work on the medical haptic master, which has 6 DOFs controlled by MR actuators. This geometry can decouple rotational motion from translational motion, making this a simple structure with decoupled dynamics. The proposed mechanism and design process of medical haptic systems can be applied to several types of high-DOF mechanisms. Lastly, to demonstrate the effectiveness of the haptic system, the cutting surgery operations of the porcine specimens are performed by eliminating a tumor section of the specimens, which is marked by dyeing it in black color.

2. Hardening Sponge Featuring Magneto-Rheological (MR) Fluid

In this work, a novel tactile device featuring a hardening sponge is devised for RMIS. As a classical RMIS system cannot provide the viscoelastic properties of biological tissues or organs, the realization of tactile sensation can be helpful for surgeons, as shown in Figure 1. We introduce an MR fluid, sponge, and film into the proposed tactile device to mimic biological tissues or organs.

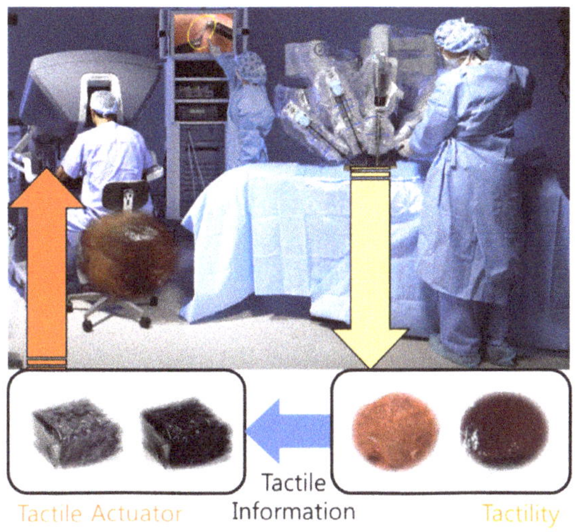

Figure 1. Concept of tactile sensation by the surgeon.

2.1. Force Measurement of Human-Like Tissues

When realizing the tactile sensation of organs or tissues, reaction force, texture, temperature, depth of impression, and weight can be considered. Among several indices, it is known that the reaction force between an organ and a human finger is an important factor for tactile sensation [18]. In this work, as the beginning research stage, only reaction force is selected as the tactile recognition index. As shown in Figure 2, a gantry-type robot with a force sensor is utilized to quantify the reaction force of organs and tissues. When the surgeon touches the organ, the operation speed is very slow. So, that process is considered to be the quasi-static process and the travel speed of end-effector is restricted to 0.4 mm/s in this study. For more information about the gantry robot, please refer to our previous study [24]. Two lead screws with step motors are placed in the vertical direction. A guide rail is horizontally placed between the lead screws. A timing belt, a step motor, and an end effector are mounted on the guide rail. Owing to this mechanism, the position of the end effector moves along the z-axis and x-axis directions. When the end effector compresses the specimen, palpation force is measured by a force sensor (ATI Corp., Nano17, Apex, NC, USA). The resolution of the force sensor is 0.0125 N. The maximum moving speed of the end effector is maintained low (0.4 mm/s) to consider only the quasi-static process. After the specimen is deformed up to 1 mm, the end effector is stopped and the force history is measured during 30 s. This is because a surgeon requires a specific time to feel unknown objects. As shown in Figure 3, the porcine heart and liver are selected as specimens. It should be noted that the features of porcine organs are largely similar to those of human tissues. The dimensions of the specimens are $25 \times 25 \times 10$ mm^3 (width × height × thickness). Generally, the stiffness and damping of deformable objects are mainly determined by the reaction force along the normal direction [20]. Thus, the end effector with the force sensor touches the specimen perpendicularly.

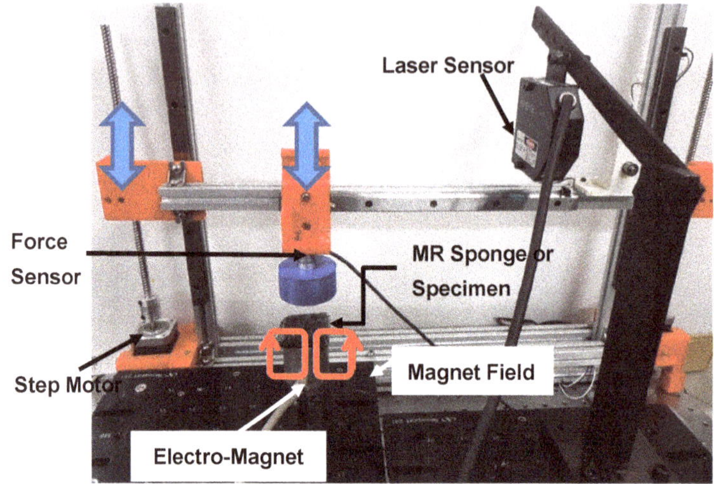

Figure 2. Photograph of force measurement system.

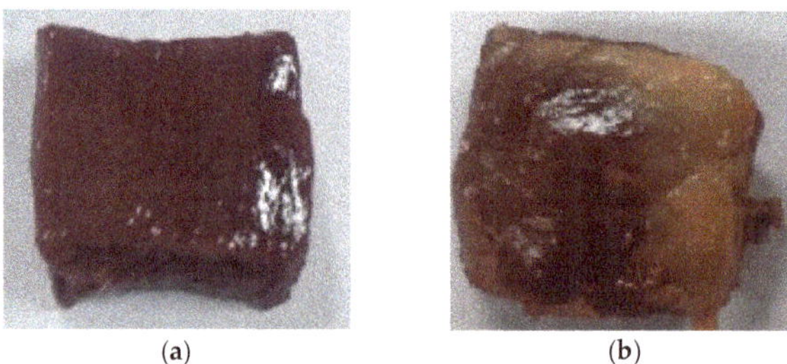

(a)   (b)

Figure 3. The specimens tested in this work; (**a**) porcine liver; (**b**) porcine heart.

Figure 4 shows the measured results of the two specimens. Reaction force increases until 2.5 s and then decreases exponentially. As mentioned earlier, the displacement of the end effector is 1 mm and travel time is 2.5 s. If the specimen is created from a purely elastic material, reaction force is proportional to deformation length. Thus, the behavior of a purely elastic material cannot represent the reduction in reaction force. In addition, the reaction force of a purely viscous material is almost zero for constant deformation. Accordingly, it is considered that the physical features of the specimens are similar to that of a viscoelastic material. It is known that viscoelastic materials show viscous and elastic characteristics [25]. It is observed from Figure 4 that the stiffness of the porcine heart is the largest based on the magnitude of reaction force. Additionally, the rate of reduction in force and the gap between the maximum and equilibrium values are different for each specimen.

2.2. Material Property of Sponge and MR Fluid

In order to mimic the tactile feeling of an organ, a material that has similar viscoelastic material properties should be used. Sponge is generally known as a viscoelastic material but the material property is constant. Since human organs and tissues have diverse reaction force characteristics as mentioned above, several sponge materials are required for surgery. During RMIS, it is very difficult to change the sponge according to the organ or tissue. Accordingly, a hardening sponge device featuring MR fluid is devised to realize several force characteristics. The phase of the MR fluid which consists of micron-sized iron particles and carrier fluid can be changed under a magnetic field. This phenomenon is due to the polarization generated in the iron particles. Owing to the chain structures, the reaction force of the MR sponge is tuned according to the intensity of the magnetic field.

Open-celled polyurethane (PU) foam is selected as the material of the sponge. Foam is a cellular structure and open-celled foams consist of numerous pores in an interconnected network. The foam has 25 pores per inch (ppi) and the reaction force of the polyurethane foam is measured with the same test conditions and gantry robot system. Figure 5a shows the measured force of the sponge only. The maximum force is 1.68 N which is smaller than that of heart and its force characteristic shows the viscoelastic behavior. From the test results, it is known that the open-celled PU foam can be classified as a soft viscoelastic material.

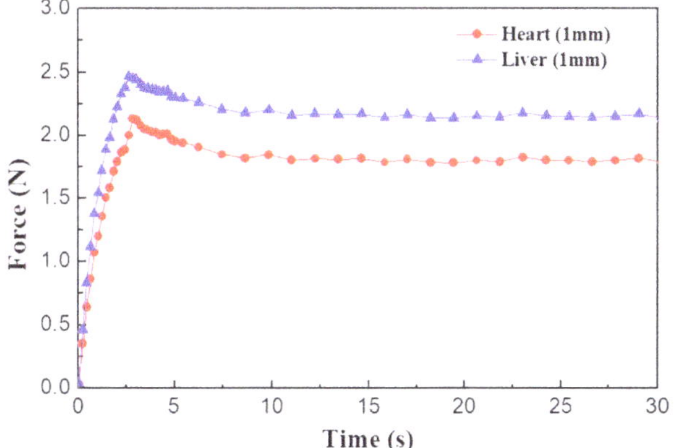

Figure 4. Force levels of two specimens.

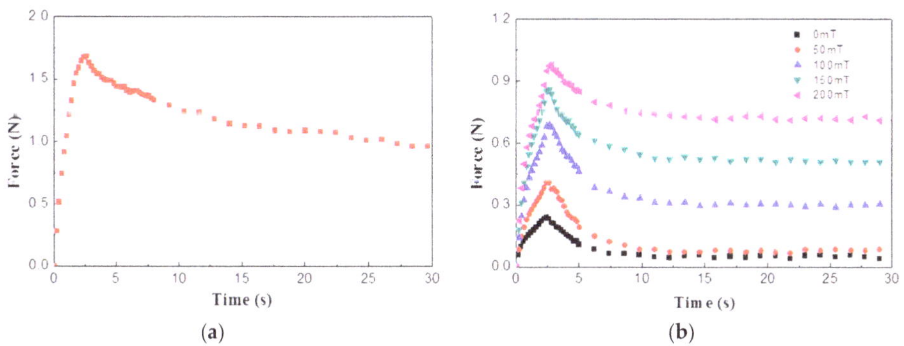

Figure 5. Measured reaction force results; (**a**) polyurethane foam; (**b**) magneto-rheological (MR) fluid.

MRF-132DG fluid provided by Lord Corp. (Cary, NC, USA) is used in this study and its reaction force during squeeze mode is measured. Figure 5b shows the measured force with several magnetic field inputs and deformations. Since the remaining force value is almost zero without magnetic field input, the behavior of MR fluid is similar to that of viscous material. However, the behavior of MR fluid is changed to that of viscoelastic material according to the magnetic field input. It is generally known that MR fluid behaves like a Newtonian fluid without a magnetic field. If a magnetic field is applied to an MR fluid, the iron particles are aligned with the direction of the magnetic field and form chain structures. It is inferred from test results that the chain structure of MR fluid is related to the measured elastic force characteristics.

2.3. Fabrication of MR Sponge

In order to take advantages of MR fluid and polyurethane foam, an MR sponge device is devised. Since, the MR fluid sinks into the polyurethane foam, the mechanical configuration of the MR sponge can be illustrated by spring and dashpot elements as shown in Figure 6. The dashpot and spring mean the viscous and elastic components of the material, respectively. k is the spring coefficient of the each materials and c is its damping coefficient. The MR fluid and polyurethane foam are connected in parallel. From the mechanical model, the following equations are derived:

$$\begin{aligned}
F_1 &= k_1 x_1 = c_2 \dot{x}_2 = F_2 \\
F_e &= k_e x_{total} \\
x_{total} &= x_1 + x_2 \\
F_{total} &= (F_e + F_1)_{MR} + (F_e + F_1)_{sponge}
\end{aligned} \tag{1}$$

where x_1, x_2 are deformed displacements of spring and damping elements in series connection. F_1 and F_2 are spring and damping forces in series connection. The magnitudes of the two forces are the same due to the law of action and reaction. From Equation (1), it can be inferred that the total reaction force of the MR sponge device is expected to be the sum of reaction forces of MR fluid and polyurethane foam. By using the proposed force model, the MR sponge device can be designed to meet the required force magnitudes. It is noted that the spring and damping constants in Equation (1) are not obtained in this work. From the measured reaction force results in Figure 4, the required peak force range is 2–2.5 N.

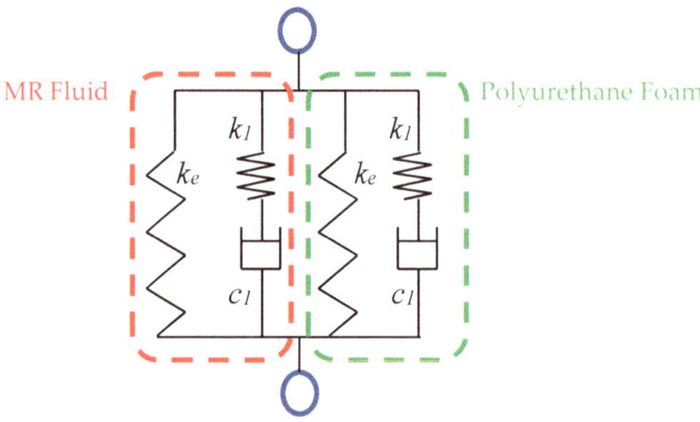

Figure 6. Mechanical model of MR sponge device.

The size of the polyurethane foam is $25 \times 25 \times 10$ mm^3. A slippery film clings to the sponge to prevent leakage of the MR fluid. Figure 7 shows the components and assembly of the single MR sponge cell. After the MR sponge is deformed, its original shape is recovered owing to its elasticity. During deflection and restoration, the repulsive force between the MR sponge and a surgeon is controlled by a magnetic field. Based on this mechanism of the MR sponge, the surgeon can distinguish between the tactile sensations of various cells and biological organs. Because the surgeon remotely manipulates the slave robot via the haptic master, the mechanism of the tactile device is extremely helpful to the surgeon in accomplishing accurate surgery.

The reaction force measurement test of the MR sponge is conducted to mimic these features of the specimens. An electromagnet (JL-4A, JL Magnet Corp., Seoul, Korea) is used to apply magnetic fields of several intensities to the MR sponge. The reaction force of the MR sponge can be tuned because the MR fluid is affected by the magnetic field. As shown in Figure 8, the maximum reaction force increases according to the magnetic field. It can be inferred that the MR sponge is a viscoelastic material and the reaction force curves can be controlled by magnetic inputs. Hence, the devised MR tactile device can realize a wide range of the tactile sensations of organs or tissues. For instance, when a magnetic field of 50 mT is applied to the MR sponge, then the reaction force curve is almost similar to that of the porcine liver. Also, the predicted maximum forces according to magnetic field inputs are 1.91 N, 2.08 N, 2.35 N, 2.52 N, and 2.65 N, respectively. The mean error percentage between measured and predicted maximum forces is 1.15%. These results show the effectiveness of the proposed force model. But this model is only valid for transient response such as peak force. In order to predict the force characteristics in a steady state response, the time constant and compression region should be considered. For more detailed information, please refer to our previous study [24].

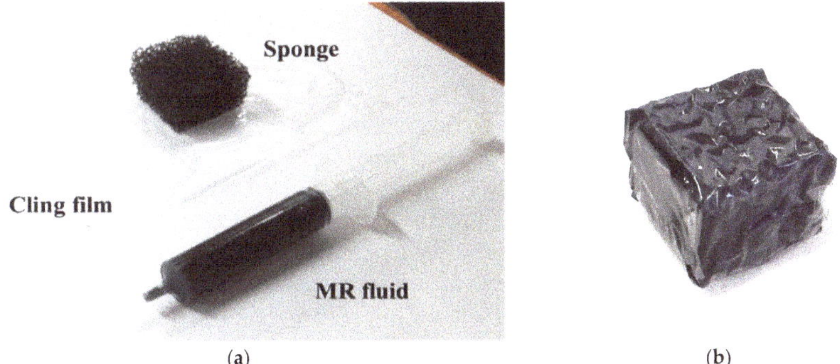

Figure 7. MR sponge tactile device; (a) components; (b) assembly.

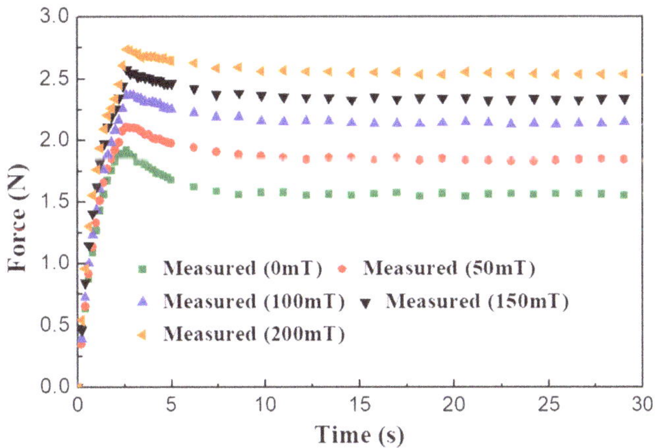

Figure 8. The field-dependent force of MR sponge.

3. 6 Degrees-of-Freedom (DOF) Haptic Master with MR Actuators

The shear stress of MRF-132DG can be expressed based on the Bingham model:

$$\tau = \eta\dot{\gamma} + \tau_y(B) \tag{2}$$

where η and γ are viscosity constant and shear rate, respectively. The nominal value of the viscosity constant is 0.092. $\tau_y(B)$ is yield stress of MR fluid which is changed according to the magnitude of the magnetic field, B. Lord Corp. presents the yield stress properties of MRF-132DG [26]. When MRF-132DG is shearing between two surfaces, the shear stress is generated and measured. The MR fluid's particles align with the direction of the magnetic field, thereby restricting the fluid's rotational motion within the gap in proportion to the strength of the magnetic field. From Figure 9, the relation between yield stress and magnetic field is obtained by using the curve-fitting method.

$$\tau_y(B) = m_{MR0}B^5 + m_{MR1}B^4 + m_{MR2}B^3 + m_{MR3}B^2 + m_{MR4}B^1 + m_{MR5}$$
$$\textit{where}$$
$$m_{MRi} = \begin{bmatrix} -4.1319 \times 10^{-11} & 4.1437 \times 10^{-08} & -1.39 \times 10^{-05} & 1.1 \times 10^{-03} & 2.886 \times 10^{-01} & -1.178 \times 10^{-01} \end{bmatrix}$$

$$(3)$$

It is noted that the above equation for the field-dependent property of MR fluid is to be used to determine appropriate dimensions of the haptic master.

It is generally known that RMIS requires 6-DOF motions (3-DOF surgical instrument motions and 3-DOF end effector motions) [14,15]. A 6-DOF MR haptic master is proposed to realize 6-DOF surgical motions and reaction force/torque, as shown in Figure 10. When a surgeon grips and manipulates the handle, the handle's 6-DOF motion command is transferred to the surgical robot, and the reaction force between the surgical instrument and organ/tissue should be provided by the haptic master. Thus, the MR actuator (clutch/brake) for each motion is required to supply reaction force to the surgeon, and several MR actuators are integrated with the motion mechanism of the haptic master. Based on the parallel robot mechanism, 3-DOF translational motions are realized using a moving platform, 6 links, and rotary/universal joints [27]. The handle and moving platform are connected to each other. Hence, when the handle moves along the X, Y, and Z axes, the moving platform and links are moved but the MR clutch rotates without moving. It is noted that the torque induced from the MR clutch is transferred to the surgeon as reaction force via the handle. In addition, 3-DOF rotational motions (pitching, rolling, and yawing) and reaction torque are realized at the handle. When the handle is rotated, the MR brake generates reaction torque against rotational motion. An additional counter mass is attached to the other side of the MR brake. The purpose of the counter mass is to prevent the handle from rotating owing to its mass. Figure 11 shows the configuration of the MR brake, which mainly consists of an inner rotor, an MR fluid, a coil, and an outer casing. When the inner rotor rotates, fluid friction, which is determined by the shear stress of the MR fluid, is induced. Because this fluid friction is converted to reaction torque, the shear stress of the MR fluid can determine the reaction torque generated by the MR brake.

The reaction torque of the MR brake can be expressed as follows:

$$
\begin{aligned}
T_b &= T_c + T_\eta + T_f \\
&= \left[2\frac{\pi D_{b2}{}^2 h_b \tau_y(B)}{2} + \frac{\pi\left(D_{b2}{}^3 - D_{b1}{}^3\right)\tau_y(B)}{12} \right] \\
&\quad + \left[2\frac{\pi\eta D_{b2}{}^3 h_b |\dot{\gamma}|}{4t_b} + \frac{\pi\eta\left[\left(\frac{D_{b2}}{2}\right)^4 - \left(\frac{D_{b1}}{2}\right)^4\right]|\dot{\gamma}|}{2t_b} \right] + T_f
\end{aligned}
$$

$$(4)$$

where, T_c is the controllable torque induced from yield stress, T_η is the fluid friction torque induced from viscosity of the MR fluid. T_f is mechanical friction torque induced from oil seal. Contrary to the MR brake mechanism, the MR clutch mechanism requires a driving actuator to generate reaction torque. Thus, two shafts and two rotors are connected to each other. When one shaft rotates, one rotor rotates and the torque of the one rotor is transferred to the outer housing via the MR fluid. Thus, if the rotational directions of the shafts are different, then the rotational direction of the rotor is different. In addition, the rotational direction of the outer housing is determined by the difference between the torques transmitted from each rotor. Similar to the mechanism of the MR brake, the magnitude of transmitted torque is determined by the shear stress of the MR fluid. Based on the configuration of the MR clutch, reaction torque can be expressed as follows:

$$T_{total} = \left| \vec{T}_1 - \vec{T}_2 \right|, \; T_i = T_{ci} + T_{\eta i} + T_{fi}, \; i = 1,2 \tag{5}$$

where T_1 and T_2 are the generated torque between each rotor and outer housing. From Equation (5) and configuration of the MR clutch, the total torque model is derived as follows:

$$T_i = 2\pi \left(\frac{D_{bR}}{2} \right)^2 \int_0^{h_b} \tau \, dz + 2\pi \int_{\frac{D_{bs2}}{2}}^{\frac{D_{bR}}{2}} r^2 \tau dr + T_f \tag{6}$$

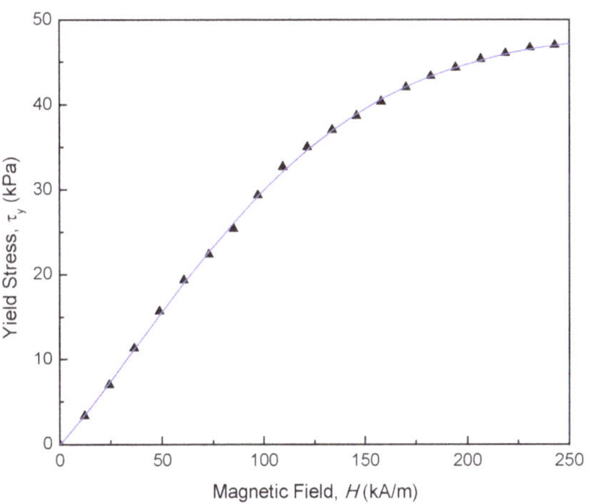

Figure 9. MRF-132 DG τ-H curve.

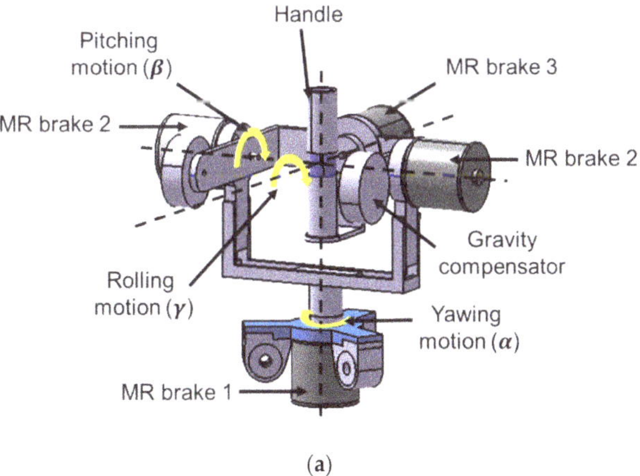

Figure 10. *Cont.*

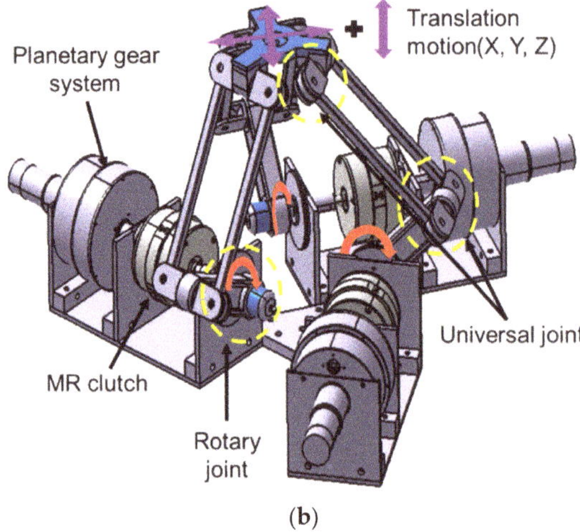

Figure 10. Overall configurations of the haptic master; (**a**) rotational part; (**b**) translational part.

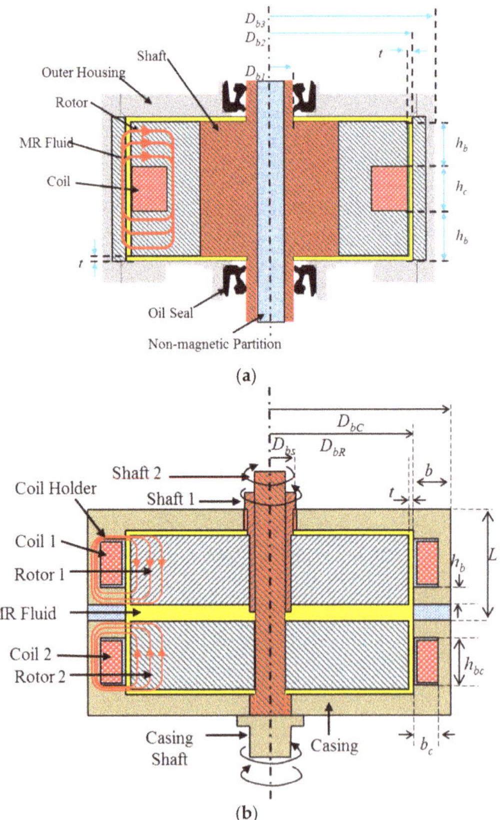

Figure 11. Internal configuration of MR devices; (**a**) MR brake; (**b**) MR clutch.

It is noted here that all variables are denoted in Figure 11. On the other hand, a DC motor and a planetary gear system are employed to supply driving power to the MR clutch. When the DC motor rotates along one direction, rotation in the opposite direction is generated via the planetary gear mechanism. The two output shafts are connected with each shaft of the MR clutch. If the MR brake is replaced with the MR clutch and planetary gear, then the weight becomes extremely high. When the reaction force along the upper direction is necessary, a higher mass of the handle requires a large MR clutch. The MR brake mechanism is applied at the handle to obtain a compact size of the MR haptic master system. Based on multiple studies [28,29], the maximum reaction torque and force were set at 0.5 Nm and 12 N, respectively. The reaction torque of the MR clutch is transformed to the reaction force along the translation motions at the handle. Hence, the objective torques of the MR brake and clutch are set to be 0.75 and 2.5 Nm, respectively. The design parameters calculated based on the generated torque model are listed in Table 1. Please refer to our previous study [19,30] for more details about the process. The diameters of the outer housing for the MR clutch and brake are 50 mm and 34 mm, respectively. Finally, the MR haptic master is manufactured as shown in Figure 12. In addition, an encoder (E40H; Autonics, Incheon, Korea) and an IMU sensor (MPU-9250, InvenSense, Sunnyvale, CA, USA) are used to measure position information; a 6-axis force/torque sensor and a 1-DOF torque sensor (SDS-100; Senstech Corp, Busan, Korea) are used for torque and force measurement.

Table 1. Design parameters of the haptic master actuators.

MR Clutch			MR Brake		
Parameter	Explanation	Value	Parameter	Explanation	Value
D_{bR}	Diameter of the MR clutch rotor	50 mm	D_{b1}	Internal diameter of the MR brake rotors	4 mm
D_{bc}	Diameter of the MR clutch's outer housing	61 mm	D_{b2}	External diameter of the MR brake rotors	34 mm
D_{bs}	Diameter of the MR clutch's shaft	10 mm	h_b	Height of the rotor	7 mm
t	Gap of the MR fluid	1 mm	h_c	Height of the coil	10 mm

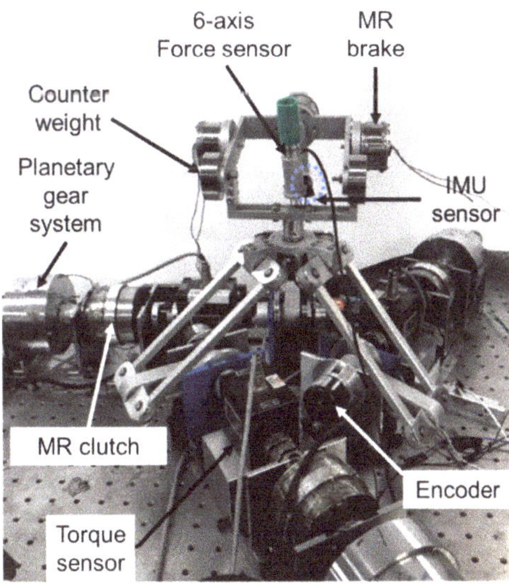

Figure 12. Photograph of the haptic master system featuring MR clutches and brakes.

4. Robot Surgery Experiment

4.1. Integration of Haptic Master and Slave Robot

As mentioned earlier, the slave robot implements the surgical command transferred from the master. The mechanism of the 6-DOF slave robot is proposed as shown in Figure 13. The proposed slave robot consists of three robot arms, servomotors, and a surgical instrument. Because the surgical instrument is inserted into a small incision, the instrument is attached to the end point of the third robot arm. In addition, a surgical tool such as forceps or a knife is placed at the opposite end of the instrument. As the wire and motor actuator are connected with the surgical tool, the surgical tool can rotate in three directions, i.e., rolling, pitching, and spinning. Each robot arm can rotate using the servomotor, and the end position of the third arm moves along the X, Y, and Z axes. The position of the end point is determined according to the translation motion of the moving platform of the haptic master. Based on the forward kinematics of the 3-DOF manipulator, the position of the moving platform, P_{mx}, P_{my}, P_{mz}, can be expressed as follows [27]:

$$
\begin{aligned}
P_{mx} &= \tfrac{1}{h_{22}h_{33}-h_{23}h_{32}}\left((h_{32}h_{43}-h_{33}h_{42})P_{mz}+h_{23}h_{42}-h_{22}h_{43}\right)\\
P_{my} &= \tfrac{1}{h_{22}h_{33}-h_{23}h_{32}}\left((h_{13}h_{32}-h_{12}h_{33})P_{mz}+h_{12}h_{23}-h_{13}h_{22}\right)\\
P_{mz} &= l_1\sin\theta_{11}+l_4\sin\theta_{21}=l_2\sin\theta_{12}+l_5\sin\theta_{22}=l_3\sin\theta_{13}+l_6\sin\theta_{23}\\
h_{1j} &= -2\cos\phi_j(l_1\cos\theta_{1j}+a-b)+2\cos\phi_j(l_2\cos\theta_{2j}+a-b)\\
h_{2j} &= -2\cos\phi_j(l_1\cos\theta_{1j}+a-b)+2\sin\phi_j(l_2\cos\theta_{2j}+a-b)\\
h_{3j} &= -2l_1\sin\theta_{1j}+2l_2\sin\theta_{2j}\\
h_{4j} &= (l_1\cos\theta_{1j}+a-b)^2+l_1{}^2\sin^2\theta_{1j}-(l_2\cos\theta_{2j}+a-b)^2-l_2{}^2\sin^2\theta_{2j}\\
\text{where} \quad & j=1,2,3
\end{aligned}
\tag{7}
$$

In Equation (7), l is the length of the 6 lower legs attached to the moving platform. ϕ_j is the angle of each lower leg on the XY plane. θ_{1j}, θ_{2j} are the angles between each lower leg. a is the displacement between the center and each lower leg, and b is the radius of the moving platform. When the position of the moving platform is measured by the encoder, the rotation angle of the servomotor for the slave robot is obtained based on inverse kinematics. In addition, the rotation information of the haptic master handle along the pitching, rolling, and spinning directions is directly transferred to the surgical tool part of the slave robot. Based on the transferred rotation command, the surgical tool is rotated using the wire actuator. Please refer to the details about the transformation in [31,32]. The rotational information of the handle is measured using the inertial measurement unit (MPU-9250).

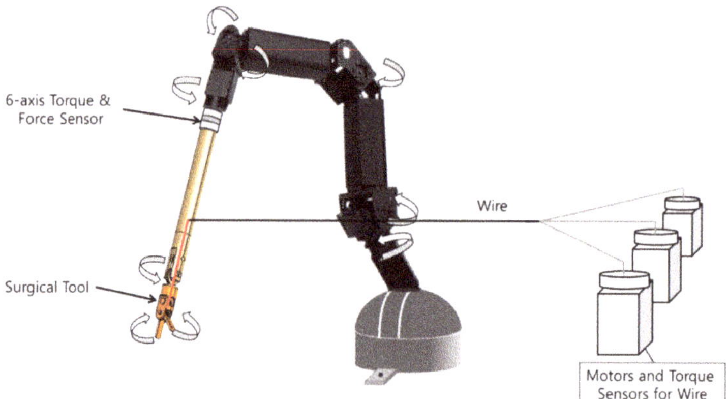

Figure 13. Overall configuration of the slave robot.

4.2. Experimental Results and Discussion

A tumor cutting experiment was performed to evaluate the performance of the proposed system. As it is extremely important to clearly distinguish between a tumor and normal tissue, an efficient method, such as haptic feedback and tactile sensation, is required. Two testing conditions are adopted to maximize the performance of the MR tactile device. In the first condition, participants cut tumors using only the haptic feedback from the MR haptic master, whereas in the second condition, they use the haptic feedback and tactile sensation from the MR tactile device. Figure 14 shows the experimental apparatus of the overall cutting surgery system, which is integrated with the MR haptic master, MR tactile device, and slave robot. When a participant manipulates the MR haptic master, translational and rotational commands are transferred to the slave robot via data acquisition boards (PXIe-6363, NI Corp., Austin, TX, USA) and A/D and D/A boards (PXIe-1082, NI Corp., Austin, TX, USA). The torques induced from the MR clutch are measured by the torque sensors (SDS-100). While implementing surgical tasks according to the commands, the slave robot transmits the reaction force between the specimen and surgical tool. This force is measured by the force sensor (Nano 17). The movement of the haptic master is measured by the encoders (E40H) and an IMU. Then, the desired angle for the slave robot is obtained based on Equation (4) and inverse kinematics. The servo motor of the slave robot rotates according to the calculated command and measured actual rotation angle. In order to reduce the position tracking error, a simple proportional-integral-derivative (PID) controller is implemented.

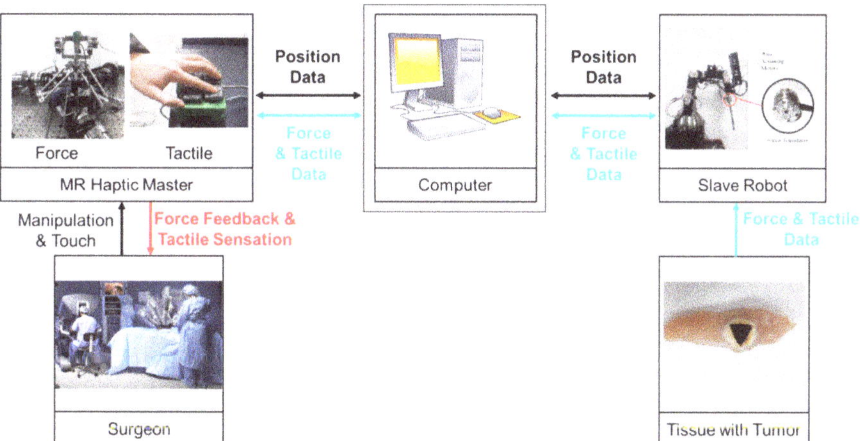

Figure 14. Experimental configuration for the robot surgery integrated with MR haptic master and tactile device.

Prior to cutting surgery, the tracking control of the end-point of the slave robot by the haptic master is evaluated. The maximum error during the experiment is 4.02 mm. Based on the error results, the position tracking controller shows excellent performance. While implementing the surgical tasks according to the commands, the slave robot transmits the reaction force between specimen and surgical tool, which is measured by the force sensor. This force sensor is attached at the end point of the slave robot arm. The generated torque/force induced from the MR haptic master are measured by torque sensors and 6-axis torque/force sensor. In order to realize the measured force of the specimen, the torque/force tracking controller featuring the PID controller is proposed. Also, the input magnetic field for the MR tactile device is determined based on the relation between the reaction force and magnetic field.

In order to undertake simple cutting surgery, a tumor is marked by dyed black in the specimen, as shown in Figure 15. The objective of the cutting experiment is to remove the black tumor from the normal tissue. It is generally known that the stiffness of the tumor and normal tissue are different. In order to emulate this, the forces of the porcine liver and heart tissues are used to replace the tumor and normal tissues in this experiment, respectively. When a magnetic field of 50 mT and 100 mT are applied to the MR sponge, then the reaction force curve is almost similar to that of the porcine liver and heart. During the cutting experiment, the haptic master generates the reaction torque/force from the MR brake/clutch by applying a magnetic field as input. At the same time, the position tracking controller of the slave robot is implemented, followed by a certain surgical operation. The operator feels the reaction force through the solidification of the MR fluid by the MR haptic master. In addition, the MR tactile device mimics the stiffness of the surgery specimen using the MR fluid effect. So, the operator can distinguish the tumor and normal tissues precisely. It can be clearly seen from Figure 15 that the use of the tactile device can increase surgical accuracy since the operator (surgeon) can feel the same stiffness of the surgical tissues from the proposed MR sponge tactile device. During cutting surgery with the haptic feedback force and the tactile sensation, the torque/force tracking control results of the MR actuators are measured and plotted, as shown in Figure 16. From the control results along three translation directions and 3 rotational directions, it can be clearly observed that the accuracy of the proposed control haptic master system is acceptable. Note that 20 specimens are used for the removal of the tumor in this surgical cutting experiment.

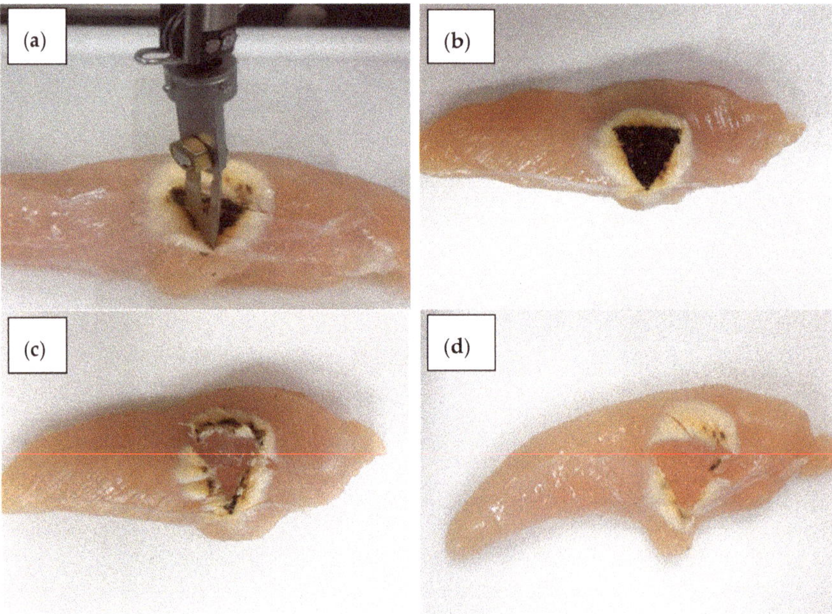

Figure 15. Surgical results; (**a**) cut operation; (**b**) tumor mark; (**c**) cut result with the haptic force only; (**d**) cut result with the haptic force and tactile sensation.

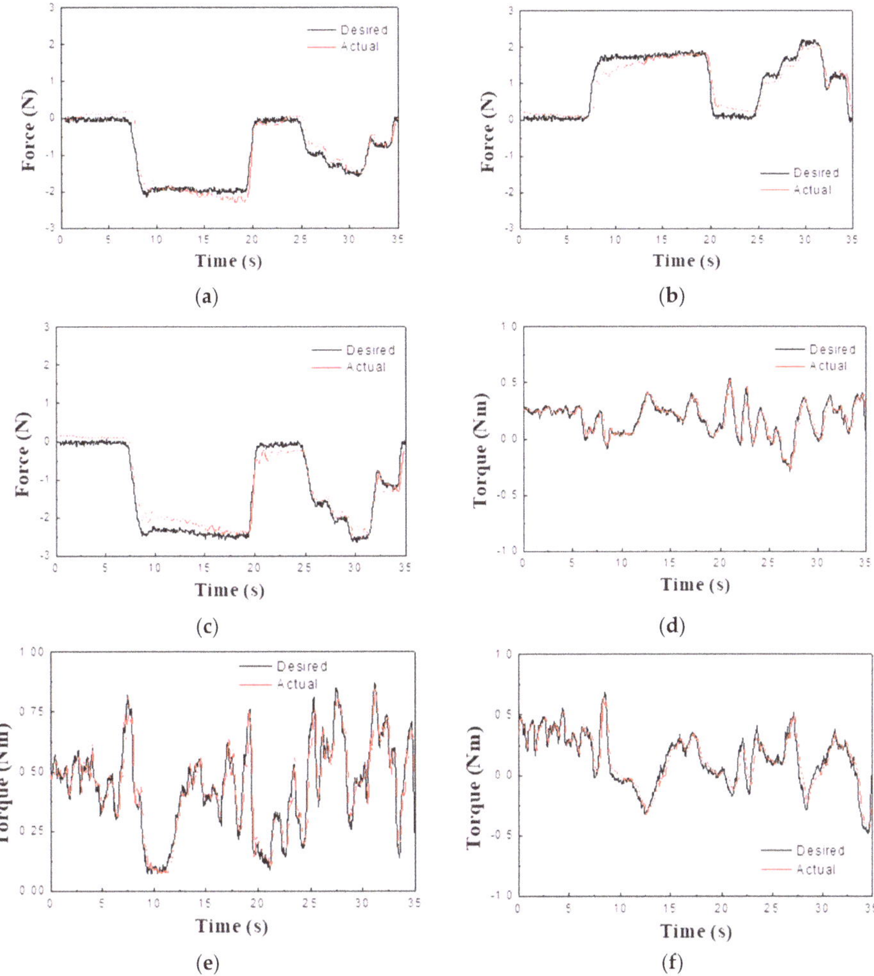

Figure 16. Force/torque tracking control results of the haptic master; (**a**) *x*-axis force; (**b**) *y*-axis force, (**c**) *z*-axis force; (**d**) pitching torque; (**e**) rolling torque; (**f**) spinning torque.

Since the usefulness of the tactile device cannot be measured simply, a psychological test is implemented to evaluate the effectiveness of the proposed tactile device for cutting surgery. Before cutting the tumor, the stiffness of tissues is measured via the force sensor of the slave robot. Then, the stiffness of the touched tissue is realized by the proposed tactile device and the participants are requested to touch the MR sponge and rate the following question on a five point scale.

- Question: Can you distinguish the different stiffness represented by the tactile device?
- Answer: (Negative) 1 ——— 2 ——— 3 (Mean) ——— 4 ——— 5 (Positive)

The number of participants was 30 and the age of participants ranged from 20 to 30. The mean value of answers was 4.5 and the standard deviation was 0.56. This result indicates that the proposed tactile device is very helpful in distinguishing the different stiffness of human organ/tissue for robot-assisted surgery.

5. Conclusions

In this work, novel types of smart devices are devised based on the control ability of MR fluid. As for the haptic master, MR clutches and MR brakes are used as actuators for the three translational motions and the three rotational motions, respectively. On the other hand, as for the tactile device, an MR sponge which can emulate the similar stiffness and/or damping property of human organs is devised. The overall network for the robotic surgery has been established by integrating the haptic master, MR sponge tactile device, slave robot and microprocessor including the PID control algorithm. Then, in order to achieve interactive motion between the command from the master followed by the slave robot, kinematic analysis was performed. As a final step, a tumor removal experiment using the porcine specimens was performed to demonstrate the superior performance of the proposed robot-assisted surgical system. It has been demonstrated via the experiment that the addition of the tactile device from which the operator (surgeon) can feel the same stiffness of the surgical object can enhance surgical accuracy. This has also been confirmed by a psychological test. Finally, it can be remarked that proposed haptic master and tactile device showed superior control performance due to the MR fluid and also the other control device can be designed based on the control ability of the MR fluid.

Author Contributions: Investigation and writing—original draft, J.S.O.; conceptualization and resources, J.W.S.; project administration and writing—review and editing, S.-B.C.

Funding: This work was supported by the National Research Foundation of Korea(NRF) grant funded by the Korea government(MSIP; Ministry of Science, ICT & Future Planning) (No. 2017R1C1B5018204).

Conflicts of Interest: The authors declare no conflict of interest.

References

1. Pierrot, F.; Dombre, E.; Dégoulange, E.; Urbain, L.; Caron, P.; Boudet, S.; Mégnien, J.L. Hippocrate: A safe robot arm for medical applications with force feedback. *Med. Image Anal.* **1999**, *3*, 285–300. [CrossRef]
2. Nguyen, Q.H.; Choi, S.B. Optimal design of a vehicle magnetorheological damper considering the damping force and dynamic range. *Smart Mater. Struct.* **2008**, *18*, 015013. [CrossRef]
3. Nguyen, Q.H.; Choi, S.B. Optimal design of an automotive magnetorheological brake considering geometric dimensions and zero-field friction heat. *Smart Mater. Struct.* **2010**, *19*, 115024. [CrossRef]
4. Shafer, A.S.; Kermani, M.R. Design and validation of a magneto-rheological clutch for practical control applications in human-friendly manipulation. In Proceedings of the 2011 IEEE International Conference on Robotics and Automation (ICRA), Shanghai, China, 9–13 May 2011; pp. 4266–4271.
5. Senkal, D.; Gurocak, H. Compact MR-brake with serpentine flux path for haptics applications. In Proceedings of the World Haptics 2009—Third Joint EuroHaptics Conference and Symposium on Haptic Interfaces for Virtual Environment and Teleoperator Systems, Salt Lake City, UT, USA, 18–20 March 2009; pp. 91–96.
6. Nguyen, P.B.; Choi, S.B. A new approach to magnetic circuit analysis and its application to the optimal design of a bi-directional magnetorheological brake. *Smart Mater. Struct.* **2011**, *20*, 125003. [CrossRef]
7. Tusset, A.M.; Janzen, F.C.; Piccirillo, V.; Rocha, R.T.; Balthazar, J.M.; Litak, G. On nonlinear dynamics of a parametrically excited pendulum using both active control and passive rotational (MR) damper. *J. Vib. Control* **2018**, *24*, 1587–1599. [CrossRef]
8. Ulasyar, A.; Lazoglu, I. Design and analysis of a new magneto rheological damper for washing machine. *J. Mech. Sci. Technol.* **2018**, *32*, 1549–1561. [CrossRef]
9. Li, W.H.; Liu, B.; Kosasih, P.B.; Zhang, X.Z. A 2-DOF MR actuator joystick for virtual reality applications. *Sens. Actuators A Phys.* **2007**, *137*, 308–320. [CrossRef]
10. Senkal, D.; Gurocak, H. Spherical brake with MR fluid as multi degree of freedom actuator for haptics. *J. Intell. Mater. Syst. Struct.* **2009**, *20*, 2149–2160. [CrossRef]
11. Ahmadkhanlou, F.; Washington, G.N.; Bechtel, S.E. Modeling and control of single and two degree of freedom magnetorheological fluid-based haptic systems for telerobotic surgery. *J. Intell. Mater. Syst. Struct.* **2009**, *20*, 1171–1186. [CrossRef]

12. Song, B.K.; Oh, J.S.; Choi, S.B. Design of a new 4-DOF haptic master featuring magnetorheological fluid. *Adv. Mech. Eng.* **2014**, *6*, 843498. [CrossRef]
13. Oh, J.S.; Choi, S.H.; Choi, S.B. Design of a 4-DOF MR haptic master for application to robot surgery: Virtual environment work. *Smart Mater. Struct.* **2014**, *23*, 095032. [CrossRef]
14. Qiu, T.; Hamel, W.R.; Lee, D. Design and control of a low cost 6 DOF master controller. In Proceedings of the 2014 IEEE International Conference on Robotics and Automation (ICRA), Hong Kong, China, 31 May–7 June 2014; pp. 5313–5318.
15. Chen, X.; Xin, X.; Zhao, B.; He, Y.; Hu, Y.; Liu, S. Design and analysis of a haptic master manipulator for minimally invasive surgery. In Proceedings of the 2017 IEEE International Conference on Information and Automation (ICIA), Macau, China, 18–20 July 2017; pp. 260–265.
16. Takei, M.; Shiraiwa, H.; Omata, S.; Motooka, N.; Mitamura, K.; Horie, T.; Ookubo, T.; Sawada, S. A new tactile skin sensor for measuring skin hardness in patients with systemic sclerosis and autoimmune Raynaud's phenomenon. *J. Int. Med. Res.* **2004**, *32*, 222–231. [CrossRef] [PubMed]
17. Zhang, L.; Ju, F.; Cao, Y.; Wang, Y.; Chen, B. A tactile sensor for measuring hardness of soft tissue with applications to minimally invasive surgery. *Sens. Actuators A Phys.* **2017**, *266*, 197–204. [CrossRef]
18. Han, Y.-M.; Oh, J.-S.; Kim, J.-K.; Choi, S.-B. Design and experimental evaluation of a tactile display featuring magnetorheological fluids. *Smart Mater. Struct.* **2014**, *23*, 77001. [CrossRef]
19. Oh, J.S.; Kim, J.K.; Lee, S.R.; Choi, S.B.; Song, B.K. Design of tactile device for medical application using magnetorheological fluid. *J. Phys. Conf. Ser.* **2013**, *412*, 12047. [CrossRef]
20. Kim, S.; Kim, P.; Park, C.-Y.; Choi, S.-B. A new tactile device using magneto-rheological sponge cells for medical applications: Experimental investigation. *Sens. Actuators A Phys.* **2016**, *239*, 61–69. [CrossRef]
21. Groenen, M.A.M.; Archibald, A.L.; Uenishi, H.; Tuggle, C.K.; Takeuchi, Y.; Rothschild, M.F.; Rogel-Gaillard, C.; Park, C.; Milan, D.; Megens, H.-J.; et al. Analyses of pig genomes provide insight into porcine demography and evolution. *Nature* **2012**, *491*, 393–398. [CrossRef] [PubMed]
22. Nava, A.; Mazza, E.; Furrer, M.; Villiger, P.; Reinhart, W.H. In vivo mechanical characterization of human liver. *Med. Image Anal.* **2008**, *12*, 203–216. [CrossRef] [PubMed]
23. Carter, F.J.; Frank, T.G.; Davies, P.J.; McLean, D.; Cuschieri, A. Measurements and modelling of the compliance of human and porcine organs. *Med. Image Anal.* **2001**, *5*, 231–236. [CrossRef]
24. Cha, S.W.; Kang, S.R.; Hwang, Y.H.; Oh, J.S.; Choi, S.B. A controllable tactile device for human-like tissue realization using smart magneto-rheological fluids: Fabrication and modeling. *Smart Mater. Struct.* **2018**, *27*, 065015. [CrossRef]
25. Bland, D.R. *The Theory of Linear Viscoelasticity*; Courier Dover Publications: New York, NY, USA, 2016.
26. MRF-132DG. Available online: http://www.lordmrstore.com/lord-mr-products/mrf-132dg-magneto-rheological-fluid (accessed on 22 July 2018).
27. Tsai, L.W.; Walsh, G.C.; Stamper, R.E. Kinematics of a novel three DOF translational platform. In Proceedings of the IEEE International Conference on Robotics and Automation, Minneapolis, MN, USA, 22–28 April 1996; pp. 3446–3451.
28. Tavakoli, M. *Haptics for Teleoperated Surgical Robotic Systems*; World Scientific: Singapore, 2008; Volume 1.
29. Kim, P.; Kim, S.; Park, Y.D.; Choi, S.B. Force modeling for incisions into various tissues with MRF haptic master. *Smart Mater. Struct.* **2016**, *25*, 035008. [CrossRef]
30. Han, Y.M.; Oh, J.S.; Kim, S.; Choi, S.B. Design of multi-degree motion haptic mechanisms using smart fluid-based devices. *Mech. Based Des. Struct. Mach.* **2017**, *45*, 135–144. [CrossRef]
31. Kang, S.R. Design and Control of 6-DOF Haptic Master Using MR Fluid for Robot Surgery. Master's Thesis, Inha University, Incheon, Korea, 2018.
32. Hwang, Y.H. Design of 7-DOF Slave Robot Integrated with Magneto-Rheological Haptic Master. Master's Thesis, Department of Mechanical Engineering, Inha University, Incheon, Korea, 2018.

 *materials*

Article

Electroactive Hydrogels Made with Polyvinyl Alcohol/Cellulose Nanocrystals

Tippabattini Jayaramudu [1,2], **Hyun-U Ko** [1], **Hyun Chan Kim** [1], **Jung Woong Kim** [1], **Ruth M. Muthoka** [1] **and Jaehwan Kim** [1,*]

1   Center for Nanocellulose Future Composites, Department of Mechanical Engineering, Inha University, 100 Inha-Ro, Nam-Gu, Incheon 22212, Korea; mr.jayaramudu@gmail.com (T.J.); lostmago@naver.com (H.-U.K.); Kim_HyunChan@naver.com (H.C.K.); jw6294@naver.com (J.W.K.); mwongelinruth@gmail.com (R.M.M.)
2   Laboratory of Material Sciences, Instituto de Quimica de Recursos Naturales, Universidad de Talca, Talca 747, Chile
*   Correspondence: jaehwan@inha.ac.kr; Tel.: +82-32-831-7325

Received: 28 May 2018; Accepted: 3 September 2018; Published: 4 September 2018

Abstract: This paper reports a nontoxic, soft and electroactive hydrogel made with polyvinyl alcohol (PVA) and cellulose nanocrystal (CNC). The CNC incorporating PVA-CNC hydrogels were prepared using a freeze–thaw technique with different CNC concentrations. Fourier transform infrared spectroscopy, thermogravimetric analysis, X-ray diffraction and scanning electron microscopy results proved the good miscibility of CNCs with PVA. The optical transparency, water uptake capacity and mechanical properties of the prepared hydrogels were investigated in this study. The CNC incorporating PVA-CNC hydrogels showed improved displacement output in the presence of an electric field and the displacement increased with an increase in the CNC concentration. The possible actuation mechanism was an electrostatic effect and the displacement improvement of the hydrogel associated with its enhanced dielectric properties and softness. Since the prepared PVA-CNC hydrogel is nontoxic and electroactive, it can be used for biomimetic soft robots, actively reconfigurable lenses and active drug-release applications.

Keywords: electroactive hydrogel; polyvinyl alcohol; cellulose nanocrystals; freeze–thaw method; actuation

---

1. Introduction

Hydrogels are hydrophilic three-dimensional network structures that are cross-linked physically or chemically and which maintain their structural integrity during formation [1]. They can hold large amounts of water molecules/biological solutions, which turn them into soft and viscoelastic materials. The soft, flexible, elastic and wet features of the hydrogels promote them as potential candidates for various biomedical and pharmaceutical applications including diapers, contact lenses, membranes, tissue engineering, drug delivery systems and biosensors [2–5]. Stimuli–response hydrogels change their structure (especially volume and shape) due to such conditions as pH, ionic strength, temperature and electric field [6,7]. Several studies revealed that acrylic acid and its polymers, as well as other hydrogels based on polymeric materials, are electric or pH responsive. However, acrylic acid is known to be toxic in nature [8]. This toxicity problem can be overcome by blending or reinforcing natural polymers into synthetic polymers. Accordingly, natural polymer-based hydrogels can show stimuli-responsive behavior, resulting in their high number of potential applications including biomimetic soft robots, haptic actuators, artificial muscles, active tunable lenses and active drug release. Thus far, many natural polymers have been used to develop hydrogels such as chitosan, cellulose, whey protein and carboxymethyl cellulose [9–13]. Among them, cellulose has merits in terms of renewability, biocompatibility, abundance, low price, superior mechanical properties and easy chemical modification.

Cellulose consists of crystal and amorphous parts connected in a row. Cellulose nanocrystal (CNC), a rod-like shaped nanocrystal, can be isolated from cellulose resources including wood pulp, tunicates, bacterial cellulose, cotton, ramie, hemp as well as other agricultural residues by treating them with acid hydrolysis [14–16]. CNC has a high degree of crystallinity, mechanical properties and a specific surface area [17,18]. The typical width of CNCs is in the range of 5–50 nm, but their length and width depend on the source and the process conditions. CNC produced by sulfuric acid hydrolysis is electrostatically stable and easily dispersed in polar aqueous suspensions due to the sulfate ester groups on their surfaces [19–22]. Based on their attractive characteristics, CNCs have been used as reinforcing agents for a wide range of applications in packaging films, nanocomposites, microchips, tissue engineering, actuators and sensors [23–27].

In hydrogels, reinforcement technology is playing a key role [21,26–29]. Cellulose can easily interact with various polar and water-soluble polymer materials. Thus, blending of CNC with hydrogels can reinforce the hydrogels in terms of mechanical properties and electromechanical properties. Especially the integration of CNC in hydrogels can increase their dielectric constant so as to improve its electroactive properties. With this background, this study aims to improve the transparency and electroactive properties of hydrogels by incorporating CNC into polyvinyl alcohol (PVA) to develop nontoxic electroactive hydrogels. PVA hydrogel was reported as an electroactive material [30] and showed higher transparency when the hydrogels were prepared using the solvent mixture of dimethyl sulfoxide (DMSO) and water (80:20 wt.%) [31]. PVA and CNC are known to be nontoxic. The basic physical properties of the prepared hydrogels including the swelling behavior, transparency and surface morphology were investigated using the water uptake capacity test, UV-vis spectroscopy and scanning electron microscopy (SEM). To study the CNC interaction and its structural and thermal characteristics, the prepared hydrogels were tested using Fourier transform infrared (FTIR) spectroscopy, X-ray diffraction (XRD) and thermogravimetric analysis (TGA). The mechanical properties of the prepared hydrogels were characterized using a universal testing machine. Furthermore, the actuation properties of the prepared hydrogels were tested by applying actuation voltage.

2. Materials and Methods

2.1. Materials

Cellulose cotton pulp (MVE, DPw-4580) of 98% purity was obtained from Buckeye Technology Inc. Poly (vinyl alcohol) (Mw = 85,000~124,000 g/mole, 99% hydrolyzed), sulfuric acid (H_2SO_4) and sodium hydroxide were purchased from Sigma-Aldrich Korea, Gyeonggi-do, South Korea. Dimethyl sulfoxide (DMSO) was purchased from Dae Jung chemicals & Metals Co. Ltd. (Gyeonggi-do, South Korea) Deionized (DI) water was used throughout the experiments.

2.2. Preparation of CNC

In this study, CNC was prepared using acid hydrolysis treatment. The preparation of CNC was described in Reference [19,20]; following is a brief explanation. The cotton pulp (20.0 g), a source of cellulose, was dispersed in H_2SO_4 (175 mL of 30% (v/v) aqueous) under mechanical stirring with 200 rpm and 6 h at 60 °C. An alkaline (NaOH, 1 M) pre-treatment was carried out on the cotton pulp to remove the non-cellulosic components and to prior obtain the high yield of CNC. The acid hydrolysis resulted in a suspension, and it was diluted (Ph = 7) by adding excessive deionized (DI) water, followed by centrifugation (11,000 rpm and 10 min). After this, the CNC suspension was homogenized and dialyzed overnight. A certain amount of homogenized CNC was dispersed in 20 mL of solvent mixture of DMSO and DI water (80:20 wt.%) by sonication for 1 h. Finally, 1% of CNC suspension was obtained and stored at room temperature until use.

2.3. Preparation of PVA-CNC Hydrogels

For the preparation of the PVA-CNC hydrogels, 9 wt.% PVA solution and 1 wt.% CNC suspension were used. The transparent PVA solution was prepared by dissolving 9 g of PVA in a 91 g solvent mixture of DMSO and DI water (80:20 wt.%) by continuous stirring at 80 °C for 8 h under a nitrogen atmosphere. To the PVA solution, different amounts of 1 wt.% CNC suspension were added, while the weight of the PVA-CNC mixtures was kept constant at 20 g. After adding the CNC suspension, the PVA-CNC mixtures were sonicated for 20 min and then subjected to magnetic stirring for another 2 h (200 rpm) at 80 °C to obtain a homogeneous mixture. Finally, PVA-CNC hydrogels were obtained via a freeze–thaw process. The PVA-CNC mixtures were poured into a petri dish and subjected to three freeze–thaw cycles consisting of a 12 h freezing step at −20 °C, followed by a 6 h thawing step at room temperature. After finishing the three freeze and thaw steps, PVA-CNC hydrogels were formed. The prepared hydrogels were immersed in 100 mL of DI water in order to remove solvents and water-soluble/unreacted materials [32,33]. The DI water was changed every 8 h up to 3 days. The thickness of the prepared hydrogels was 4 ± 0.05 mm. The prepared PVA-CNC hydrogels were kept in DI water until use. The sample codes of the PVA-CNC hydrogels were designated as PVA-CNCx according to the amount of CNC suspension used in the hydrogels. Table 1 provides the feed composition ratio of PVA to CNC.

Table 1. Feed composition ratio of PVA-CNC hydrogels.

Hydrogels	Weight of 9 wt.% PVA (g)	Weight of 1% wt. CNCs (g)
PVA	20	0
PVA-CNC1	17	3
PVA-CNC2	15	5
PVA-CNC3	13	7

3. Characterization

3.1. Physical Properties

A water uptake capacity test of the prepared PVA-CNC hydrogels was carried out. The hydrogels were dried in an oven for 24 h at 60 °C until their weight reached saturation. The weight of the dried hydrogels were noted and immersed in a 100 mL beaker containing 50 mL distilled water at room temperature to equilibrate for up to 48 h. Then the samples were taken out and blotted with wiper paper to remove water on their surface and again reweighed using an analytical balance (GH-200, A&D weighing, Tokyo, Japan). The water uptake ratio, W.U., can be represented using the following equation:

$$W.U._{(g/g)} = (W_{wet} - W_{dry})/W_{dry}, \tag{1}$$

where W_{wet} and W_{dry} denote the weight of the equilibrated hydrogel at 48 h and initial weight of the dried hydrogel, respectively.

The optical transparency of the prepared PVA-CNC hydrogels was measured using a UV-visible spectrophotometer (HP8452A, Agilent, Santa Clara, CA, USA). For the measurement, the hydrogels were cut into the desired shape and the spectra range of 200–800 nm wavelengths were recorded.

A Scanning electron microscope (SEM, S-4000, Hitachi, Tokyo, Japan) was used to observe morphologies of the prepared hydrogels. To prepare specimens, the prepared hydrogels were freeze-dried and coated with platinum. The images were taken using the SEM, at 15 kV accelerating voltage.

3.2. FTIR, XRD and TGA

FTIR spectroscopy was used to study the transmission of light and the interaction of CNCs of the prepared PVA-CNC hydrogels. For the FTIR analysis, the samples were completely dried in a vacuum oven at 60 °C for 6 h. The FTIR spectra were recorded on a FTIR spectrometer (Bruker Optics, Billerica, MA, USA) with the range of 400–4000 cm^{-1} using the KBr disk pellet method and averaging 16 scans.

XRD patterns of the prepared CNC, PVA, and PVA-CNC hydrogels were recorded using an X-ray diffractometer (DMAX 2500, Rigaku, Japan), with Cu Kα radiation source (λ = 0.1542 nm) at 40 kV and 300 mA. The scan speed was 2° per min and the spectra of 2θ (Bragg angle) ranged from 2.5 to 60°. The thermal stability of the prepared CNC, PVA and PVA-CNC hydrogels was studied using a TGA (STA 409 PC, NETZSCH , Selb, Germanay) at a constant heating rate of 10 °C/min in the range of 30–600 °C under a constant nitrogen flow (20 mL/min).

3.3. Mechanical Testing

The compression test of the PVA-CNC hydrogels was conducted at a fully-hydrated stage and was followed by the ASTM D-882-97 test method using a universal test machine (Won Shaft Jeong Gong, Gyeonggi-do, South Korea) under the ambient condition with compression rate of 0.0005 mm/s. The size of the specimens was 20 × 20 × 5 mm^3. The specimen was kept between two parallel plates and the upper plate pressed the specimen until it reached the maximum value.

3.4. Actuation Test

The actuation test was carried out using a laser displacement sensor (Keyance LK-G85, Tokyo, Japan), a high voltage amplifier (Model 10/10, Trek, Lockport, NY, USA) and a function generator (33220A, Agilent, Santa Clara, CA, USA). Figure 1 shows the schematic setup of the actuation test. Before conducting the actuation test, the hydrogel specimens (10 × 10 × 4 mm^3) were equilibrated in DI water for 24 h and kept between two electrodes (polyimide tape attached to indium tin oxide coated glass (ITO glass)). A high voltage was applied on the electrodes via the function generator and the high voltage amplifier. The displacement of the hydrogel specimen was measured using the laser displacement sensor along with a data acquisition system (Pulse, B&K, Nærum, Denmark) connected to personal laptop. The actuation test was conducted at a constant environmental condition (25 °C, 95% RH) using an environmental chamber.

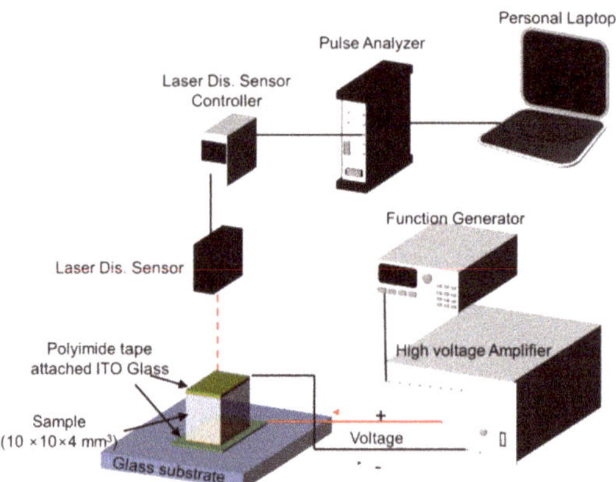

Figure 1. Schematic setup of the actuation test.

4. Results and Discussion

4.1. Physical Properties

The water uptake ratios of the pure PVA and PVA-CNC hydrogels were calculated using Equation (1). Figure 2A shows the results for the pure PVA and the hydrogels. As the CNC concentration increased, the water uptake ratio consistently increased from 220% to 250%. This might be due to the fact that the hydrogen bonds between the CNC and PVA chains decreased the residual hydrogen bonds in the PVA chains, which resulted in increased water uptake [34].

The optical transparency of the prepared hydrogels was measured using the UV-vis spectroscopy at 300 to 700 nm. Figure 2B shows the optical transparency of the pure PVA and PVA-CNC hydrogels. The optical transparency taken at 500 nm of pure PVA was 92.4% and it decreased to 91.0, 77.7 and 75.9%, as the CNC concentration in the PVA-CNC hydrogels increased. Increasing the CNC concentration reduced the transparency due to CNC aggregation, which enhanced the turbidity of PVA hydrogel so as to decrease its transparency [20].

The SEM images of PVA-CNC3 hydrogel were taken to observe the morphology of the hydrogel. Figure 3A,B shows the surface morphologies of the pure PVA and PVA-CNC3 hydrogel, respectively. The pure PVA showed a smooth surface morphology, meanwhile the CNC-incorporated PVA-CNC3 hydrogel exhibited a uniform but rough surface. The CNCs were shown to be well-dispersed in the hydrogel. The cross-sectional SEM image of the PVA-CNC3 hydrogel (Figure 3C) showed that the rod-shaped CNCs were dispersed in the cross-sectional image of the hydrogel. This uniform dispersion might be associated with the interaction between the PVA and the CNC.

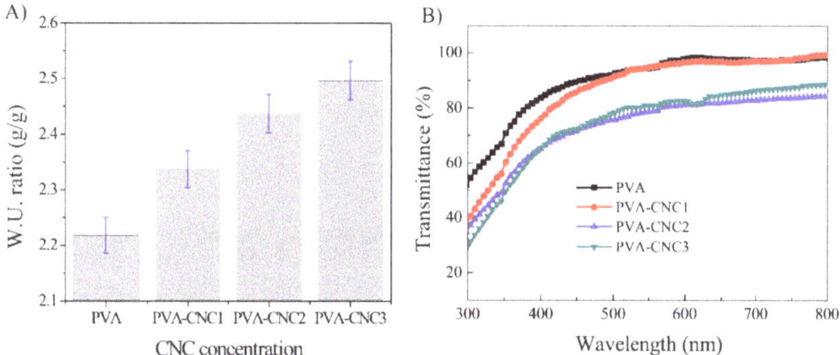

Figure 2. (**A**) Water uptake capacity of pure PVA and PVA-CNC hydrogels and (**B**) Optical transparency of the pure PVA and PVA-CNC hydrogels.

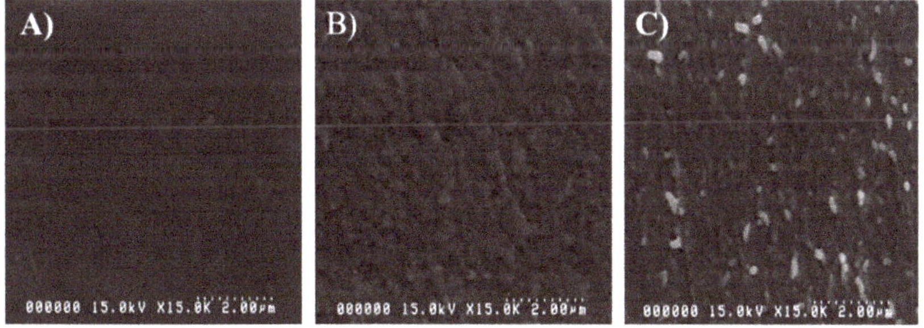

Figure 3. SEM images: (**A**) the pure PVA, (**B**) surface of PVA-CNC3 hydrogel and (**C**) cross-section of PVA-CNC3 hydrogel.

4.2. FTIR, XRD and TGA

To study the transmission of light and the interactions of CNC in the prepared PVA-CNC hydrogels, FTIR spectroscopy analysis was performed. Figure 4 shows the FTIR spectra of the prepared CNC, PVA, and PVA-CNC3 hydrogel. The O-H stretching vibration of the pure PVA is shown at 3422 cm^{-1}. The characteristic peaks at 1070 cm^{-1} and 2901 cm^{-1} are related to stretching vibrations of the C-O and C-H. The peak at 1628 cm^{-1} is related to an acetyl group (C=O), which is induced from the preparation of PVA. A bending vibration related to CH$_2$ groups is observed in the region of 1430–1446 cm^{-1} [10]. The FTIR spectra of the prepared CNC indicates the characteristic peaks assigned to cellulose I structures: Peaks are shown at 3374 cm^{-1} (O-H region), 2900 cm^{-1} (C-H stretching vibration), 1430 cm^{-1} (CH$_2$ symmetric bending) and 1320 cm^{-1} (CH$_2$ wagging at C-6). Peaks at 1065 cm^{-1}, 1124 cm^{-1} and 1160 cm^{-1} demonstrate the presence of sulfate ester bonds, which are induced by the sulfuric acid hydrolysis for CNC preparation [35]. In the case of the PVA-CNC3 hydrogel, the O-H stretching vibration peak shifted to 3402 cm^{-1} due to the overlap of intermolecular hydrogen bonded O-H peaks from PVA-PVA and CNC-CNC. There is a new peak at 3201 cm^{-1} on the FTIR spectrum of PVA-CNC3. This peak might correspond to intermolecular hydrogen bonded between PVA and CNC. The result shows that CNC and PVA are well-interacted. This gives rise to the increase in the water uptake capacity of PVA-CNC hydrogels.

Figure 5A represents XRD patterns that give crystalline information for the pure PVA, CNC and PVA-CNC3 hydrogel. PVA is known to be semi-crystalline in nature and the pure PVA shows the main strong diffraction intensity characteristic peak at 2θ = 19.5° [36] and the CNC sample exhibits four well-defined diffraction peaks at 2θ = 14.6, 16.2, 22.5, and 34.4°, which correspond to a typical cellulose I structure [22]. Note that the PVA-CNC3 hydrogel exhibits a similar diffraction peak at 2θ = 19.6°, with decreased intensity, and a shoulder peak at 2θ = 22.4°, with increased intensity, which suggests the physical interaction of PVA and CNC. This observation indicates that the incorporation of CNCs in PVA does not affect the crystalline structure of the PVA matrix. This means that the CNCs are well-dispersed in the PVA matrix so as to form the PVA-CNC hydrogel [37]. This fact was also confirmed for the FTIR spectra shown in Figure 4.

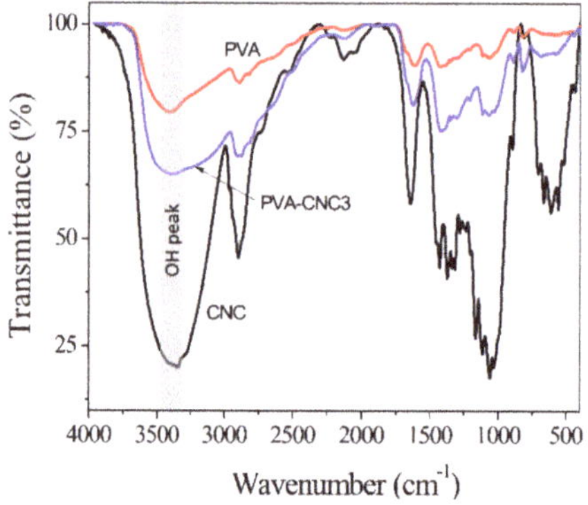

Figure 4. FTIR spectra of CNC, PVA, and PVA-CNC3 hydrogel.

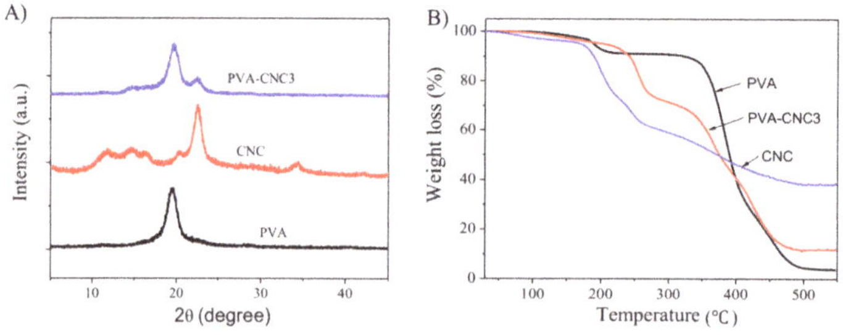

Figure 5. (**A**) X-ray diffraction patterns of CNC, PVA, and PVA-CNC3 hydrogel, (**B**) TGA curves of the pure PVA, CNC and PVA-CNC3 hydrogel.

TGA measures the weight changes as a function of temperature. As the temperature increases, the weight of the sample decreases, indicating the continuous decomposition of the sample. Figure 5B shows the TGA curves of the pure PVA, CNC and PVA-CNC3 hydrogel. Below 150 °C, a minor weight loss occurred in all samples near 89 °C, which is associated with the evaporation of the absorbed water molecules. The pure PVA hydrogel showed mainly two weight-loss steps. The first weight loss started from 179 °C and finished at 216 °C (the weight loss was 8.9%), and was mainly associated with the dehydration of the hydroxyl groups by applying heat. The second weight loss started at 345 °C and degraded rapidly up to 500 °C (the weight loss was 95.4%), which was due to the degradation of the main chain. The PVA-CNC3 hydrogel showed two weight loss steps. The weight loss started from 212 °C and continually decreased up to 500 °C and a maximum 88.4% of weight loss was observed. Note that the PVA-CNC3 hydrogel showed higher thermal stability than the pure PVA, which might be due to the formation of the intermolecular bond between the CNC and the PVA. A similar observation has been reported previously [38,39]. In addition, the TGA spectrum of the CNC sample showed that the starting degradation temperature (160 °C) was lower than that of the pure PVA and the PVA-CNC3 hydrogel because CNC has many hydroxyl groups on its surface [40].

4.3. Mechanical Testing

The mechanical properties of the pure PVA and PVA-CNC hydrogels were studied using the universal testing machine. The test specimens were fully hydrated in distilled water. Figure 6A shows the compressional stress–strain curves of the pure PVA and PVA-CNC hydrogels with various CNC concentrations. Figure 6B shows the compressive modulus values of the hydrogels with different CNC concentrations. The modulus decreased with the increasing CNC concentration. The compressive modulus decrease of the PVA-CNC hydrogels can be surmised from the water uptake results. The mechanical property was inversely proportional to the water uptake capacity in the hydrogels: Increasing the water uptake capacity decreased the compressive modulus, due to the softening of the hydrogel structures. The CNC concentration plays an important role in the successful dispersion and formation of strong interfaces within the PVA polymer matrix. When the CNC concentration is above a critical value, the compressive strength of the hydrogels could be significantly decreased due to poor dispersion of the CNC as well as limited interfacial interaction between the CNC and PVA [41]. The compressive modulus of the pure PVA hydrogel was 82 kPa, and as the CNC concentration increased, it gradually decreased to 7 kPa for the PVA-CNC3 hydrogel.

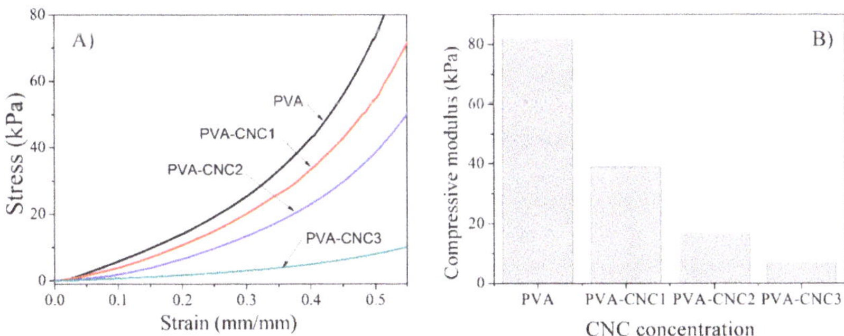

Figure 6. (**A**) Compressive stress–strain curves of PVA, PVA-CNC composite hydrogels and (**B**) compressive modulus.

Figure 7 shows the formation of PVA-CNC hydrogels. The hydroxyl groups of the PVA as well as on the surface of the CNC can interact with each other to form hydrogen bonds. However, as a large amount of CNC is dispersed in the hydrogel, for example PVA-CNC3, it seems that CNC aggregation occurs due to the hydrophilic nature of CNC, resulting in the evenly rough surface of the hydrogel.

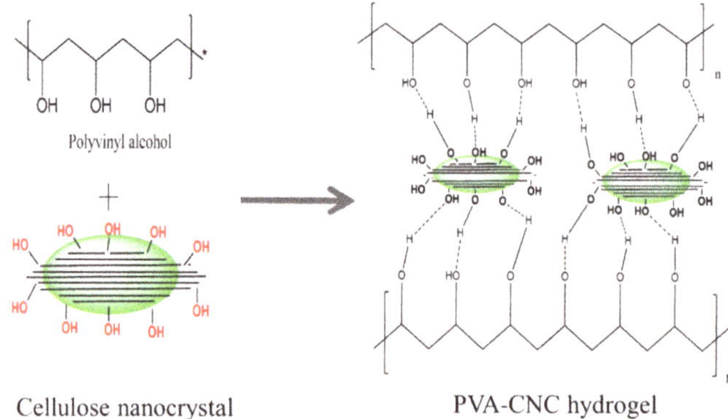

Figure 7. Formation of the PVA-CNC hydrogel.

4.4. Actuation Test

A soft actuator deforms in the presence of an external electric field. In this study, the deformation was defined as displacement and investigated in terms of the applied electric field and frequency. The actuation test of the prepared hydrogels was carried out in an aqueous swollen state in the DI water of the PVA-CNC hydrogels. Figure 8 shows the displacement of the hydrogels in terms of CNC concentration with the applied voltage and the frequency. Figure 8A shows the displacement of the hydrogels with a voltage change at 0.1 Hz. The displacement increased with increased voltage as well as increased CNC concentration. The maximum displacement of 14.4 µm was shown from the PVA-CNC3 hydrogel at 1.6 kV. This displacement corresponded to 3600 ppm strain and the applied voltage does corresponded to 0.4 V/µm electric field strength. This actuation performance is better than the cellulose hydrogel case (1800 ppm at 0.25 V/µm) [10]. Figure 8B displays the frequency-dependent displacement of the hydrogels under a constant voltage of 1.6 kV. The displacement output decreased with an increasing frequency. The CNC concentration played a significant role in the electroactive behavior of the PVA-CNC hydrogels, and the higher CNC concentration exhibited larger displacement than the pure PVA case. The PVA-CNC3 hydrogel showed higher displacement than the other hydrogels. This result is associated with the interfacial polarization between CNCs and the PVA polymer matrix [42]. It is a known fact that dispersed CNCs in a polymer matrix increases its dielectric properties, which is beneficial to improving its electroactive behavior.

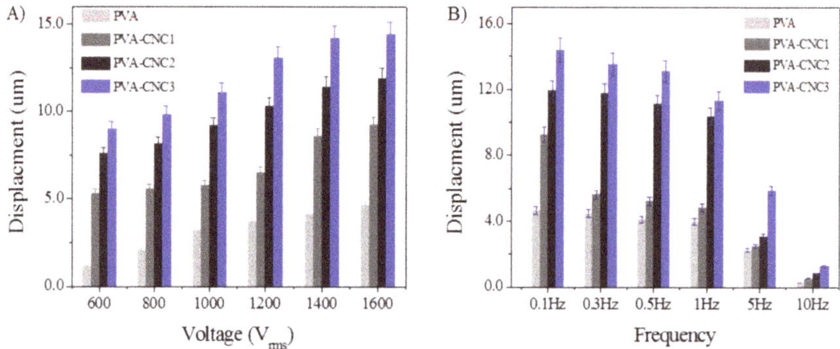

Figure 8. Actuation results for the pure PVA and PVA-CNC hydrogels: (**A**) actuation voltage variation at 0.1 Hz and (**B**) actuation frequency at 1.0 kV.

Generally, an electroactive response occurs through a combination of piezoelectricity, electrostriction, flexoelectricity, electrostatic effect (Coulombic force), electrophoretic effect, electrochemical effect (ion migration) and electroosmotic interaction [5]. In the PVA/DMSO hydrogel, electrostrictive interaction was claimed as the mechanism causing its electroactive response since the displacement was proportional to the square of the electrical actuation voltage [30]. In some electroactive materials, it is hard to clarify the mechanism because although the same materials are used, different morphologies or physical/chemical properties could result in different mechanisms of electroactive behavior.

The prepared PVA-CNC hydrogel is not an ionic hydrogel. Although a sulfate ester bond peak was observed from the FTIR, this peak was caused by the sulfuric acid hydrolysis for the CNC preparation. During the CNC preparation, sulfuric acid was applied to isolate CNC from the pulp and some of the sulfuric ions remained on the surface of the CNCs. However, the remnant sulfuric ions were not significant because CNCs were dialyzed and washed with DI water several times. Thus, we believe that the prepared hydrogel was a nonionic hydrogel and the ion migration effect was not significant in the prepared hydrogel.

When the displacement and voltage curves are considered, as shown in Figure 8A, the displacements increased linearly or quadratically with the actuation voltage, as well as CNC concentration. This may give piezoelectricity in the hydrogel. However, it is very hard to possess dipole domains in the hydrogel. Thus, the prepared hydrogel was far from a piezoelectric material. On the other hand, as increasing the CNC concentration causes the compressive modulus to decrease, which is beneficial in terms of increasing the electrostatic effect associated with the Coulombic force. Under Coulombic force, a large strain can be produced when the electroactive material is soft. This is a well-known fact for dielectric elastomer electroactive polymers. In summary, since the PVA-CNC hydrogel is a soft, nonionic hydrogel, and it has high dielectric properties, the electrostatic effect may be its dominant actuation mechanism.

5. Conclusions

In this research, nontoxic, soft and electroactive PVA-CNC hydrogels were prepared using the freeze–thaw process with different CNC concentrations, and their characteristics were analyzed. The water uptake capacity of the hydrogels increased and the compressive modulus decreased as the CNC concentration increased. The optical transparency of the prepared hydrogels was inversely proportional to the CNC concentration. The thermogravimetric analysis and scanning electron microscopy results showed good miscibility of CNC with PVA. The CNC incorporated PVA-CNC hydrogels showed an improved displacement output in the presence of the electric field and with the increasing CNC concentration. The maximum 3600 ppm strain was produced under 0.4 V/µm electric field strength from the PVA-CNC3 hydrogel. The displacement improvement of the PVA-CNC hydrogels is associated with their enhanced dielectric properties and reduced compressive modulus. Since the PVA-CNC hydrogel is nonionic, soft, and it possible has strong dielectric properties, the electrostatic effect may be its dominant actuation mechanism. Since the prepared PVA-CNC hydrogel is nontoxic and electroactive, it may be a promising material for biomimetic soft robots, actively reconfigurable lenses and active drug-release.

Author Contributions: Conceptualization, T.J. and J.K.; Methodology, T.J. and H.K.; Validation, J.K.; Formal Analysis, T.J. and H.K.; Investigation, T.J., H.C.K and J.W.K; Resources, H.C.K.; Data Curation, T.J.; Writing-Original Draft Preparation, T.J.; Writing-Review & Editing, R.M. and J.K.; Visualization, T.J. and H.K.; Supervision, J.K.; Project Administration, J.W.K.; Funding Acquisition, J.K.

Funding: This research was supported by Creative Research Initiatives Program through the National Research Foundation of Korea (NRF) funded by the Ministry of Science, Technology and ICT (NRF-2015R1A3A2066301).

Conflicts of Interest: The authors declare no conflicts of interest.

References

1. Du, X.; Zhou, J.; Shi, J.; Xu, B. Supramolecular hydrogelators and hydrogels: From soft matter to molecular biomaterials. *Chem. Rev.* **2015**, *115*, 13165–13307. [CrossRef] [PubMed]
2. Zolfagharian, A.; Kouzani, A.Z.; Khoo, S.Y.; Moghadam, A.A.A.; Gibson, I.; Kaynak, A. Evolution of 3D printed soft actuators. *Sens. Actuator A Phys.* **2016**, *250*, 258–272. [CrossRef]
3. Kwon, I.C.; Bae, Y.H.; Kim, S.W. Characteristics of charged networks under an electric stimulus. *J. Polym. Sci. Part B Polym. Phys.* **1994**, *32*, 1085–1092. [CrossRef]
4. Sun, S.; Wong, Y.W.; Yao, K.; Mak, A.F. A study on mechano-electro-chemical behavior of chitosan/poly (propylene glycol) composite fibers. *J. Appl. Polym. Sci.* **2000**, *76*, 542–551. [CrossRef]
5. Koetting, M.C.; Peters, J.T.; Steinchen, S.D.; Peppas, N.A. Stimulus-responsive hydrogels: Theory, modern advances, and applications. *Mater. Sci. Eng. R* **2015**, *93*, 1–49. [CrossRef] [PubMed]
6. Kim, S.J.; Lee, K.J.; Kim, S.I.; Lee, Y.M.; Chung, T.D.; Lee, S.H. Electrochemical behavior of an interpenetrating polymer network hydrogel composed of poly (propylene glycol) and poly (acrylic acid). *J. Appl. Polym. Sci.* **2003**, *89*, 2301–2305. [CrossRef]
7. Wandera, D.; Wickramasinghe, S.R.; Husson, S.M. Stimuli-responsive membranes. *J. Membr. Sci.* **2010**, *357*, 6–35. [CrossRef]
8. Jayaramudu, T.; Li, Y.; Ko, H.U.; Shishir, I.R.; Kim, J. Poly (acrylic acid)-Poly (vinyl alcohol) hydrogels for reconfigurable lens actuators. *Int. J. Precis. Eng. Manuf. Green Technol.* **2016**, *3*, 375–379. [CrossRef]
9. Yuan, N.; Xu, L.; Zhang, L.; Ye, H.; Zhao, J.; Liu, Z.; Rong, J. Superior hybrid hydrogels of polyacrylamide enhanced by bacterial cellulose nanofiber clusters. *Mater. Sci. Eng. C Mater. Biol. Appl.* **2016**, *67*, 221–230. [CrossRef] [PubMed]
10. Jayaramudu, T.; Ko, H.U.; Zhai, L.; Li, Y.; Kim, J. Preparation and characterization of hydrogels from polyvinyl alcohol and cellulose and their electroactive behavior. *Soft Matter* **2017**, *15*, 64–72. [CrossRef]
11. Kamal, M.R.; Khoshkava, V. Effect of cellulose nanocrystals (CNC) on rheological and mechanical properties and crystallization behavior of PLA/CNC nanocomposites. *Carbohydr. Polym.* **2015**, *123*, 105–114. [CrossRef] [PubMed]
12. Khoshkava, V.; Kamal, M.R. Effect of cellulose nanocrystals (CNC) particle morphology on dispersion and rheological and mechanical properties of polypropylene/CNC nanocomposites. *ACS Appl. Mater. Interfaces* **2014**, *6*, 8146–8157. [CrossRef] [PubMed]
13. Jayaramudu, T.; Raghavendra, G.M.; Varaprasad, K.; Sadiku, R.; Raju, K.M. Development of novel biodegradable Au nanocomposite hydrogels based on wheat: For inactivation of bacteria. *Carbohydr. Polym.* **2013**, *92*, 2193–2200. [CrossRef] [PubMed]
14. Kim, J.H.; Shim, B.S.; Kim, H.S.; Lee, Y.J.; Min, S.K.; Jang, D.; Kim, J. Review of nanocellulose for sustainable future materials. *Int. J. Precis. Eng. Manuf. Green Technol.* **2015**, *2*, 197–213. [CrossRef]
15. Azizi Samir, M.A.S.; Alloin, F.; Dufresne, A. Review of recent research into cellulosic whiskers, their properties and their application in nanocomposite field. *Biomacromolecules* **2005**, *6*, 612–626. [CrossRef] [PubMed]
16. Mueller, S.; Weder, C.; Foster, E.J. Isolation of cellulose nanocrystals from pseudostems of banana plants. *RSC Adv.* **2014**, *4*, 907–915. [CrossRef]
17. Gao, F. *Advances in Polymer Nanocomposites: Types and Applications*; Woodhead Publishing: Cambridge, UK, 2012; pp. 131–155, ISBN 978-1-84569-940-6.
18. George, J.; Sabapathi, S.N. Cellulose nanocrystals: Synthesis, functional properties, and applications. *Nanotechnol. Sci. Appl.* **2015**, *8*, 45–54. [CrossRef] [PubMed]
19. Gao, X.; Sadasivuni, K.K.; Kim, H.C.; Min, S.K.; Kim, J. Designing pH-responsive and dielectric hydrogels from cellulose nanocrystals. *J. Chem. Sci.* **2015**, *127*, 1119–1125. [CrossRef]
20. Sadasivuni, K.K.; Ponnamma, D.; Ko, H.U.; Zhai, L.; Kim, H.C.; Kim, J. Electroactive and optically adaptive bionanocomposite for reconfigurable microlens. *J. Phys. Chem. B* **2016**, *120*, 4699–4705. [CrossRef] [PubMed]
21. Domingues, R.M.; Silva, M.; Gershovich, P.; Betta, S.; Babo, P.; Caridade, S.G.; Gomes, M.E. Development of injectable hyaluronic acid/cellulose nanocrystals bionanocomposite hydrogels for tissue engineering applications. *Bioconjug. Chem.* **2015**, *26*, 1571–1581. [CrossRef] [PubMed]
22. Kumar, A.; Negi, Y.S.; Choudhary, V.; Bhardwaj, N.K. Characterization of cellulose nanocrystals produced by acid-hydrolysis from sugarcane bagasse as agro-waste. *J. Mater. Phys. Chem.* **2014**, *2*, 1–8. [CrossRef]

23. Favier, V.; Chanzy, H.; Cavaille, J.Y. Polymer nanocomposites reinforced by cellulose whiskers. *Macromolecules* **1995**, *28*, 6365–6367. [CrossRef]
24. Capadona, J.R.; Van Den Berg, O.; Capadona, L.A.; Schroeter, M.; Rowan, S.J.; Tyler, D.J.; Weder, C. A versatile approach for the processing of polymer nanocomposites with self-assembled nanofibre templates. *Nat. Nanotechnol.* **2007**, *2*, 765–769. [CrossRef] [PubMed]
25. Einchhorn, S.J.; Dufresne, A.; Aranguren, M.M.; Capadona, J.R.; Rowan, S.J.; Weder, C.; Veigel, S. Review: Current international research into cellulose nanofibres and composites. *J. Mater. Sci.* **2010**, *45*, 1–33. [CrossRef]
26. McKee, J.R.; Hietala, S.; Seitsonen, J.; Laine, J.; Kontturi, E.; Ikkala, O. Thermoresponsive nanocellulose hydrogels with tunable mechanical properties. *ACS Macro Lett.* **2014**, *3*, 266–270. [CrossRef]
27. Yang, J.; Zhang, X.M.; Xu, F. Design of cellulose nanocrystals template-assisted composite hydrogels: Insights from static to dynamic alignment. *Macromolecules* **2015**, *48*, 1231–1239. [CrossRef]
28. Ooi, S.Y.; Ahmad, I.; Amin, M.C.I.M. Cellulose nanocrystals extracted from rice husks as a reinforcing material in gelatin hydrogels for use in controlled drug delivery systems. *Ind. Crops Prod.* **2016**, *93*, 227–234. [CrossRef]
29. De France, K.J.; Chan, K.J.; Cranston, E.D.; Hoare, T. Enhanced mechanical properties in cellulose nanocrystal–poly (oligoethylene glycol methacrylate) injectable nanocomposite hydrogels through control of physical and chemical cross-linking. *Biomacromolecules* **2016**, *17*, 649–660. [CrossRef] [PubMed]
30. Liang, S.; Xu, J.; Weng, L.; Zhang, L.; Guo, X.; Zhang, X. Biologically Inspired Path-Controlled Linear Locomotion of Polymer Gel in Air. *J. Phys. Chem. B* **2007**, *111*, 941–945. [CrossRef] [PubMed]
31. Hou, Y.; Chen, C.; Liu, K.; Tu, Y.; Zhang, L.; Li, L. Preparation of PVA hydrogel with high-transparence and investigations of its transparent mechanism. *RSC Adv.* **2015**, *5*, 24023–24030. [CrossRef]
32. Jayaramudu, T.; Raghavendra, G.M.; Varaprasad, K.; Sadiku, R.; Ramam, K.; Raju, K.M. Iota-Carrageenan-based biodegradable Ag^0 nanocomposite hydrogels for the inactivation of bacteria. *Carbohydr. Polym.* **2013**, *95*, 188–194. [CrossRef] [PubMed]
33. Jayaramudu, T.; Raghavendra, G.M.; Varaprasad, K.; Raju, K.M.; Sadiku, E.R.; Kim, J. 5-Fluorouracil encapsulated magnetic nanohydrogels for drug-delivery applications. *J. Appl. Polym. Sci.* **2016**, *133*, 43921. [CrossRef]
34. Kudo, S.; Otsuka, E.; Suzuki, A. Swelling Behavior of Chemically Crosslinked PVA Gels in Mixed Solvents. *J. Polym. Sci. Part B Polym. Phys.* **2010**, *48*, 1978–1986. [CrossRef]
35. Xu, X.; Shen, Y.; Wang, W.; Sun, C.; Li, C.; Xiong, Y.; Tu, J. Preparation and in vitro characterization of thermosensitive and mucoadhesive hydrogels for nasal delivery of phenylephrine hydrochloride. *Eur. J. Pharm. Biopharm.* **2014**, *88*, 998–1004. [CrossRef] [PubMed]
36. Bajpai, A.K.; Gupta, R. Synthesis and characterization of magnetite (Fe_3O_4)—Polyvinyl alcohol-based nanocomposites and study of superparamagnetism. *Polym. Compos.* **2010**, *31*, 245–255. [CrossRef]
37. Kakroodi, A.R.; Cheng, S.; Sain, M.; Asiri, A. Mechanical, thermal, and morphological properties of nanocomposites based on polyvinyl alcohol and cellulose nanofiber from Aloe vera rind. *J. Nanomater.* **2014**, *2014*, 903498. [CrossRef]
38. Abitbol, T.; Johnstone, T.; Quinn, T.M.; Gray, D.G. Reinforcement with cellulose nanocrystals of poly (vinyl alcohol) hydrogels prepared by cyclic freezing and thawing. *Soft Matter* **2011**, *7*, 2373–2379. [CrossRef]
39. Kaneko, D.; Shimoda, T.; Kaneko, T. Preparation methods of alginate micro-hydrogel particles and evaluation of their electrophoresis behavior for possible electronic paper ink application. *Polym. J.* **2010**, *42*, 829–833. [CrossRef]
40. Han, J.; Lei, T.; Wu, Q. High-water-content mouldable polyvinyl alcohol-borax hydrogels reinforced by well-dispersed cellulose nanoparticles: Dynamic rheological properties and hydrogel formation mechanism. *Carbohydr. Polym.* **2014**, *102*, 306–316. [CrossRef] [PubMed]
41. Tanpichai, S.; Oksman, K. Cross-linked nanocomposite hydrogels based on cellulose nanocrystals and PVA: Mechanical properties and creep recovery. *Compos. Part A Appl. Sci. Manuf.* **2016**, *88*, 226–233. [CrossRef]
42. Ko, H.U.; Kim, H.C.; Kim, J.W.; Zhai, L.; Jayaramudu, T.; Kim, J. Fabrication and characterization of cellulose nanocrystal based transparent electroactive polyurethane. *Smart Mater. Struct.* **2017**, *26*, 085012. [CrossRef]

Article

Spray Deposition of Ag Nanowire–Graphene Oxide Hybrid Electrodes for Flexible Polymer–Dispersed Liquid Crystal Displays

Yumi Choi [1], Chang Su Kim [2] and Sungjin Jo [1,*]

[1]   School of Architectural, Civil, Environmental, and Energy Engineering, Kyungpook National University, Daegu 41566, Korea; yums1026@gmail.com
[2]   Advanced Functional Thin Films Department, Korea Institute of Materials Science (KIMS), Changwon 51508, Korea; cskim1025@kims.re.kr
*   Correspondence: sungjin@knu.ac.kr; Tel.: +82-53-950-8971

Received: 18 October 2018; Accepted: 6 November 2018; Published: 9 November 2018

Abstract: We investigated the effect of different spray-coating parameters on the electro-optical properties of Ag nanowires (NWs). Highly transparent and conductive Ag NW–graphene oxide (GO) hybrid electrodes were fabricated by using the spray-coating technique. The Ag NW percolation network was modified with GO and this led to a reduced sheet resistance of the Ag NW–GO electrode as the result of a decrease in the inter-nanowire contact resistance. Although electrical conductivity and optical transmittance of the Ag NW electrodes have a trade-off relationship, Ag NW–GO hybrid electrodes exhibited significantly improved sheet resistance and slightly decreased transmittance compared to Ag NW electrodes. Ag NW–GO hybrid electrodes were integrated into smart windows based on polymer-dispersed liquid crystals (PDLCs) for the first time. Experimental results showed that the electro-optical properties of the PDLCs based on Ag NW–GO electrodes were superior when compared to those of PDLCs based on only Ag NW electrodes. This study revealed that the hybrid Ag NW–GO electrode is a promising material for manufacturing the large-area flexible indium tin oxide (ITO)-free PDLCs.

Keywords: silver nanowire; graphene oxide; polymer-dispersed liquid crystal; smart window; hybrid transparent conductive electrode

---

1. Introduction

Smart windows are controllable windows whose optical properties can be altered by applying an electric field. They are used for various applications including switchable privacy glasses, vehicle windows, and energy-saving windows [1–3]. Smart windows have recently attracted significant attention since they can minimize heating and cooling energies in buildings and transportation systems. Among the different electrooptically switchable active components available for smart windows such as polymer-dispersed liquid crystals (PDLCs), chromic materials, and suspended particles, PDLCs are an excellent candidate due to their high transmittance, wide viewing angle, high switching speed, and a relatively simple fabrication process [4,5].

The PDLCs consist of birefringent liquid crystal (LC) droplets that are uniformly dispersed in a solid polymer matrix. In order to fabricate smart windows based on PDLCs, the PDLC film is positioned between two transparent conductive electrodes (TCEs). The PDLC film can be switched from an opaque to a transparent state since the electric field between the TCEs aligns the directors of the LCs along the same direction. Therefore, TCEs with low sheet resistances and high transmittances are necessary in order to minimize the voltage drop across the electrode and ensure fast switching as well as smaller power consumption. Indium tin oxide (ITO) has been commonly used as a TCE for typical smart windows based on PDLCs. Although ITO has high conductivity and transmittance, its relatively high cost and fragile characteristics make it unsuitable for fabrication of large-area flexible smart windows.

There are several potential alternatives to ITO including conductive polymers, graphene, Ag nanowires (NWs), metal grids, and carbon nanotubes. Recently, conductive polymers, graphene, and Ag NWs have been successfully integrated into PDLC-based smart windows [6–8]. Although these emerging candidate materials are of primary interest, they suffer from lower electrical conductivities, complex fabrication processes, lower transmittances, and uneven distributions of the electric currents. In order to overcome the disadvantages of these individual TCE materials, hybrid TCEs such as Ag NW–graphene, Ag NW–conductive polymer, Ag NW–metal oxide, Ag NW–carbon nanotubes, and Ag NW–metal grids [9–13] have also been investigated. Even though these hybrid TCEs have been investigated for applications in various electronic devices including solar cells, organic light emitting diodes, flexible heaters, flexible sensors, and touch panels [14–18], no extensive studies have been reported on the application of hybrid electrodes for PDLC-based smart windows.

In this study, the hybrid Ag NW–GO electrode was used as a substitute for an ITO electrode in a PDLC. To the best of our knowledge, we report for the first time Ag NW–GO hybrid electrodes integrated into smart windows based on PDLCs. Because of the inverse relationship between optical transmittance and electrical resistivity of TCE, the optical properties of TCE deteriorate with increasing electrical conductivity. However, Ag NW–GO hybrid structure can reduce the resistances of Ag NW networks without affecting their transmittances. In this case, we have described the preparation of highly conductive and transparent Ag NW–GO hybrid electrodes by a simple spray-coating technique for the fast production of large-area flexible smart windows with a decreased production cost.

2. Materials and Methods

2.1. Ag NW–GO Hybrid Electrode Fabrication

A hybrid electrode was fabricated by using Ag NWs (Nanopyxis) dispersed in isopropyl alcohol (IPA) and GO (Uninanotech) dispersed in IPA. In order to fabricate the GO suspension, a 6.2 g/L GO suspension was diluted with IPA to 0.2 g/L and sonicated for 20 min. The polyethylene terephthalate (PET) substrates were cleaned with acetone, IPA, and deionized water. After drying, the PET films were treated with ultraviolet-ozone (UVO) for 5 min. Immediately after the UVO treatment, the Ag NW suspension was uniformly spray coated and annealed at 65 °C for 1 min. Lastly, in order to form Ag NW–GO hybrid electrodes, the GO suspension was spray coated onto the Ag NWs and dried at 65 °C for 5 min. For spray deposition, a fully automated spray coater was utilized.

2.2. PDLC Fabrication

The PDLCs, which is commercially available from Qingdao Liquid Crystal, were mixed with spacers at a weight ratio of 100:1 and stirred for 4 h to obtain a uniform thickness of the PDLC layer. The fully mixed solution was drop-dispersed on the substrate coated with Ag NW–GO. Subsequently, it was covered with another Ag NW–GO-coated substrate, which produced a sandwich structure (Figure S1). Lastly, the PDLC layer was photo-polymerized by using a mask aligner for 5 min.

3. Results and Discussion

In order to produce highly uniform, transparent, and conductive Ag NW networks by spray coating, various processing parameters such as dispensing pressure, spray pressure, nozzle–to–sample distance, scan speed, and substrate temperature should be simultaneously controlled [19]. We investigated the influence of each of these parameters on the morphology of the Ag NW network, which is known to affect their electro-optical properties. First, Ag NW networks were prepared at three different dispensing pressures: 0.1 psi, 0.5 psi, and 1 psi. Their optical transmittances (T) and sheet resistances (R_s) were measured and their morphologies were characterized by using scanning electron microscopy (SEM). Figure 1a shows a decrease in the T and R_s values with the increasing dispensing pressure. A higher dispensing pressure led to a higher flow rate of the Ag NW suspension, which affected the density of the Ag NW networks. As shown in Figure 2a,b and Figure S2a,b, the Ag NW density increased with the dispensing pressure. However, a high dispensing pressure could clog the nozzle, which leads to an irregular flow of the Ag NW suspension through the nozzle and, thus, to an irregular coating of Ag NWs. Ag NW-deficient areas were observed in Figure S2c due to the irregular deposition of Ag NWs at a dispensing pressure of 1 psi. In order to avoid this phenomenon, the dispensing pressure was maintained at 0.5 psi. Second, the nozzle–to–sample distance, which controls the mass of Ag NWs sprayed per unit area, was optimized to obtain uniform Ag NW networks [20]. The T and R_s values in Figure 1b show that the density of Ag NWs increased with a decreasing nozzle–to–substrate distance. The nozzle–to–substrate distance was chosen to be 6 cm since a smaller distance led to nonuniform Ag NW networks while a larger distance led to low-density Ag NW networks, which is shown in Figure 2d–f and Figure S2d–f. Lastly, we investigated the effect of the nozzle pressure on the electro-optical performance of the Ag NW network. The nozzle pressure was chosen to be as high as 35 psi because an increased nozzle pressure increased the Ag NW density, as shown in Figure 2g–i. The relationship between T and R_s of the Ag NW networks (Figure 1c) was consistent with the SEM results. A higher nozzle pressure led to large shear forces, which promoted the Ag NW suspension into smaller droplets that were beneficial for the deposition of more uniform films since small droplets tend to produce less prominent "coffee-staining" effects [21].

In order to enhance the electro-optical properties of the Ag NW networks, they were spray-coated with the GO suspension. Electrical conduction in an Ag NW network is dominated by the resistances at junctions of Ag NWs due to the percolative nature of conduction. The GO tends to bond with Ag NWs due to the strong electrostatic adhesion caused by the large number of oxygen-containing groups in GO [22]. The GO sheets adhered and wrapped around the Ag NWs, which led to the soldering of the inter-nanowire junctions and caused a significant reduction in the contact resistance of these junctions [23]. The optimized spray-coating parameters for the formation of Ag NW networks were also used for the spray coating of the GO suspension. As shown in Figure 3b and Figure S3, the Ag NW junctions were wrapped by the GO sheets and a small number of GO sheets were deposited in the optical pathway and unblocked by Ag NWs, which enhanced the electro-optical properties of the Ag NW–GO hybrid networks by reducing R_s while minimizing the loss of T. The transmittances, sheet resistances, and figures of merit (FoMs) of the Ag NW and Ag NW–GO hybrid networks are summarized in Table 1. The spray-coated Ag NW–GO hybrid networks exhibited excellent properties. A representative film had $T = 90.7\%$ and $R_s = 15.6\ \Omega/\text{sq}$. Table 1 shows that the FoM was enhanced upon the formation of the Ag NW–GO hybrid network when compared to that of the Ag NW network.

Table 1. Sheet resistances, transmittances, and FoMs of the Ag NW and Ag NW–GO films.

Electrode	Sheet Resistance (Ω/sq)	Transmittance (%) (at 550 nm)	FoM ($10^{-3}\ \Omega^{-1}$)
Ag NW	22.8	92.0	19.1
Ag NW–GO	15.6	90.7	24.2

Furthermore, the uniformity of R_s of the Ag NW networks, which is one of the most important quality factors for large-area transparent conductive films, was improved upon the formation of the Ag NW–GO hybrid networks. The uniformities of large-area Ag NW and Ag NW–GO networks (200 mm $\times$ 200 mm) were estimated by measuring their R_s values at 81 points with intervals of 20 mm in the horizontal and vertical directions. Figure 4 shows the distribution of the R_s values of the Ag NW and Ag NW–GO networks. The standard deviation of R_s of the Ag NW and Ag NW–GO networks were 4.21 Ω/sq and 1.48 Ω/sq, respectively. Therefore, the improved electro-optical properties and high uniformity of the Ag NW–GO network fabricated by spray coating made it suitable for large-area smart window applications.

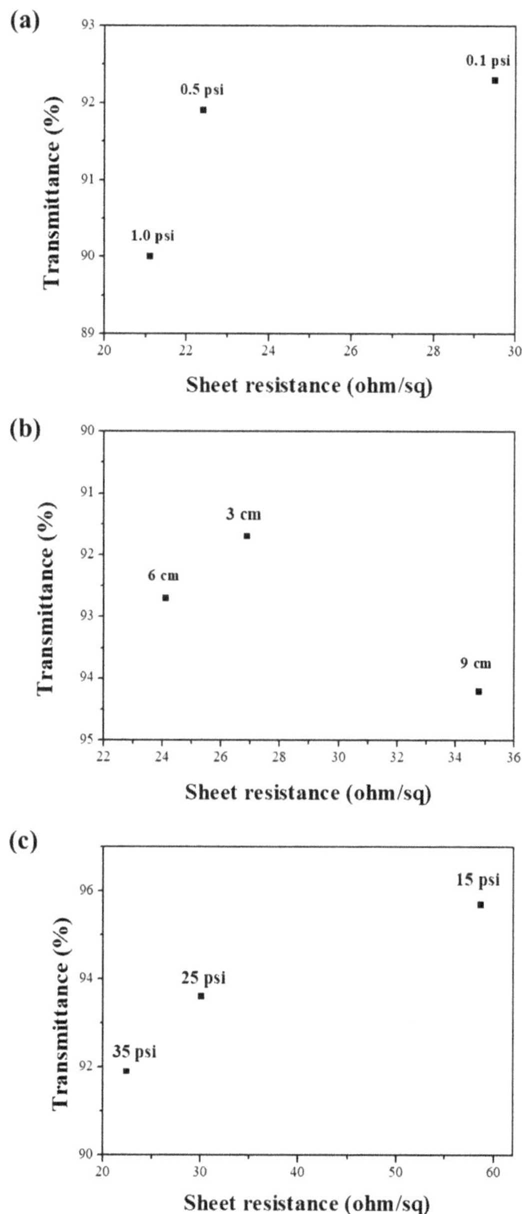

Figure 1. Transmittances at 550 nm and sheet resistances of the spray-coated Ag NW films as a function of the (**a**) dispensing pressure, (**b**) nozzle–to–substrate distance, and (**c**) nozzle pressure.

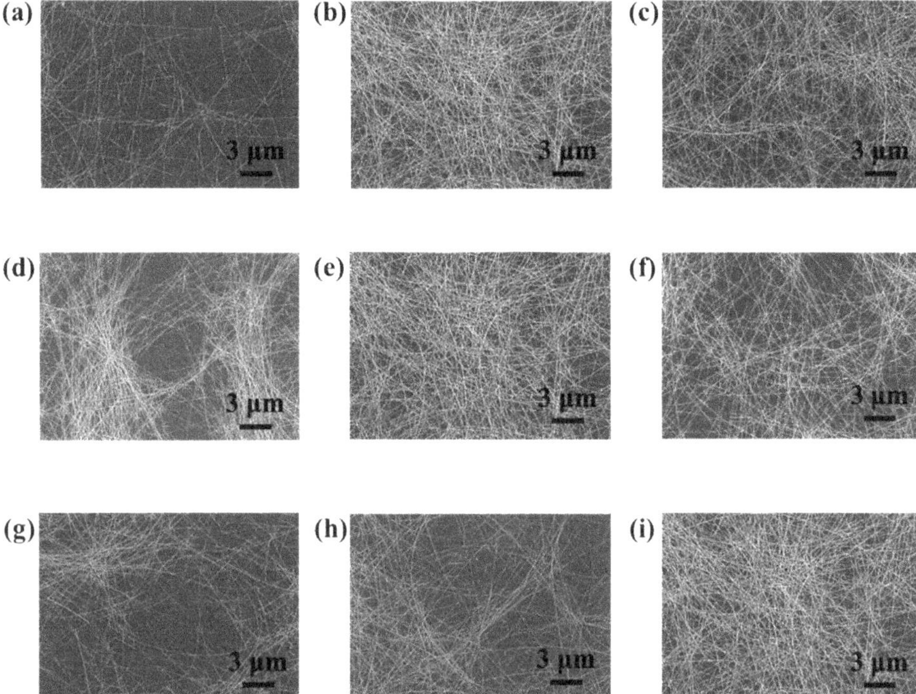

Figure 2. SEM images of Ag NW networks obtained with different spray-coating parameters including dispensing pressures of (**a**) 0.1 psi, (**b**) 0.5 psi, and (**c**) 1 psi, nozzle–to–substrate distances of (**d**) 3 cm, (**e**) 6 cm, and (**f**) 9 cm, and nozzle pressures of (**g**) 15 psi, (**h**) 25 psi, and (**i**) 35 psi.

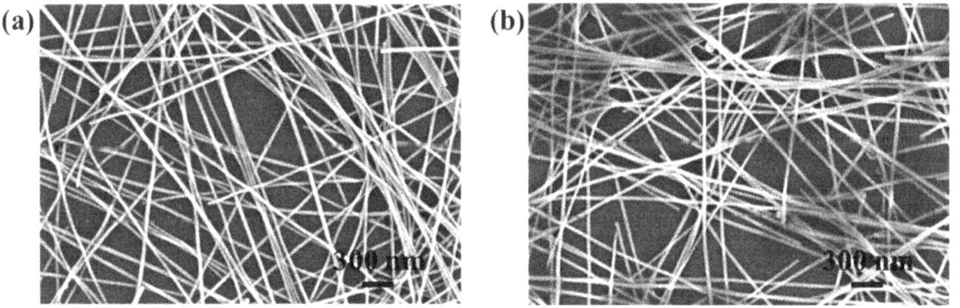

Figure 3. SEM images of the spray-coated (**a**) Ag NW network and (**b**) Ag NW network covered by GO nanosheets.

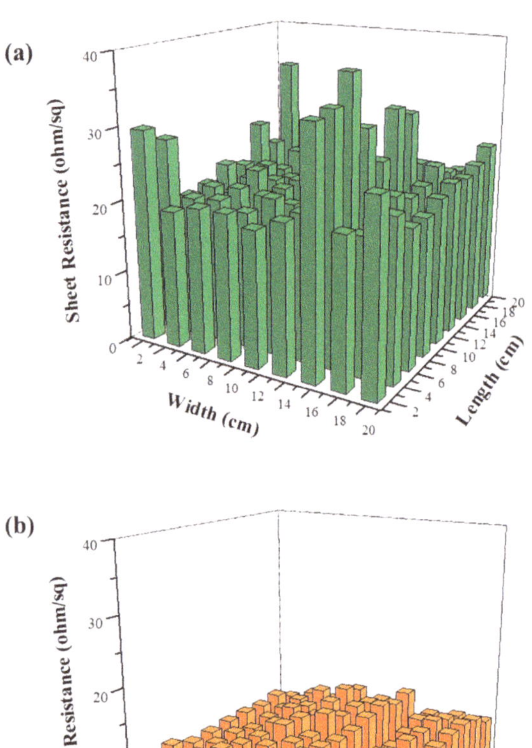

Figure 4. Histograms of the sheet resistances of the large-area, (**a**) Ag NW, and (**b**) Ag NW–GO hybrid electrodes.

The transmittances of the PDLC were measured at applied voltage amplitudes in the range of 0 to 80 V. Voltage–transmittance characteristics of PDLCs fabricated with Ag NW and Ag NW–GO hybrid electrodes are shown in Figure 5a. The difference in driving voltage between these two sets of PDLCs was clear. The driving voltage of the PDLC with Ag NW electrodes was larger than that of the PDLC fabricated by using Ag NW–GO electrodes. The transmittance of the PDLC based on the Ag NW–GO hybrid electrodes could reach 63% at 40 V while that of the Ag NW-based PDLC was only 53% at 40 V and the maximum transmittance was only 61% at 80 V. This result implied that the lower R_s of the Ag NW–GO electrodes than that of the Ag NW electrodes led to a lower driving voltage and lower energy consumption. Figure 5b shows the transmittances of the Ag NW and Ag NW–GO-based PDLCs in the on- (80 V) and off- (0 V) states as a function of the wavelength. The transmittance differences at 550 nm between the on-states and off-states for the Ag NW and Ag NW–GO-based PDLCs were 57% and 65%, respectively (Figure S4). Therefore, the Ag NW–GO-based PDLC was more desirable for use as a smart window since it was not only more transparent in the on state but could also be operated at a lower voltage.

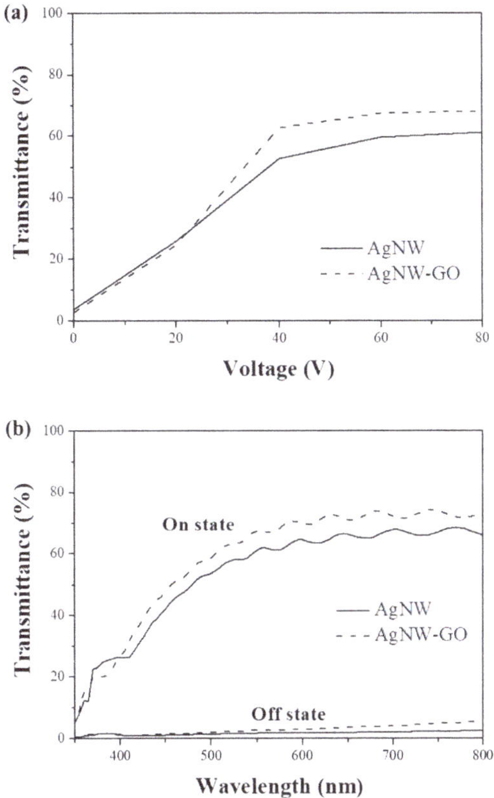

Figure 5. (**a**) Transmittances of the Ag NW and Ag NW–GO PDLCs under applied voltages in the range of 0 to 80 V at a wavelength of 550 nm. (**b**) Transmittance spectra of the Ag NW and Ag NW–GO PDLCs in the on- (80 V) and off- (0 V) states.

Photographs of the fabricated PDLCs in the on-states and off-states are shown in Figure 6. The difference in transmittance between the on-states and off-states was evident and in agreement with the results in Figure 5b. The fabricated PDLC was opaque in the off-state at which the printed text below the PDLC could hardly be observed. On the contrary, in the on-state, the printed text underneath the PDLC could be clearly observed since the PDLC was transparent. In the case of the Ag NW–GO-based PDLC, the printed text appears clearer than in the case of the Ag NW-based PDLC due to its high transmittance in the on-state. The high contrast between the on-states and off-states made the PDLC with Ag NW–GO suitable for smart windows.

Lastly, we aimed to fabricate a flexible large-area PDLC by utilizing the advantage of the PDLC based on the Ag NW–GO electrodes that were obtained by using the spray-coating technique. In order to fabricate a flexible large-area PDLC, Ag NW–GO was spray-coated on a PET substrate rather than on a glass substrate [24–26]. As shown in Figure 7a, the flexible PDLC operated perfectly and no damage was observed even though the PDLC was bent. The images of the large-area PDLC with the Ag NW–GO electrodes in Figure 7b,c show that the large-area PDLC operated uniformly in both on-states and off-states. This implied that the electric field between the Ag NW–GO electrodes was sufficiently uniform, which is consistent with the results illustrated in Figure 4b.

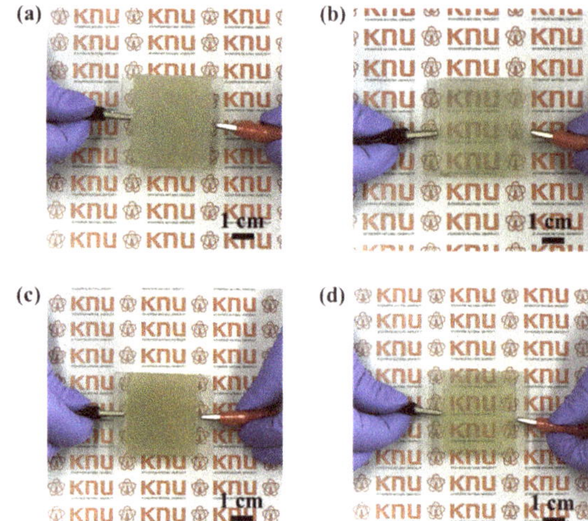

Figure 6. Photographs of the Ag NW PDLC in the (**a**) off-states and (**b**) on-states. Photographs of the Ag NW–GO PDLC in the (**c**) off-states and (**d**) on-states.

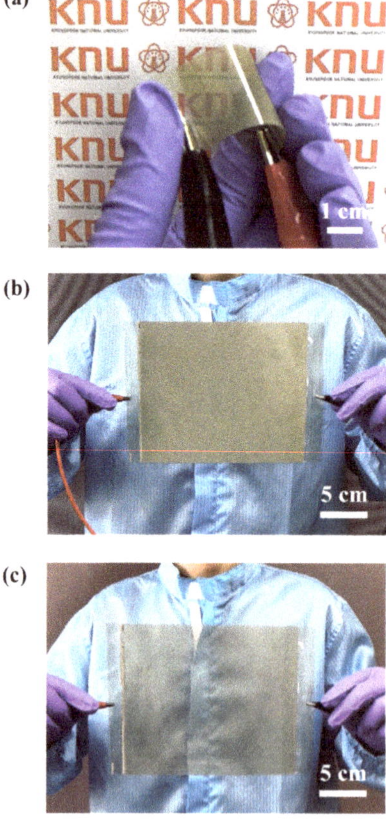

Figure 7. (**a**) Photograph of the flexible PDLC in the on-state. Photographs of the large-area PDLC in the (**b**) off-states and (**c**) on-states.

4. Conclusions

The spray-coating conditions were optimized for the fabrication of Ag NW–GO hybrid TCEs with high T and low R_s values, which can be used as superior alternatives to ITO TCEs. The PDLCs based on the Ag NW–GO electrodes exhibited a high transmittance of 63% in the on-state under a low driving voltage (40 V) due to the low R_s. The new class of flexible large-area PDLCs presented in this study demonstrate the promising potential of the spray-coated Ag NW–GO hybrid electrodes for PDLCs. This study provides a significant advancement toward the realization of flexible smart windows for future flexible optoelectronic applications.

Supplementary Materials: The following are available online at http://www.mdpi.com/1996-1944/11/11/2231/s1, Figure S1: Schematic illustration of the steps involved in fabricating PDLC, Figure S2: SEM images of Ag NW networks obtained with different spray-coating parameters, including dispensing pressures of (a) 0.1, (b) 0.5, and (c) 1 psi, nozzle–to–substrate distances of (d) 3, (e) 6, and (f) 9 cm, and nozzle pressures of (g) 15, (h) 25, and (i) 35 psi., Figure S3: The Raman spectra of Ag NW–GO, Figure S4: Transmittance spectra of the ITO, Ag NW and Ag NW–GO PDLCs in the on-state (80 V) and off-state (0 V).

Author Contributions: Y.C., C.S.K., and S.J. conceived and designed the experiments; Y.C., C.S.K., and S.J. performed the experiments and analyzed the data; Y.C. and S.J contributed to the manuscript preparation.

Funding: This study was supported by the Basic Science Research Program through the National Research Foundation of Korea (NRF) and funded by the Ministry of Science, ICT, and Future Planning (2017R1A2B4011499).

Conflicts of Interest: The authors declare no conflict of interest.

References

1. Lampert, C.M. Large-area smart glass and integrated photovoltaics. *Sol. Energ. Mater. Sol. Cells* **2003**, *76*, 489–499. [CrossRef]
2. Sol, J.A.; Timmermans, G.H.; van Breugel, A.J.; Schenning, A.P.; Debije, M.G. Multistate Luminescent Solar Concentrator "Smart" Windows. *Adv. Energy Mater.* **2018**, *8*, 1702922. [CrossRef]
3. Granqvist, C.G. Electrochromics for smart windows: Oxide-based thin films and devices. *Thin Solid Films* **2014**, *564*, 1–38. [CrossRef]
4. Cupelli, D.; Nicoletta, F.P.; Manfredi, S.; Vivacqua, M.; Formoso, P.; De Filpo, G.; Chidichimo, G. Self-adjusting smart windows based on polymer-dispersed liquid crystals. *Sol. Energ. Mater. Sol. Cells* **2009**, *93*, 2008–2012. [CrossRef]
5. Kim, Y.; Jung, D.; Jeong, S.; Kim, K.; Choi, W.; Seo, Y. Optical properties and optimized conditions for polymer dispersed liquid crystal containing UV curable polymer and nematic liquid crystal. *Curr. Appl. Phys.* **2015**, *15*, 292–297. [CrossRef]
6. Kim, Y.; Kim, K.; Kim, K.B.; Park, J.; Lee, N.; Seo, Y. Flexible polymer dispersed liquid crystal film with graphene transparent electrodes. *Curr. Appl. Phys.* **2016**, *16*, 409–414. [CrossRef]
7. Khaligh, H.H.; Liew, K.; Han, Y.; Abukhdeir, N.M.; Goldthorpe, I.A. Silver nanowire transparent electrodes for liquid crystal-based smart windows. *Sol. Energ. Mater. Sol. Cells* **2015**, *132*, 337–341. [CrossRef]
8. Chou, T.; Chen, S.; Chiang, Y.; Chang, T.; Lin, C.; Chao, C. Highly conductive PEDOT: PSS film by doping p-toluenesulfonic acid and post-treatment with dimethyl sulfoxide for ITO-free polymer dispersed liquid crystal device. *Org. Electron.* **2017**, *48*, 223–229. [CrossRef]
9. Lim, J.; Lee, S.; Kim, S.; Kim, T.; Koo, H.; Kim, H. Brush-paintable and highly stretchable Ag nanowire and PEDOT: PSS hybrid electrodes. *Sci. Rep.* **2017**, *7*, 14685. [CrossRef] [PubMed]
10. Ricciardulli, A.G.; Yang, S.; Wetzelaer, G.A.; Feng, X.; Blom, P.W. Hybrid Silver Nanowire and Graphene-Based Solution-Processed Transparent Electrode for Organic Optoelectronics. *Adv. Funct. Mater.* **2018**, *28*, 1706010. [CrossRef]
11. Chen, D.; Liang, J.; Liu, C.; Saldanha, G.; Zhao, F.; Tong, K.; Liu, J.; Pei, Q. Thermally stable silver nanowire–polyimide transparent electrode based on atomic layer deposition of zinc oxide on silver nanowires. *Adv. Funct. Mater.* **2015**, *25*, 7512–7520. [CrossRef]
12. Kim, C.; Jung, C.; Oh, Y.; Kim, D. A highly flexible transparent conductive electrode based on nanomaterials. *NPG Asia Mater.* **2017**, *9*, e438. [CrossRef]

13. Jang, J.; Im, H.; Jin, J.; Lee, J.; Lee, J.; Bae, B. A Flexible and Robust Transparent Conducting Electrode Platform Using an Electroplated Silver Grid/Surface-Embedded Silver Nanowire Hybrid Structure. *ACS Appl. Mater. Interfaces* **2016**, *8*, 27035–27043. [CrossRef] [PubMed]

14. Zhang, Q.; Di, Y.; Huard, C.M.; Guo, L.J.; Wei, J.; Guo, J. Highly stable and stretchable graphene–polymer processed silver nanowires hybrid electrodes for flexible displays. *J. Mater. Chem. C* **2015**, *3*, 1528–1536. [CrossRef]

15. Lee, J.; Lee, P.; Lee, H.B.; Hong, S.; Lee, I.; Yeo, J.; Lee, S.S.; Kim, T.; Lee, D.; Ko, S.H. Room-temperature nanosoldering of a very long metal nanowire network by conducting-polymer-assisted joining for a flexible touch-panel application. *Adv. Funct. Mater.* **2013**, *23*, 4171–4176. [CrossRef]

16. Zilberberg, K.; Gasse, F.; Pagui, R.; Polywka, A.; Behrendt, A.; Trost, S.; Heiderhoff, R.; Görrn, P.; Riedl, T. Highly Robust Indium-Free Transparent Conductive Electrodes Based on Composites of Silver Nanowires and Conductive Metal Oxides. *Adv. Funct. Mater.* **2014**, *24*, 1671–1678. [CrossRef]

17. Kim, D.; Zhu, L.; Jeong, D.; Chun, K.; Bang, Y.; Kim, S.; Kim, J.; Oh, S. Transparent flexible heater based on hybrid of carbon nanotubes and silver nanowires. *Carbon* **2013**, *63*, 530–536. [CrossRef]

18. Fan, Z.; Liu, B.; Liu, X.; Li, Z.; Wang, H.; Yang, S.; Wang, J. A flexible and disposable hybrid electrode based on Cu nanowires modified graphene transparent electrode for non-enzymatic glucose sensor. *Electrochim. Acta* **2013**, *109*, 602–608. [CrossRef]

19. Lee, J.; Shin, D.; Park, J. Fabrication of silver nanowire-based stretchable electrodes using spray coating. *Thin Solid Films* **2016**, *608*, 34–43. [CrossRef]

20. Scardaci, V.; Coull, R.; Lyons, P.E.; Rickard, D.; Coleman, J.N. Spray deposition of highly transparent, low-resistance networks of silver nanowires over large areas. *Small* **2011**, *7*, 2621–2628. [CrossRef] [PubMed]

21. Deegan, R.D.; Bakajin, O.; Dupont, T.F.; Huber, G.; Nagel, S.R.; Witten, T.A. Contact line deposits in an evaporating drop. *Phys. Rev. E* **2000**, *62*, 756. [CrossRef]

22. Ha, B.; Jo, S. Hybrid Ag nanowire transparent conductive electrodes with randomly oriented and grid-patterned Ag nanowire networks. *Sci. Rep.* **2017**, *7*, 11614. [CrossRef] [PubMed]

23. Liang, J.; Li, L.; Tong, K.; Ren, Z.; Hu, W.; Niu, X.; Chen, Y.; Pei, Q. Silver nanowire percolation network soldered with graphene oxide at room temperature and its application for fully stretchable polymer light-emitting diodes. *ACS Nano* **2014**, *8*, 1590–1600. [CrossRef] [PubMed]

24. Singh, A.; Salmi, Z.; Joshi, N.; Jha, P.; Decorse, P.; Lecoq, H.; Lau-Truong, S.; Jouini, M.; Aswal, D.; Chehimi, M. Electrochemical investigation of free-standing polypyrrole–silver nanocomposite films: A substrate free electrode material for supercapacitors. *RCS Adv.* **2013**, *3*, 24567–24575. [CrossRef]

25. Kim, Y.; Hong, J.; Lee, S. Fabrication of a highly bendable LCD with an elastomer substrate by using a replica-molding method. *J. Soc. Inf. Disp.* **2006**, *14*, 1091–1095. [CrossRef]

26. Kim, I.; Kim, T.; Lee, S.; Kim, B. Extremely Foldable and Highly Transparent Nanofiber-Based Electrodes for Liquid Crystal Smart Device. *Sci. Rep.* **2018**, *8*, 11517. [CrossRef] [PubMed]

 *materials*

Article

Soft-Material-Based Smart Insoles for a Gait Monitoring System

Changwon Wang [1], Young Kim [2] and Se Dong Min [1,*

[1] Department of Medical IT Engineering, Soonchunhyang University, Asan 31538, Korea; changwon@sch.ac.kr
[2] Wellness Coaching Service Research Center, Soonchunhyang University, Asan 31538, Korea; ykim02@sch.ac.kr
* Correspondence: sedongmin@sch.ac.kr; Tel.: +82-41-530-4871

Received: 26 October 2018; Accepted: 26 November 2018; Published: 30 November 2018

Abstract: Spatiotemporal analysis of gait pattern is meaningful in diagnosing and prognosing foot and lower extremity musculoskeletal pathologies. Wearable smart sensors enable continuous real-time monitoring of gait, during daily life, without visiting clinics and the use of costly equipment. The purpose of this study was to develop a light-weight, durable, wireless, soft-material-based smart insole (SMSI) and examine its range of feasibility for real-time gait pattern analysis. A total of fifteen healthy adults (male: 10, female: 5, age 25.1 $\pm$ 2.64) were recruited for this study. Performance evaluation of the developed insole sensor was first executed by comparing the signal accuracy level between the SMSI and an F-scan. Gait data were simultaneously collected by two sensors for 3 min, on a treadmill, at a fixed speed. Each participant walked for four times, randomly, at the speed of 1.5 km/h (C1), 2.5 km/h (C2), 3.5 km/h (C3), and 4.5 km/h (C4). Step count from the two sensors resulted in 100% correlation in all four gait speed conditions (C1: 89 $\pm$ 7.4, C2: 113 $\pm$ 6.24, C3: 141 $\pm$ 9.74, and C4: 163 $\pm$ 7.38 steps). Stride-time was concurrently determined and R2 values showed a high correlation between the two sensors, in both feet ($R^2 \geq 0.90$, $p < 0.05$). Bilateral gait coordination analysis using phase coordination index (PCI) was performed to test clinical feasibility. PCI values of the SMSI resulted in 1.75 $\pm$ 0.80% (C1), 1.72 $\pm$ 0.81% (C2), 1.72 $\pm$ 0.79% (C3), and 1.73 $\pm$ 0.80% (C4), and those of the F-scan resulted in 1.66 $\pm$ 0.66%, 1.70 $\pm$ 0.66%, 1.67 $\pm$ 0.62%, and 1.70 $\pm$ 0.62%, respectively, showing the presence of a high correlation ($R^2 \geq 0.94$, $p < 0.05$). The insole developed in this study was found to have an equivalent performance to commercial sensors, and thus, can be used not only for future sensor-based monitoring device development studies but also in clinical setting for patient gait evaluations.

Keywords: conductive textile; capacitive pressure sensor; gait; monitoring; phase coordination index

---

1. Introduction

Human gait is an essential means of locomotion for daily life and the most important function necessary for quality of life [1]. Walking dysfunctions can lead to falling, fracture, progression of disease, decreased mobility, and depression; all limiting the performances in daily activities. Early diagnosis of gait-related impairments is important in preventing symptom aggravation and irreversible deformities. Reliable yet practical diagnostic medical devices with high resolution sensors need to be continuously advanced.

Substantial evidence on cognitive neuroscience and motor control suggests that gait parameters can reveal important factors that determine the overall health and well-being [2,3]. However, present gait analysis systems and the results reported are based on costly equipment, bulky and complicated set-up, require multiple types of sensors, and are limited to indoors [4–6]. Therefore, development of high-performance, wearable smart analytic systems with affordable prices, is in demand for real-time daily activity monitoring and analysis. With practical methodologies, neurological, musculoskeletal, kinematic, and sports-related problems associated with gait and foot pathologies, can be more efficiently addressed.

Meaningful gait parameters include gait speed, step-count, stride-time, center of pressure (CoP), and phase coordination index (PCI) [7–12]. Walking speed is the product of step-length and step-frequency (step-count/time) and is considered the sixth vital sign, because it has been validated as a marker of frailty and mortality [13,14]. Wearable smart insole that can monitor the changes in walking speed, during various types of activities, is expected to enhance the quality of gait-related research and therapy.

For walking improvements in the elderly or rehabilitation patients, mobility skill practice is essential; use of a smart insole feedback system could accelerate the procedure. Step-count in a controlled environment is known to be a reliable and valid indicator in quantifying the temporal frequency of gait [14]. It is used to evaluate current motor control functions and set future goals in rehabilitation therapy. Stride-time and stride-length present one's gait cycle, and the coordination patterns of limb segments can be used to identify the joint mechanics [15]. Understanding the coordination patterns, in movement analysis, can be used to diagnose and prognose neuro-cognitive functions. Plantar pressure analyses are considered meaningful in examining the biomechanical characteristics of the foot, because related sports injuries to functional deformities can be diagnosed [16].

Among these parameters, PCI is reported to be a relatively more sensitive measure in analyzing the bilateral coordination or asymmetry of locomotion and balance, which are significant variables, especially in rehabilitative medicine [17–19]. PCI is an indicator evaluating the coordination of left–right stepping phase, and the PCI value closer to 0% means that the two feet are moving in a higher coordination. Balance and coordination are fundamental motor control functions for normal gait. PCI analysis can also be extensively applied to diagnose the severity of scoliosis, hemiparesis, and aging. Despite the significance, most of the current insole sensors, used for gait analysis, are not designed to detect and analyze PCI. The need for development of a cost-effective wearable sensor that can measure PCI in real-life conditions, is prominent.

A number of scientists have developed wearable sensors for gait analysis, but most of them are equipped with multiple bulky measurement devices, including dual 3-axis accelerometer, gyroscope, torque, ground reaction force sensor, and pressure sensor, making the measuring procedure complicated and inconvenient [20–22]. In recent studies by Wu et al. (2015) and Park et al. (2018), insole-type pressure sensor and a smart shoes system were developed for gait analysis and smart phone applications-enabled real-time monitoring of the activities have been carried out [23,24]. However, many of these research-based, newly developed wearable devices are rather expensive and the bio-signals collected and analyzed are not as accurate, compared to that of commercial sensors [25].

With the aim to overcome and complement the aforementioned limitations and fulfill the needs of a clinical field, this study used conductive textile, a type of soft material reported to be user-friendly, inexpensive, and easily transformable, to develop a practical sensor. A light-weight, durable, wireless, soft-material-based smart insole (SMSI) sensor was developed for accurate and affordable real-time gait analysis system and its range of feasibility was examined.

2. Materials and Methods

2.1. Textile Capacitive Pressure Insole

To obtain gait data, a parallel capacitance-based pressure sensor, using conductive textile, was developed. For the sensor, a W-290-PCN model (A-jin Electron, Busan, Korea) was used as shown in Table 1. This model is made of polyester, sequentially-plated with nickel, copper, and nickel.

Table 1. Specifications of the W-290-PCN.

Parameter	W-290-PCN	Textile Structure
Based material	Polyester	
Type	Woven	
Width (mm)	1100 ± 5	
Weight (g/m^2)	81 ± 5	
Thickness (mm)	0.1 ± 0.01	
Density (g/m^3)	188 ± 5	

Capacitance is a physical quantity that indicates the ability of an object to accumulate electrical charges. The unit is F, and 1 F is equal to the capacitance of the capacitor charged at 1 C, when a voltage of 1 V is applied. The parallel capacitance of a capacitor can be calculated by Equation (1).

$$C = \frac{Q}{V} = \varepsilon\frac{A}{d} \tag{1}$$

where, d is distance between the plates, A is the area of plates, ε is the permittivity material between the plates. Figure 1 shows the structure of parallel capacitances, where C is inversely proportional to the distance of the two plates and is proportional to the area of the material and the dielectric constant between the plates.

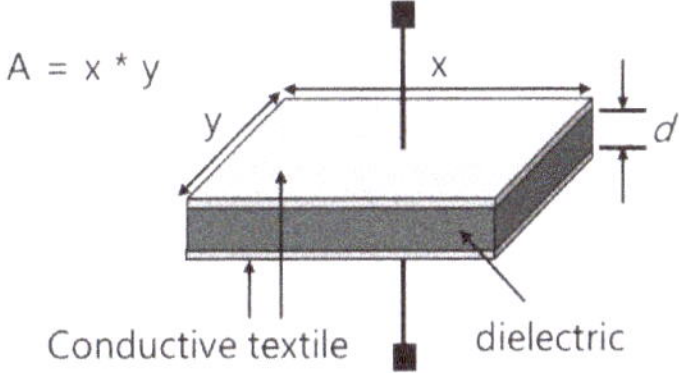

Figure 1. The structure of a parallel capacitor.

Figure 2 shows the structure of the soft-material-based smart insole (SMSI). The insole sensor was developed with two plates of conductive textile (W-290-PCN) and a non-conductive rubber, with a thickness of 3 mm, placed between the two sensor layers. The size of each sensor was 2×2 cm^2, embedded with ten channels for each foot. Figure 3 shows the location of each channel in the insole for sizes of 270 mm and 240 mm.

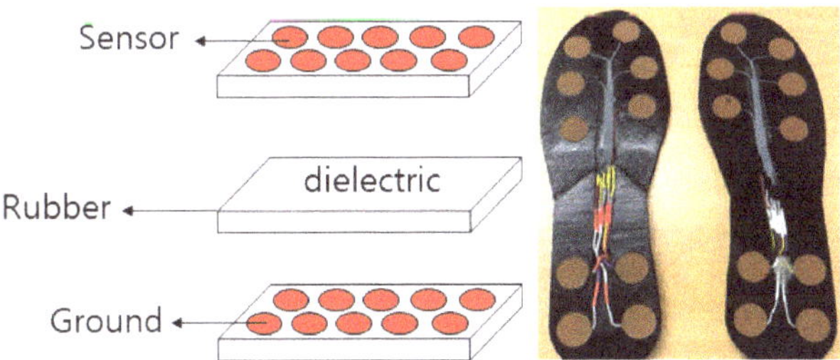

Figure 2. The structure of the proposed sensor—the soft-material-based smart insole (SMSI).

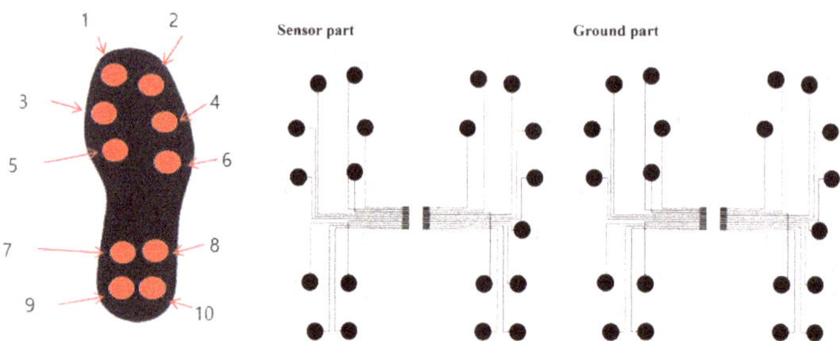

Figure 3. The sensor location of SMSI.

2.2. Gait Data Measurement and Monitoring System

Figure 4 shows the block diagram of our proposed gait measurement and monitoring system. The proposed system is divided into hardware and software division. The hardware collects the gait data from the feet and transmits the data to the software, using Bluetooth communication. The software saves the data and displays raw data from each foot.

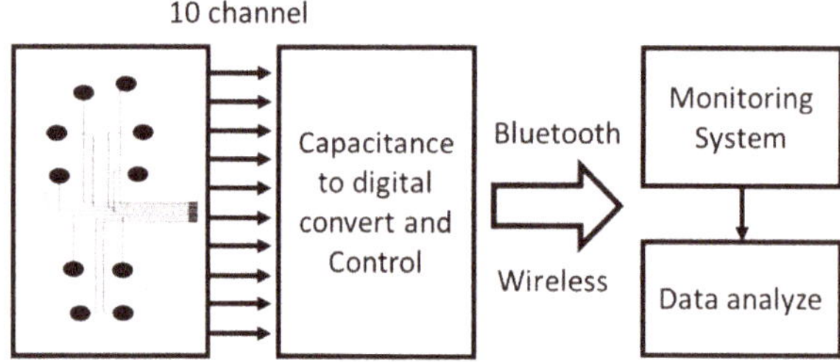

Figure 4. Block diagram of the proposed gait measurement and monitoring system.

2.2.1. Hardware Design

Figure 5 shows the schematic of hardware design of gait measurement system and Figure 6 shows the structure of the capacitance-measuring printed circuit board (PCB). Our proposed board size was 2.3×3.3 cm^2 and the operation power was 3.7 V. To convert an analog signal to a digital signal of capacitance, the sensor MPR121QR2 (Freescale Inc., Austin, TX, USA) was used. It has a measurement range of 10 pF to 2000 pF, and has a resolution of 0.01 pF. A micro controller unit (MCU), developed by STMicroelectronics (Geneva, Switzerland), was used to measure the capacitance and the PCB was developed. Data from the ten channels were sampled at 100 Hz.

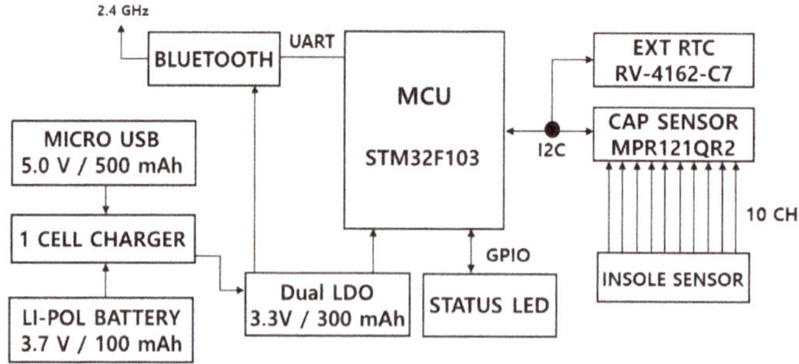

Figure 5. Schematic of the gait measurement system.

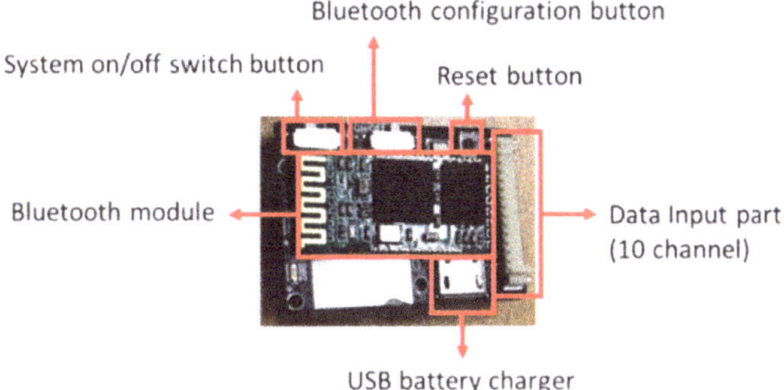

Figure 6. The structure of the capacitance-measuring printed circuit board (PCB).

2.2.2. Software Design

The gait monitoring system was developed in C# language, as illustrated in Figure 7. It was developed to transmit data between the PCB board and a monitoring system via Bluetooth communication. Baud rate was set at 115,200, non-parity bit was 0, and stop bit was 1. Since the monitoring system used separate PCBs for each side of the foot, a total of two Bluetooth devices were paired up at the same time. Our monitoring system was developed to present real-time graphs to confirm the raw gait data being collected from twenty channels. It also contains functions to save the data as a text file, as well as an Excel file, from a desired point in time.

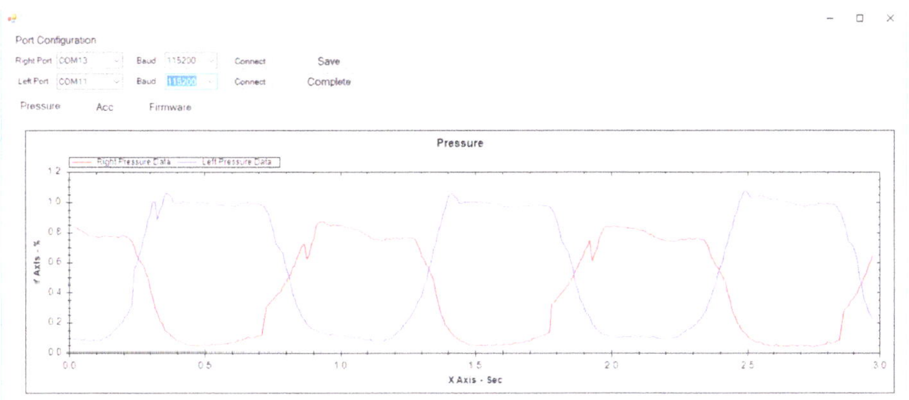

Figure 7. Gait monitoring system.

2.3. Data Acquisition

Spatiotemporal data were detected by the SMSI, during gait. The temporal moments of heel strike, midstance, and toe-off, during walking, were collected by ten insole sensor channels, in each foot. Summation of all collected data from the ten channels were analyzed. Channels 9 and 10, specifically, detected the pressure distribution area at the time of heel strike, and Channels 1 and 2 detected the toe-off moment as shown in Figure 8.

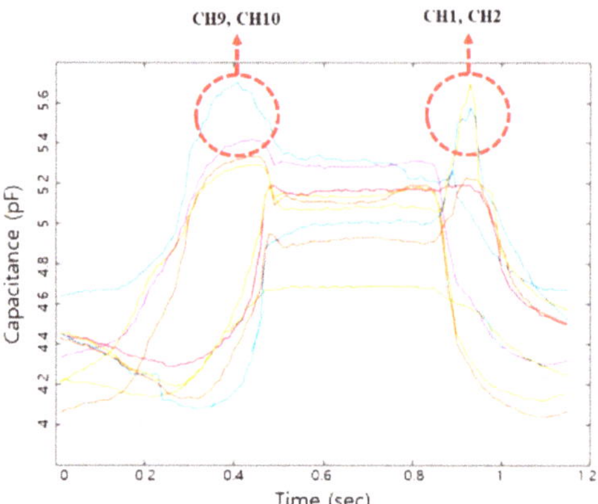

Figure 8. The ten channel data from the SMSI.

2.4. Signal Processing

To analyze gait features, we implemented a peak detection algorithm by reflecting the method of Pan-Tomkins algorithm [26], as shown in Figure 9. A low-pass filter was first applied with a cut-off frequency of 3 Hz. Then, a moving average filter of five points was applied. The low-pass filter and the moving-average filter were used for smoothing the data. The first-order differential filter was applied to make the slope of the original signal larger, as the change value of the Y-axis increased. We developed an algorithm to detect the highest peak at 300 ms intervals, and the heel strike and toe off points were calculated, based on the local maxima algorithm.

Each heel strike was calculated for the step-count, and the stride-time was defined by the time between two consecutive heel strikes, in the same foot [27].

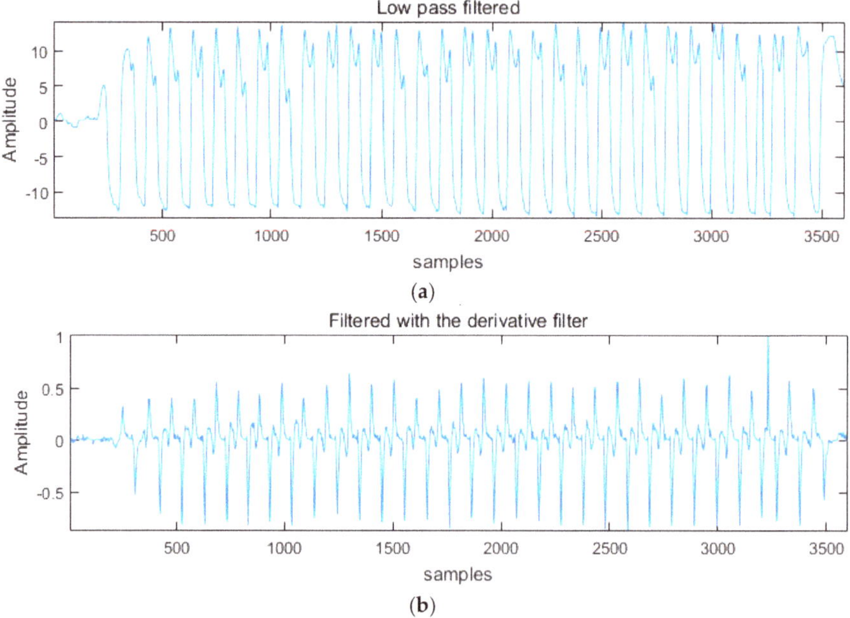

Figure 9. *Cont.*

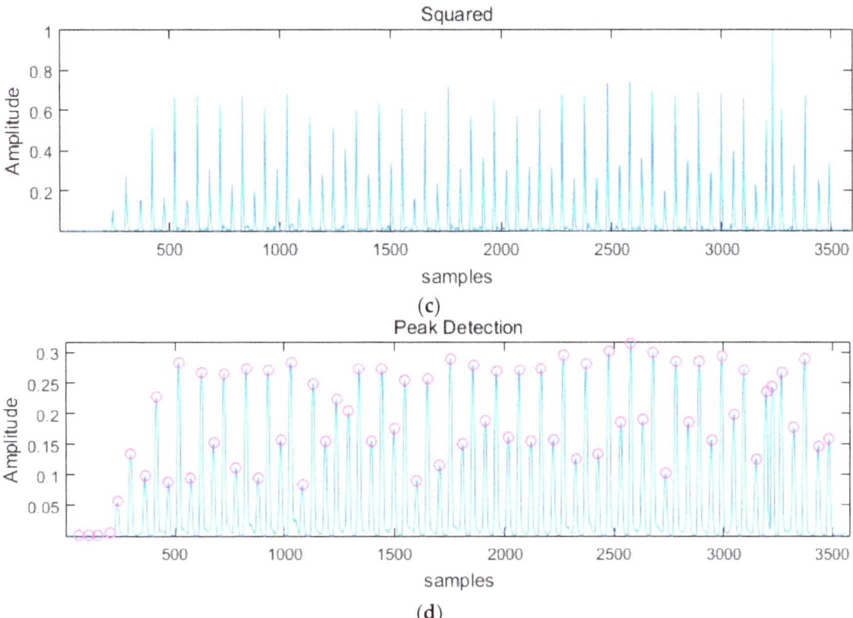

Figure 9. The procedure of signal processing: (**a**) low pass filtered gait signal; (**b**) derivative filtered gait signal; (**c**) squared gait signal; (**d**) the result of peak detection (heel strike and toe off).

PCI was calculated by φ_i, φ_ABS and φ_CV, as shown below in Equations (2)–(6) [28]. φ_i was an index that evaluated the symmetry of bilateral stepping phases and it was the distance between one heel strike and the next of the opposite leg, calculated as an angle, ideally φ_i = 180° (Figure 10). As shown in the equation below, φ_ABS indicated the balance between the two feet. φ_CV referred to the coefficient of variation of φ_i, which represented the consistency of the right and left foot, during walking.

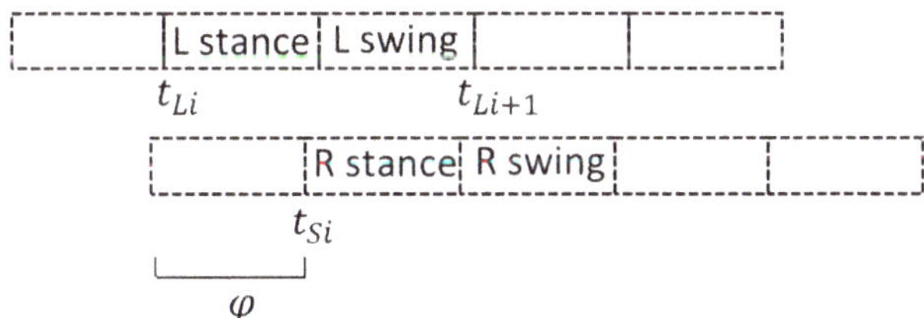

Figure 10. Left -Right stepping phase (φ) in a gait cycle.

$$\varphi_i = 360° \times \frac{t_{S_i} - t_{L_i}}{t_{Li+1} - t_{L_i}} \tag{2}$$

$$\varphi_{ABS} = |\,\varphi_i - 180°\,| \tag{3}$$

$$P\varphi_{ABS} = 100 \times (\varphi_{ABS}/180°) \tag{4}$$

$$\varphi' = \frac{1}{N} \sum_{i=1}^{n} \varphi_i, \ \delta = \sqrt{\frac{1}{N} \sum_{i=1}^{n} (\varphi' - \varphi_i)^2}, \ \varphi\_CV = \frac{\delta}{\varphi'} \tag{5}$$

$$PCI = \varphi\_CV + P\varphi\_ABS \tag{6}$$

2.5. Experimental Methods

2.5.1. Characteristics of the Subjects

A total of fifteen healthy subjects (male: 10, female: 5) participated in this experiment, as summarized in Table 2. Healthy men and women in their twenties (average age: 25.1 ± 2.64), who had no history of gait disorders, during the past six months, were recruited. The experimental protocol is shown in Table 3. All subjects were adequately informed about the experiment procedure and the experiment was conducted after obtaining written consent from all participants.

Table 2. The characteristics of the subjects.

Variables	Male	Female
Age (years)	26.10 ± 2.18	23.20 ± 2.58
Gender (M/F)	10	5
Diagnosis	N/A	N/A

Table 3. The experimental protocol on the treadmill.

Speed (km/h)	Time (min)
1.5 (C1)	3
Rest	1
2.5 (C2)	3
Rest	1
3.5 (C3)	3
Rest	1
4.5 (C4)	3
Rest	1

2.5.2. The Feasibility Test Protocol

For the feasibility test, the performance of the SMSI was evaluated by comparing its accuracy level with a commercial sensor (F-scan, Tekscan, South Boston, MA, USA). Gait data from the two sensors were simultaneously acquired, while walking on the treadmill, for 3 min, at four different pre-set speed conditions. Condition 1 (C1) was set at 1.5 km/h, Condition 2 (C2) at 2.5 km/h, Condition 3 (C3) at 3.5 km/h, and Condition 4 (C4) at 4.5 km/h (Table 3). A one-minute resting time was given between the conditions, and the order of the four speed conditions were randomly selected.

Subjects were instructed to walk on the treadmill with both the SMSI and the F-scan sensors placed inside the given sneakers. For the analysis, the step-count, the stride-time, and the PCI were calculated and compared between the two sensors. The raw data, simultaneously collected from the SMSI and the F-scan sensors, were monitored in real-time, as shown below in Figure 11.

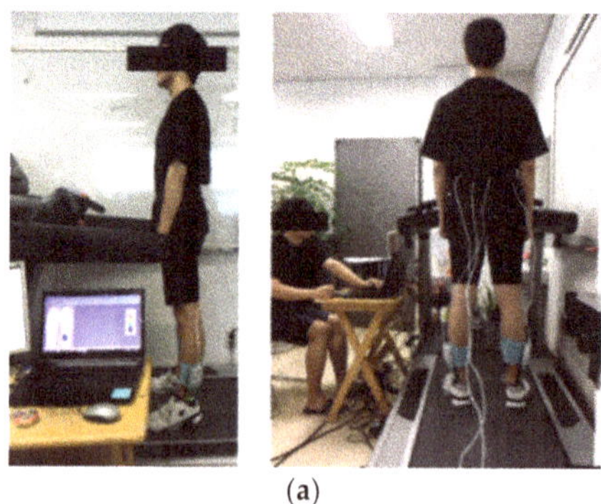

(**a**)

Figure 11. *Cont.*

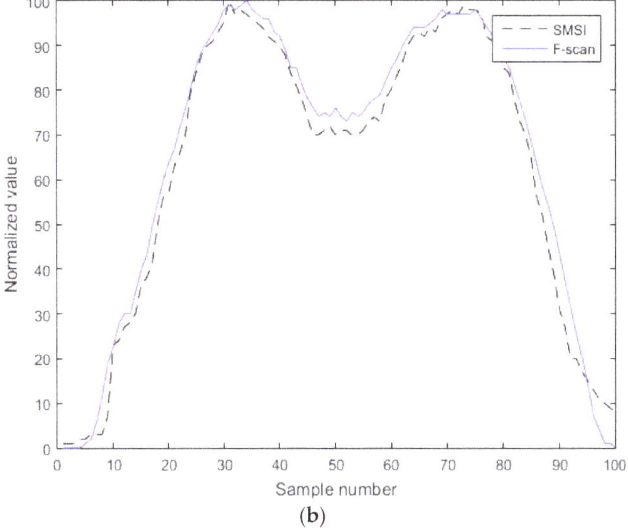

Figure 11. Experimental method, (**a**) experimental environment and (**b**) real-time raw data of the SMSI and the F-scan.

3. Results

The Results of the Sensor Performance

Table 4 shows a 100% consistency in the step-count detected by the two sensors in four different gait speed conditions (C1: 89, C2: 113, C3: 141, C4: 163).

Table 4. The results of the step-count detection between the two sensors.

Variables	C1	C2	C3	C4
SMSI	89	113	141	163
F-scan	89	113	141	163
Consistency (%)	100%	100%	100%	100%

The results of stride-time are presented below in Table 5 and refer to Figure 12. The average stride-time and standard deviation, of each subject, under the four different gait speed conditions were calculated for the left and right leg. The stride-time of only the left foot is reported here because there was a high correlation in stride-time detected from both side of the legs.

Table 5. Average stride-time of each subject for SMSI and F-scan of the left foot.

Subject No.	Sensors	C1	C2	C3	C4
1	SMSI	1.75 ± 0.12	1.54 ± 0.12	1.24 ± 0.07	1.10 ± 0.02
	F-scan	1.75 ± 0.09	1.54 ± 0.08	1.24 ± 0.05	1.09 ± 0.03
2	SMSI	1.65 ± 0.09	1.46 ± 0.08	1.21 ± 0.04	1.07 ± 0.02
	F-scan	1.65 ± 0.10	1.46 ± 0.06	1.20 ± 0.05	1.07 ± 0.02
3	SMSI	1.81 ± 0.09	1.56 ± 0.06	1.22 ± 0.07	1.08 ± 0.02
	F-scan	1.80 ± 0.09	1.55 ± 0.06	1.20 ± 0.05	1.06 ± 0.01
4	SMSI	1.88 ± 0.09	1.51 ± 0.05	1.23 ± 0.02	1.08 ± 0.01
	F-scan	1.88 ± 0.10	1.51 ± 0.04	1.23 ± 0.01	1.07 ± 0.01
5	SMSI	1.87 ± 0.13	1.53 ± 0.05	1.26 ± 0.03	1.07 ± 0.01
	F-scan	1.86 ± 0.11	1.53 ± 0.03	1.26 ± 0.01	1.05 ± 0.01

Table 5. *Cont.*

Subject No.	Sensors	C1	C2	C3	C4
6	SMSI	1.81 ± 0.11	1.49 ± 0.06	1.19 ± 0.02	1.09 ± 0.01
	F-scan	1.80 ± 0.09	1.50 ± 0.05	1.19 ± 0.01	1.07 ± 0.01
7	SMSI	1.85 ± 0.10	1.45 ± 0.05	1.22 ± 0.04	1.12 ± 0.02
	F-scan	1.85 ± 0.07	1.45 ± 0.03	1.21 ± 0.01	1.10 ± 0.01
8	SMSI	1.89 ± 0.17	1.44 ± 0.03	1.16 ± 0.02	1.06 ± 0.01
	F-scan	1.88 ± 0.15	1.44 ± 0.04	1.15 ± 0.01	1.05 ± 0.01
9	SMSI	1.82 ± 0.08	1.63 ± 0.04	1.22 ± 0.05	1.08 ± 0.02
	F-scan	1.82 ± 0.10	1.64 ± 0.06	1.21 ± 0.04	1.08 ± 0.01
10	SMSI	1.83 ± 0.06	1.51 ± 0.05	1.26 ± 0.02	1.09 ± 0.01
	F-scan	1.81 ± 0.09	1.51 ± 0.04	1.25 ± 0.03	1.10 ± 0.01
11	SMSI	1.77 ± 0.18	1.53 ± 0.06	1.21 ± 0.01	1.18 ± 0.01
	F-scan	1.78 ± 0.14	1.54 ± 0.05	1.20 ± 0.04	1.19 ± 0.03
12	SMSI	1.86 ± 0.10	1.38 ± 0.05	1.24 ± 0.04	1.12 ± 0.01
	F-scan	1.85 ± 0.11	1.37 ± 0.06	1.24 ± 0.03	1.11 ± 0.01
13	SMSI	1.81 ± 0.09	1.29 ± 0.03	1.14 ± 0.04	1.06 ± 0.01
	F-scan	1.81 ± 0.13	1.28 ± 0.03	1.13 ± 0.02	1.06 ± 0.01
14	SMSI	1.70 ± 0.09	1.46 ± 0.10	1.21 ± 0.02	1.03 ± 0.01
	F-scan	1.71 ± 0.07	1.46 ± 0.05	1.22 ± 0.03	1.05 ± 0.01
15	SMSI	1.74 ± 0.08	1.50 ± 0.07	1.22 ± 0.02	1.03 ± 0.01
	F-scan	1.74 ± 0.06	1.50 ± 0.04	1.22 ± 0.04	1.01 ± 0.02

C1: 1.5 km/h, C2: 2.5 km/h, C3: 3.5 km/h, C4: 4.5 km/h.

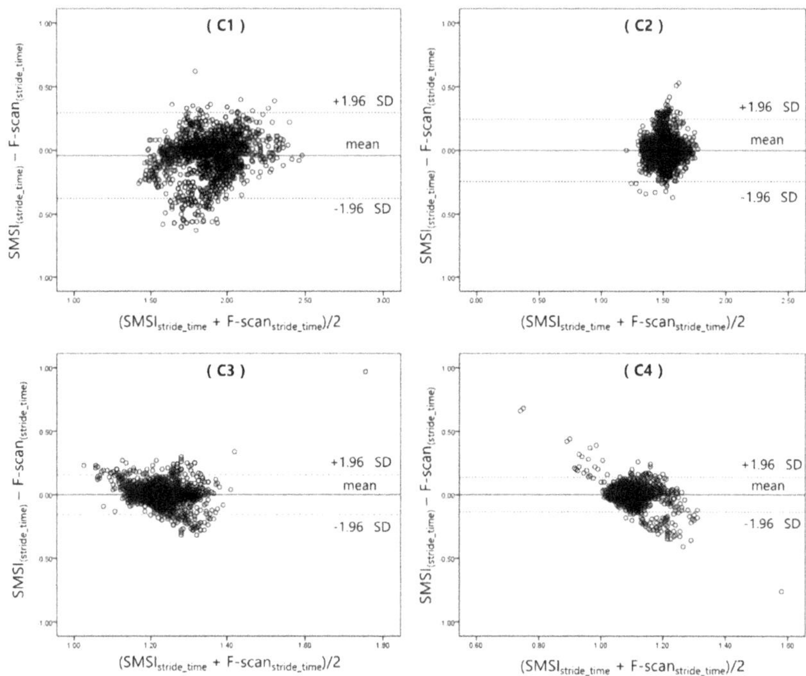

Figure 12. The Bland-Altman plot showing the difference between the SMSI and F-scan for the left foot stride-time, in each gait speed condition. The dashed line in the middle is the mean value of the differences, the lines above and below denote the standard deviation (95% CI).

Table 6 shows the results of the correlation analysis on the stride-time detected by the SMSI and the F-scan. Data from the left foot of all subjects were compared between the two sensors, and the R^2 value resulted in 0.91 ± 0.04 (C1), 0.93 ± 0.02 (C2), 0.94 ± 0.02 (C3), and 0.95 ± 0.03 (C4). The R^2 values of the right foot were 0.90 ± 0.03, 0.93 ± 0.02, 0.94 ± 0.02, and 0.93 ± 0.03, respectively. A p-value of <0.05 was set, for statistical significance.

Table 6. The results of the correlation analysis on the stride-time, between the two sensors.

Subject No.	Foot side	C1	p	C2	p	C3	p	C4	p
1	LF	0.92 *	0.00	0.94 *	0.00	0.98 *	0.001	0.93 *	0.00
	RF	0.94 *	0.00	0.92 *	0.00	0.91 *	0.00	0.91 *	0.00
2	LF	0.88 *	0.00	0.92 *	0.01	0.97 *	0.00	0.94 *	0.00
	RF	0.88 *	0.00	0.95 *	0.00	0.93 *	0.00	0.91 *	0.00
3	LF	0.88 *	0.001	0.96 *	0.00	0.91 *	0.02	0.97 *	0.00
	RF	0.93 *	0.00	0.92 *	0.00	0.95 *	0.00	0.91 *	0.00
4	LF	0.87 *	0.00	0.95 *	0.00	0.93 *	0.00	0.97 *	0.00
	RF	0.89 *	0.00	0.94 *	0.00	0.96 *	0.00	0.98 *	0.00
5	LF	0.84 *	0.00	0.90 *	0.03	0.97 *	0.00	0.97 *	0.00
	RF	0.87 *	0.001	0.97 *	0.00	0.98 *	0.00	0.96 *	0.00
6	LF	0.97 *	0.00	0.91 *	0.00	0.98 *	0.00	0.96 *	0.00
	RF	0.82 *	0.00	0.92 *	0.00	0.94 *	0.00	0.92 *	0.00
7	LF	0.98 *	0.00	0.91 *	0.00	0.91 *	0.02	0.98 *	0.00
	RF	0.84 *	0.00	0.91 *	0.00	0.98 *	0.00	0.93 *	0.00
8	LF	0.97 *	0.00	0.90 *	0.00	0.97 *	0.00	0.97 *	0.00
	RF	0.88 *	0.00	0.90 *	0.00	0.95 *	0.00	0.95 *	0.00
9	LF	0.91 *	0.00	0.94 *	0.00	0.94 *	0.00	0.95 *	0.00
	RF	0.83 *	0.00	0.92 *	0.00	0.91 *	0.00	0.90 *	0.00
10	LF	0.88 *	0.001	0.94 *	0.00	0.96 *	0.00	0.97 *	0.01
	RF	0.89 *	0.024	0.91 *	0.00	0.91 *	0.00	0.98 *	0.00
11	LF	0.98 *	0.00	0.96 *	0.00	0.96 *	0.00	0.95 *	0.00
	RF	0.87 *	0.00	0.92 *	0.01	0.99 *	0.00	0.98 *	0.00
12	LF	0.91 *	0.00	0.90 *	0.00	0.93 *	0.00	0.91 *	0.00
	RF	0.90 *	0.00	0.97 *	0.00	0.96 *	0.001	0.92 *	0.00
13	LF	0.92 *	0.01	0.93 *	0.00	0.96 *	0.00	0.95 *	0.00
	RF	0.93 *	0.00	0.98 *	0.00	0.93 *	0.00	0.90 *	0.00
14	LF	0.88 *	0.00	0.95 *	0.00	0.93 *	0.00	0.96 *	0.00
	RF	0.86 *	0.00	0.95 *	0.00	0.93 *	0.00	0.91 *	0.00
15	LF	0.97 *	0.02	0.91 *	0.00	0.92 *	0.00	0.94 *	0.00
	RF	0.89 *	0.00	0.93 *	0.00	0.91 *	0.00	0.93 *	0.00
Avg (STD)	LF	0.91 * ± 0.04		0.93 * ± 0.02		0.94 * ± 0.02		0.95 * ± 0.03	
	RF	0.90 * ± 0.03		0.93 * ± 0.02		0.94 * ± 0.02		0.93 * ± 0.03	

LF: Left Foot, RF: Right foot; * Significance level: $p < 0.05$.

The PCI was calculated to evaluate the coordination of both feet, as illustrated in Table 7. The average value of the PCI, in the SMSI, was $1.75 \pm 0.80\%$ (C1), $1.72 \pm 0.81\%$ (C2), $1.72 \pm 0.79\%$ (C3), and $1.73 \pm 0.80\%$ (C4), and the average value of the PCI, in the F-scan, was $1.66 \pm 0.66\%$, $1.70 \pm 0.66\%$, $1.67 \pm 0.62\%$, and $1.70 \pm 0.62\%$, respectively. R^2 values between the two sensors were 0.94 (C1), 0.95 (C2), 0.95 (C3), and 0.95 (C4). A p-value of <0.05 was set, for statistical significance.

Table 7. The results of the phase coordination index (PCI) detection.

Sensor	Variables	C1	C2	C3	C4
SMSI	PCI (%)	1.75% ± 0.80%	1.72% ± 0.81%	1.72% ± 0.79%	1.73% ± 0.80%
	φ_CV (%)	1.19% ± 0.17%	1.12% ± 0.11%	1.11% ± 0.15%	1.11% ± 0.15%
	φ_ABS (deg)	1.02° ± 0.12°	1.09° ± 0.10°	1.09° ± 0.09°	1.12° ± 0.09°
	φ (deg)	180.22° ± 1.14°	180.10° ± 1.01°	180.12° ± 1.05°	180.10° ± 1.08°
F-scan	PCI (%)	1.66% ± 0.66%	1.70% ± 0.66%	1.67% ± 0.62%	1.70% ± 0.62%
	φ_CV (%)	1.11% ± 0.11%	1.14% ± 0.17%	1.11% ± 0.15%	1.12% ± 0.11%
	φ_ABS(deg)	1.00° ± 0.12°	1.02° ± 0.08°	1.01° ± 0.09°	1.08° ± 0.08°
	φ (deg)	180.20° ± 1.04°	180.11° ± 1.00°	180.12° ± 1.05°	180.10° ± 1.04°
	R^2 (p)	0.94 * (0.00)	0.95 * (0.01)	0.95 * (0.00)	0.95 * (0.00)

* Significance level: $p < 0.05$.

4. Discussion

This study aimed to develop an SMSI and examine its feasibility for gait pattern analysis in healthy young adults. Gait data simultaneously collected by an SMSI and an F-scan, were compared in real-time, for analysis. Step-count was calculated by the peak detection method, and the results presented a 100% consistency in all four different gait speed conditions, showing that SMSI has an equivalent performance to the F-scan.

Considering the number of our subjects (n = 15), Spearman's rho correlation analysis was performed to analyze the stride-time detected by the two sensors. R^2 values for the left foot were higher or equal to 0.91 and those for the right foot were higher or equal to 0.90. The correlation coefficients showed to be statistically significant ($p < 0.05$), confirming the accuracy and feasibility of our sensor. However, the R^2 values of C1, on both sides of the foot, were lower than those of the other speed conditions. This may have been caused by dislocation of the insole sensors, during gait; C1 being the slowest gait speed condition, the foot and the sensor may have got detached, from time-to-time.

We calculated the PCI value to test the clinical feasibility of the SMSI. The PCI is an indicator for evaluating balance function in the lower extremity and is presented in percentage. A value closer to 0% refers to higher balance between the two feet [12,17,19,28]. PCI was originally developed to evaluate the degree of asymmetry, during walking, and many studies evaluated the gait asymmetry of the patients with Parkinson's disease and stroke [17,19,28]. In the studies that evaluated gait asymmetry in healthy subjects, the PCI value was reported to be 2.52% and 2.47% [17,28].

In our study, the average PCI value of the SMSI was between 1.75 ± 0.80% (C1) and 1.72 ± 0.79% (C3), and that of the F-scan was between 1.66 ± 0.66% (C1) and 1.70 ± 0.62% (C4), which were similar to the findings from previous studies [17,28]. In addition, the R^2 values for the PCI showed a high correlation between the two sensors in both feet ($R^2 \geq 0.94$, $p < 0.05$). These values were also similar to the past findings, in other studies. Gait pattern is highly variable from person to person, especially when gender, age, body weight, cultural background, and medical history are taken into account. The reason behind the similarity in results may be due to the subject inclusion criteria. The average age of the healthy subjects in our study was 25.1 ± 2.64 and that of the other studies were 26.3 ± 0.5 and 26.3 ± 0.19, respectively.

Our study has a few limitations, even though our SMSI sensor was tested to have a high accuracy and performance capacity even with only ten channels embedded. Generalizing the results of this study may be difficult because the number of subjects was small. Future studies need to recruit a larger number of participants, as well as different age groups, for a more accurate and diverse data analysis. In this study, performance evaluation was conducted in a laboratory and did not consider temperature and humidity factors. As the results of this study showed clinical feasibility, further study in various environments (indoor and outdoor), considering changes in the temperature and humidity, could make use of the developed sensor and monitoring system, for both healthy and unhealthy individuals.

5. Conclusions

We developed a cost-effective, user-friendly, wearable soft-material-based smart insole sensor, with a real-time monitoring system and performed a feasibility test for the gait pattern analysis, in young healthy individuals, by compensating the limitations of the existing lab-based, expensive, analytic devices. Based on the results of this study, the utilization of our developed system is expected to expand to broader clinical, biomechanical, and quality of life-related studies.

Author Contributions: Conceptualizaion, C.W.; Writing—original draft, C.W.; Writing—review & editing, Y.K.; Project administration, S.D.M.; Supervision, S.D.M.

Funding: This research was supported by a grant (NRF-2015M3A9D7067388) of the Bio & Medical Technology Development Program of the National Research Foundation (NRF) funded by the Ministry of Science and ICT, Republic of Korea and was supported by the Soonchunhyang University research fund.

Conflicts of Interest: The authors declare no conflict of interest.

References

1. Hausdorff, J.M.; Peng, C.K.; Ladin, Z.; Wei, J.Y.; Goldberger, A.L. Is walking a random walk? Evidence for long-range correlations in stride interval of human gait. *J. Appl. Physiol.* **1995**, *78*, 349–358. [CrossRef] [PubMed]

2. Avvenuti, M.; Carbonaro, N.; Cimino, M.; Cola, G.; Tognetti, A.; Vaglini, G. Smart Shoe-Assisted Evaluation of Using a Single Trunk/Pocket-Worn Accelerometer to Detect Gait Phases. *Sensors* **2018**, *18*, 3811. [CrossRef] [PubMed]

3. Li, K.Z.H.; Bherer, L.; Mirelman, A.; Maidan, I.; Hausdorff, J.M. Cognitive Involvement in Balance, Gait and Dual-Tasking in Aging: A Focused Review from a Neuroscience of Aging Perspective. *Front. Neurol.* **2018**, *9*, 913. [CrossRef] [PubMed]

4. Wang, L.; Tan, T.; Ning, H.; Hu, W. Silhouette analysis-based gait recognition for human identification. *IEEE Trans. Pattern Anal. Mach. Intell.* **2003**, *25*, 1505–1518. [CrossRef]

5. Moeslund, T.B.; Adrian, H.; Volker, K. A survey of advances in vision-based human motion capture and analysis. *Comp. Vis. Image Underst.* **2006**, *104*, 90–126. [CrossRef]

6. Zeng, W.; Shu, L.; Li, Q.; Chen, S.; Wang, F.; Tao, X.-M. Fiber-Based Wearable Electronics: A Review of Materials, Fabrication, Devices, and Applications. *Adv. Mater.* **2014**, *26*, 5310–5336. [CrossRef] [PubMed]

7. Storm, F.A.; Heller, B.W.; Mazza, C. Step Detection and Activity Recognition Accuracy of Seven Physical Activity Monitors. *PLoS ONE* **2015**, *10*, 1–13. [CrossRef] [PubMed]

8. Tudor-Locke, C.; Sisson, S.B.; Collova, T.; Lee, S.M.; Swan, P.D. Pedometer-Determined Step Count Guidelines for Classifying Walking Intensity in a Young Ostensibly Healthy Population. *Can. J. Appl. Physiol.* **2005**, *30*, 666–676. [CrossRef] [PubMed]

9. Montero-Odasso, M.; Verghese, J.; Beauchet, O.; Hausdorff, J.M. Gait and Cognition: A Complementary Approach to Understanding Brain Function and the Risk of Falling. *J. Am. Geriatr. Soc.* **2012**, *60*, 2127–2136. [CrossRef] [PubMed]

10. Rampp, A.; Barth, J.; Schulein, S.; Gabmann, K.-G.; Klucken, J.; Eskofier, B.M. Inertial Sensor-Based Stride Parameter Calculation from Gait Sequences in Geriatric Patients. *IEEE Trans. Biomed. Eng.* **2015**, *62*, 1089–1097. [CrossRef] [PubMed]

11. Studenski, S.; Perera, S.; Patel, K. Gait Speed and Survival in Older Adults. *JAMA* **2011**, *305*, 50–58. [CrossRef] [PubMed]

12. Nanhoe-Mahabier, W.; Snijders, A.H.; Delval, A.; Weerdesteyn, V.; Duysens, J.; Overeem, S.; Bloem, B.R. Walking patterns in Parkinson's disease with and without freezing of gait. *Neuroscience* **2011**, *182*, 217–224. [CrossRef] [PubMed]

13. Clark, K.; Leathers, T.; Rotich, D.; He, J.; Wirtz, K.; Daon, E.; Flynn, B.C. Gait Speed Is Not Associated with Vasogenic Shock or Cardiogenic Shock following Cardiac Surgery, but Is Associated with Increased Hospital Length of Stay. *Crit. Care. Res. Pract.* **2018**, *2018*, 1538587. [CrossRef] [PubMed]

14. Hurt, C.P.; Lein, D.H., Jr.; Smith, C.R.; Curtis, J.R.; Westfall, A.O.; Cortis, J.; Rice, C.; Willig, J.H. Assessing a novel way to measure step count while walking using a custom mobile phone application. *PLoS ONE* **2018**, *13*, e0206828. [CrossRef] [PubMed]

15. Kibushi, B.; Moritani, T.; Kouzaki, M. Local dynamic stability in temporal pattern of intersegmental coordination during various stride time and stride length combinations. *Exp. Brain Res.* **2018**. [CrossRef] [PubMed]

16. Fernando, M.; Crowther, R.G.; Cunningham, M.; Lazzarini, P.A.; Sangla, K.S.; Buttner, P.; Golledge, J. The reproducibility of acquiring three dimensional gait and plantar pressure data using established protocols in participants with and without type 2 diabetes and foot ulcers. *J. Foot Ankle Res.* **2016**, *9*, 4. [CrossRef] [PubMed]

17. Plotnik, M.; Bartsch, R.P.; Zeev, A.; Giladi, N.; Hausdorff, J.M. Effects of walking speed on asymmetry and bilateral coordination of gait. *Gait Posture* **2013**, *38*, 864–869. [CrossRef] [PubMed]

18. Olney, S.J.; Richards, C. Hemiparetic gait following stroke. Part I: Characteristics. *Gait Posture* **1996**, *4*, 136–148. [CrossRef]

19. Meijer, R.; Plotnik, M.; Zwaaftink, E.G.; van Lummel, R.C.; Ainsworth, E.; Martina, J.D.; Hausdorff, J.M. Markedly impaired bilateral coordination of gait in post-stroke patients: Is this deficit distinct from asymmetry? A cohort study. *J. Neuro Eng. Rehabil.* **2011**, *8*, 1–8. [CrossRef] [PubMed]

20. Kong, P.W.; Heer, H.D. Wearing the F-scan mobile in-shoe pressure measurement system alters gait characteristics during running. *Gait Posture* **2009**, *29*, 143–145. [CrossRef] [PubMed]

21. Lemarie, E.D.; Biswas, A.; Kofman, J. Plantar Pressure Parameter for Dynamic Gait Stability Analysis. In Proceedings of the 2006 International Conference of the IEEE Engineering in Medicine and Biology Society, New York, NY, USA, 30 August–3 September 2006.

22. Pantelopoulos, A.; Bourbakis, N.G. A Survey on Wearable Sensor-Based Systems for Health Monitoring and Prognosis. *IEEE Trans. Syst. Man Cybern Part C—Appl. Rev.* **2010**, *40*, 1–12. [CrossRef]

23. Wu, Y.; Xu, W.; Liu, J.J.; Huang, M.-C.; Luan, S.; Lee, Y. An Energy-Efficient Adaptive Sensing Framework for Gait Monitoring Using Smart Insole. *IEEE Sens. J.* **2015**, *15*, 2335–2343. [CrossRef]

24. Park, S.W.; Das, P.S.; Park, J.Y. Development of wearable and flexible insole type capacitive pressure sensor for continuous gait signal analysis. *Org. Electron.* **2018**, *53*, 213–220. [CrossRef]

25. Scilingo, E.P.; Gemignani, A.; Paradiso, R.; Taccini, N.; Ghelarducci, B.; De Rossi, D. Performance evaluation of sensing fabrics for monitoring physiological and biomechanical variables. *IEEE Trans. Inf. Technol. Biomed.* **2005**, *9*, 345–352. [CrossRef] [PubMed]

26. Pan, J.; Tomkins, W.J. A Real-Time QRS Detection Algorithm. *IEEE Trans. Biomed. Eng.* **1985**, *32*, 230–236.

27. Gimmon, Y.; Rashad, H.; Kurz, I.; Plotnik, M.; Riemer, R.; Debi, R.; Shapiro, A.; Melzer, I. Gait Coordination Deteriorates in Independent Old-Old Adults. *J. Aging Phys. Act.* **2018**, *26*, 382–389. [CrossRef] [PubMed]

28. Plotnik, M.; Giladi, N.; Hausdorff, J.M. A new measure for quantifying the bilateral coordination of human gait: Effects of aging and Parkinson's disease. *Exp. Brain Res.* **2007**, *181*, 561–570. [CrossRef] [PubMed]

Article

A Soft Polydimethylsiloxane Liquid Metal Interdigitated Capacitor Sensor and Its Integration in a Flexible Hybrid System for On-Body Respiratory Sensing

Yida Li [1,*,†], Suryakanta Nayak [1,†], Yuxuan Luo [1], Yijie Liu [1],
Hari Krishna Salila Vijayalal Mohan [1], Jieming Pan [1], Zhuangjian Liu [2], Chun Huat Heng [1]
and Aaron Voon-Yew Thean [1,*]

[1] Department of Electrical and Computer Engineering, National University of Singapore, 4 Engineering Drive 3, Singapore 117583, Singapore; suryakanta.nayak@nus.edu.sg (S.N.); elelyux@nus.edu.sg (Y.L.); liuyijie1987@outlook.com (Y.L.); harikrishnasv@nus.edu.sg (H.K.S.V.M.); e0361620@u.nus.edu (J.P.); elehch@nus.edu.sg (C.H.H.)
[2] Institute of High Performance Computing, A*STAR Research Entities, 1 Fusionopolis Way, #16-16 Connexis, Singapore 138632, Singapore; liuzj@ihpc.a-star.edu.sg
* Correspondence: elelyida@nus.edu.sg (Y.L.); aaron.thean@nus.edu.sg (A.V.-Y.T.)
† Equal contributions.

Received: 30 March 2019; Accepted: 30 April 2019; Published: 6 May 2019

Abstract: We report on the dual mechanical and proximity sensing effect of soft-matter interdigitated (IDE) capacitor sensors, together with its modelling using finite element (FE) simulation to elucidate the sensing mechanism. The IDE capacitor is based on liquid-phase GaInSn alloy (Galinstan) embedded in a polydimethylsiloxane (PDMS) microfludics channel. The use of liquid-metal as a material for soft sensors allows theoretically infinite deformation without breaking electrical connections. The capacitance sensing is a result of E-field line disturbances from electrode deformation (mechanical effect), as well as floating electrodes in the form of human skin (proximity effect). Using the proximity effect, we show that spatial detection as large as 28 cm can be achieved. As a demonstration of a hybrid electronic system, we show that by integrating the IDE capacitors with a capacitance sensing chip, respiration rate due to a human's chest motion can be captured, showing potential in its implementation for wearable health-monitoring.

Keywords: stretchable; polydimethylsiloxane; liquid-metal; capacitor

---

1. Introduction

Soft electronics that enable conformal contacts on irregular surfaces is an emerging area with increasing importance. The development of this technology is expected to enhance human–machine interfaces to cover areas such as medical and e-health applications, robotics, and communications [1–12]. Flexible and stretchable electronics are vigorously studied for the choice of materials and integration strategies [5–7,9,10,13]. However, flexible electronics is only suitable for a non-conformal substrate that is static or does not undergo significant strain in the x-y axes during operation [7,13]. Stretchable electronics, on the other hand, offers more degrees of freedom that can theoretically tolerate mechanical strain in all three axes. This leads to increasing interest in the field of wearable devices for health-monitoring. In such applications, soft sensors with a Young's modulus that matches that of skin are particularly attractive. This allows for prolonged human body attachment for continuous health monitoring [5,9,13,14]. Hence, besides the functionality of the sensor, there is a need to address the mechanical robustness and the system integration strategies with high performance integrated circuits (IC) [5,7,9,13,15–18]. This calls for novel materials and approaches, such as the use of liquid metal or stretchable conducting materials, in place of conventional metal materials that are rigid [19–21]. Several studies have reported soft-matter capacitors comprising of a microfludics channel of liquid-metal (Galinstan) and described the theoretical model of capacitance change effect under mechanical deformation [22,23]. While there are reports of sensors with similar structures being used for bio-sensing and proximity sensing, the implementation of such sensors in real systems is still lacking [24,25].

In this work, we report on the experimental characterization together with a finite element (FE) simulation model of soft interdigitated (IDE) capacitors sensors. The IDE capacitor was fabricated from serpentine microfluidics channels of GaInSn liquid alloy (Galinstan) embedded in a polydimethylsiloxane (PDMS) matrix. The soft IDE capacitor, besides responding to perpendicular strains in the x-y axes, also responded to proximity sensing of a human finger that has not been reported before. In addition, previous reports have not described the IDE capacitor using FE simulation with proper boundary conditions. We will address these in this paper. Experimentally, the opposite capacitance change in the x and y directions allows for detection of direction specific strain with a resolution of 0.02 pF/(% engineering strain) as verified by our FE simulation model. Proximity sensing is achieved via the modifications of the capacitor's fringing E-field by the surrounding dielectric medium [26]. We show that spatial detection as large as 28 cm can be achieved by varying a human's finger distance to the IDE capacitor. The described FE simulation model provides a design guideline for future implementation of such a class of sensors. Finally, as a demonstration of a hybrid electronic system, we show that when the IDE capacitor sensor attached to a human chest is integrated with a functional capacitance sensing chip, it can be used for on-body respiratory sensing by utilizing the proximity effect, thus demonstrating its potential as a soft sensor suitable for use in wearables for health monitoring.

2. Fabrication Process of Soft PDMS-Liquid Metal IDE Capacitor

The stretchable capacitor was composed of two layers of PDMS (Dow Corning Sylgard 184). The first layer consisted of a microfluidics channel in the form of an IDE design, fabricated using a soft lithography approach [9]. Special design care in providing for the inlet and outlet points were necessary to ensure the Galinstan filled the microfludics channel properly without trapped air. The master mold was made using a permanent resist (SU-8 3050, Microchem, Westborough, MA, USA), where the PDMS pre-polymer was casted over the pre-fabricated design to a thickness of ~500 μm and removed after curing by peeling. The second layer consisted of a plain piece of PDMS molded to the same thickness of ~500 μm. For the formation of the closed microfluidics channel, the two PDMS layers were surface treated using a light remote O_2 plasma treatment before contacting with each other. A post-baking step (90 °C, 2 h) ensured a permanent bond of the two layers of PDMS. Finally, the microfluidics channel was completely filled with Galinstan using a needle and syringe approach. The Galinstan was injected from the inlet point while air was extracted from the outlet point simultaneously to allow the complete fill of the microfluidics channel. The height of the Galinstan electrodes was dependent on the master mold, and was 100 μm in this case [9,27,28]. Figure 1a,b shows the fabrication process flow of the soft IDE capacitor, and a photo image of the fabricated IDE soft capacitors with various sizes, respectively.

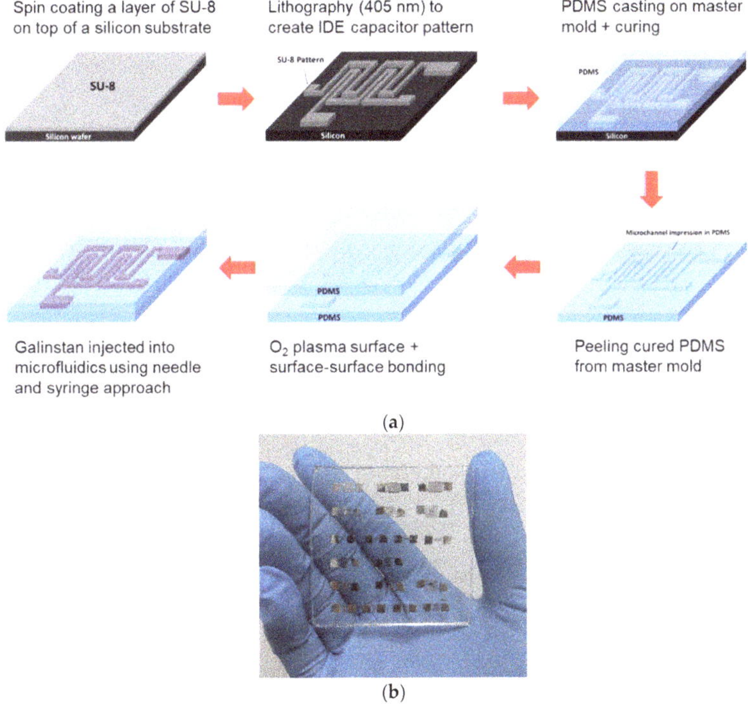

(a)

(b)

Figure 1. (**a**) Fabrication process flow of the soft IDE capacitor, and (**b**) photo image of the fabricated IDE soft capacitors of various sizes. The height of the liquid metal electrode was 100 μm.

While encapsulated in the elastomer matrix, the PDMS provided good barrier resistance to moisture as reported earlier, which otherwise would cause the oxidation of the Galinstan [9,28]. We immersed the soft IDE capacitor in water for a period of 10 min and no visible and electrical change to the capacitor were observed. Hence, this allows for its use without degradation in practical environment. For electrical contact to the IDE capacitor, short sections of tungsten (W) wires (Goodfellow, 125 μm, Huntingdon, UK) perforated the PDMS into the Galinstan reservoir and was resealed with PDMS to avoid leakage.

3. Electrical Characterization and Modelling of a Static IDE Capacitor

The capacitance of all the fabricated IDE capacitors were first measured and calibrated using a benchtop LCR meter (Keysight E4980A, Santa Rosa, CA, USA) at 1 kHz. In this set of measurements, parasitic capacitance from the connecting wires to the LCR meter caused an offset to the intrinsic capacitance by a fixed positive value. From the linear fit of the data points, the parasitic capacitance was extracted as the y-intercept when the electrode length was zero. Correcting for the parasitic capacitance, the actual capacitance of the IDE capacitors ranged from 1.05 pF to 2.40 pF. Figure 2 shows the as-measured capacitance across a total electrode length from 15 mm to 44 mm, together with the corrected capacitance values and modeled values based on empirical equations. The corrected capacitances agree very well with the capacitor model. It should be noted that the main discrepancy between actual values and measured values comes from the wires' parasitic capacitances of ~0.68 pF. In subsequent measurements, the wires' parasitic capacitances were removed from the measured values.

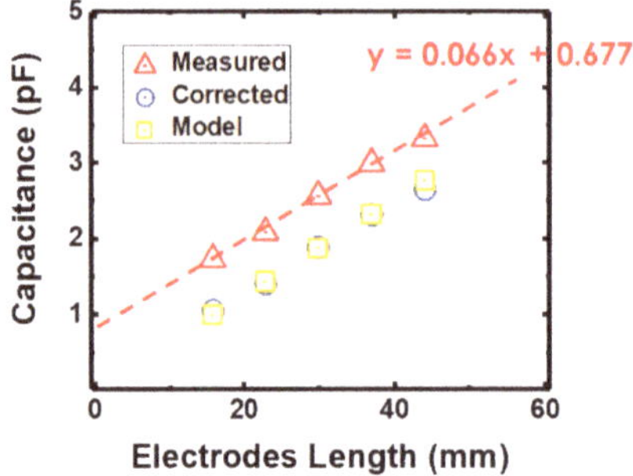

Figure 2. As-measured capacitance across a total electrode length from 15 mm to 44 mm, together with the corrected capacitance values and modeled values based on empirical equations indicated in the legends.

In our IDE capacitor design, the electrodes were of non-negligible thickness, and that was advantageous to enhance its capacitance and sensitivity. As such, a single coplanar capacitor model was not sufficient. Instead, we used a combination of co-planar capacitor and bi-planar capacitor model connected in parallel, as shown by the schematic in Figure 3.

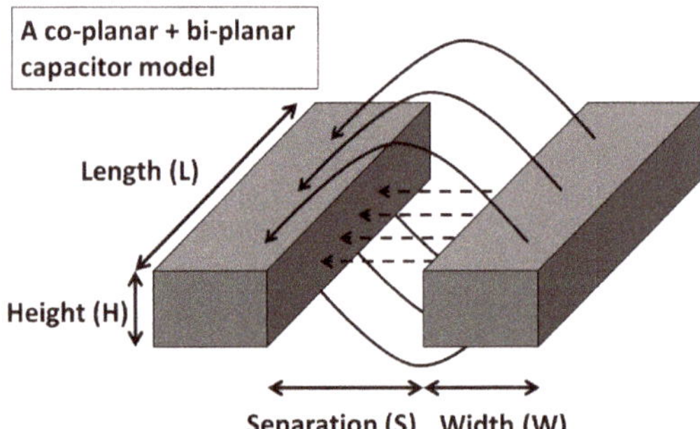

Figure 3. Schematic showing the electric field lines connecting two electrodes of the IDE capacitor. Due to the non-negligible thickness of the electrode, the two contributions of the electric field lines are the direct field lines, and the fringing field lines connected in parallel.

In this model, the electric field lines consist of those that flows directly between the sides of the two electrodes, as well as the fringing field lines that flow from the top and bottom of the electrodes. Further, the effective capacitance of such a structure is described using Equations (1) and (2) for the coplanar model, and Equation (3) for the parallel plate model.

$$C = \frac{\epsilon_r l \, \ln\left(-\dfrac{2}{\sqrt[4]{1-\frac{s^2}{(s+2w)^2}}-1}\left(\sqrt[4]{1-\frac{s^2}{(s+2w)^2}}-1\right)\right)}{377\pi V_o}, \quad for \; 0 < \frac{s}{s+2w} \le \frac{1}{\sqrt{2}} \tag{1}$$

$$C = \frac{\epsilon_r l}{377\pi V_o \, \ln\left(-\dfrac{2}{\sqrt{\frac{s}{s+2w}}-1}\left(\sqrt{\frac{s}{s+2w}}-1\right)\right)}, \quad for \; \frac{1}{\sqrt{2}} < \frac{s}{s+2w} \le 1 \tag{2}$$

$$C = \varepsilon_r \frac{A}{d} \tag{3}$$

where ε is the dielectric constant of the PDMS matrix, A is the area of the electrode given by channel length $\times$ height of the microfluidics channel, and d is the spacing (100 μm) between the electrodes. The above model for a static IDE capacitor was found to be in excellent agreement with experimental values, plotted out together with the measured capacitance in Figure 2, validating the accuracy of the model.

Fassler and co-workers have attempted to study the relationship between the deformation of IDE capacitors and the effect on its capacitance using empirical equations [23]. In their model, they assume that the electrode thickness plays a negligible role in the capacitance, and use only a single co-planar model. However, the thickness of the electrode does contribute to the effective capacitance and should be accounted for as described in Equation (3). In addition, the study is only limited to mechanical strains in the x-y axes. Although model has been reported to predict the change in such a capacitor due to the deformation of the system, it is done by setting assumptions, and does not allow for dynamic boundaries conditions. In addition, the model does not take into account the modifications of electric field lines due to an external change in dielectric material, which we termed as the proximity effect in this work. This effect is similar to reported works on electric-field sensing but has not been discussed thus far for IDE capacitors [26]. Hence, in this work, we further the study by modelling both the IDE capacitor's deformation and proximity effect using FE simulation. This allows for dynamic boundary conditions to be set. By coupling the simulated results together with experimental measurements, we elucidated the mechanism of capacitance sensing. The FE simulation was performed using Abaqus and the setup is described in Appendix A of this paper. Separately, for the following functional tests described in the following sections, a larger IDE capacitor was fabricated in order to achieve a higher sensitivity and easier handling. The new IDE capacitor was fabricated using the same process flow described earlier and the base capacitance was ≈9.6 pF after the parasitic capacitance correction.

3.1. Functional Test—Strain Effect

The mechanical deformation of the IDE capacitor led to a change in its geometry and a resultant change in capacitance. As the IDE capacitor was strained in the x-axis (space between electrodes pulled apart), the adjacent electrodes increased in distance from each other, causing a reduction in the capacitance. On the other hand, when the IDE capacitor was strained in the y-axis (electrodes elongating), the effect of the electrode elongating and a decrease in distance between adjacent electrodes resulted in an increase of the capacitance. In the FE simulation, we applied the same boundary condition at both ends of the PDMS during the stretching action. Figure 4a shows the FE simulation model where we applied a fixed constraint on one side and a stretched displacement condition on the other side, while Figure 4b,c shows the experimental measurements, FE simulated values of the capacitance change w.r.t to the two different strain directions up to 50%, and a comparison to the model described by Fassler et al. [23]. Good agreement between experimental and simulated values was obtained, validating the accuracy of the FE model. On the other hand, though showing a similar trend, there existed some error between the measured values as compared to earlier proposed model, thus showing the advantage of FE simulation in such a study [23]. Finally, the IDE capacitor demonstrated a resolution up to 0.02 pF/(% engineering strain).

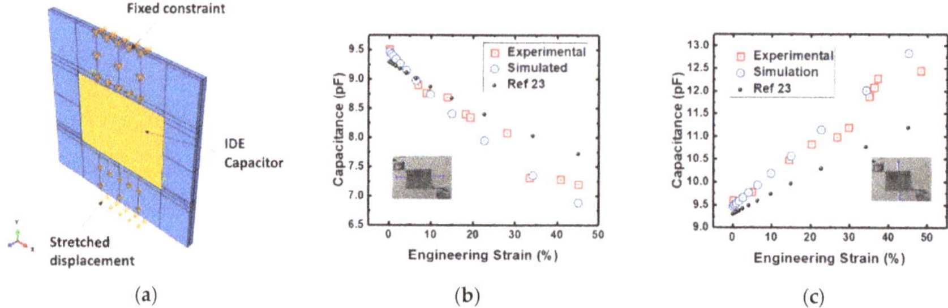

(a)   (b)   (c)

Figure 4. (**a**) FE simulation model where a fixed constraint was applied to one side and a stretched displacement condition was applied to the other side. Stretching in both x- and y-axes was simulated. (**b**) Relationship between capacitance and the x-axis strain. and (**c**) relationship between capacitance and y-axis strain comparison between experimental, FE simulated, and the earlier model proposed [23].

3.2. Proximity Effect

In this section, we describe the proximity sensing effect of the IDE capacitor through the disturbance of the E-field lines with objects placed at different distances from it. Figure 5a shows the testing setup photo images where the IDE capacitor was connected to a precision impedance analyzer (Agilent 4294A, Santa Clara, CA, USA) for capacitance measurement. The corresponding capacitance was measured as a human finger was moved between different distances from the surface of the IDE capacitor, as shown in Figure 5a. The proximity sensing mechanism is described as follows. Capacitive sensors use capacitive transducers to detect the proximity of a body, and are broadly classified in three modes: transmit mode, shunt mode, and loading mode [26]. Our setup behaved as a transmit mode where one side of the electrodes acted as a transmitter, and the other side as a receiver. Figure 5b shows an illustration of the different capacitance paths of the system as a human finger approached. The original capacitor electric field was now coupled to the receiver side through the human finger, creating two parasitic capacitances (C_1 and C_2) in series. Thus, the effective capacitance of the system increased based on the capacitance summation rule in a parallel configuration between the intrinsic capacitance C_0 and the parasitic capacitances C_1 and C_2. C_0, C_1, and C_2 were, in turn, dependent on the distance of the human finger from the capacitor due to the strength of the electric-field coupling. The further away the finger was from the capacitor, the weaker the coupling and the smaller the effective capacitance. In our experiment, a reduction of capacitance was seen as the finger moved towards the capacitor, while capacitance increased when the finger moved away from the capacitor. The proximity effect could be felt by the sensor from a distance as far as 28 cm. Compared with the dielectric property of air, human tissue is considered to be a conductive material. From the perspective of electrostatics, the boundary of human tissue was assumed to be a floating electrode, where the total electrical charge at the surface was equal to zero. The entire FE simulation model had a volumetric size of 10 cm × 5 cm × 3 cm, and the IDE capacitor, air, and the human finger model is shown in Figure 5c.

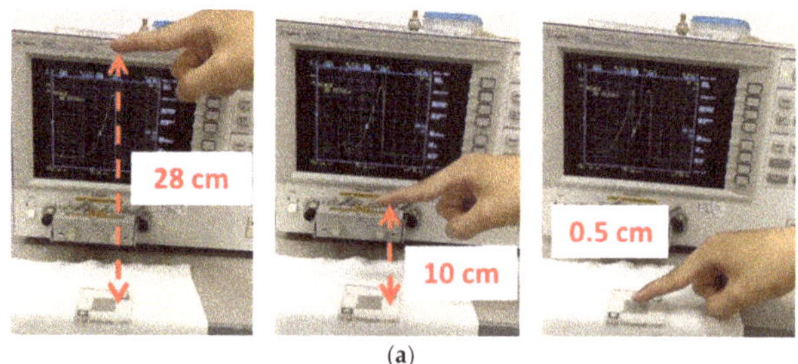

(a)

Figure 5. *Cont.*

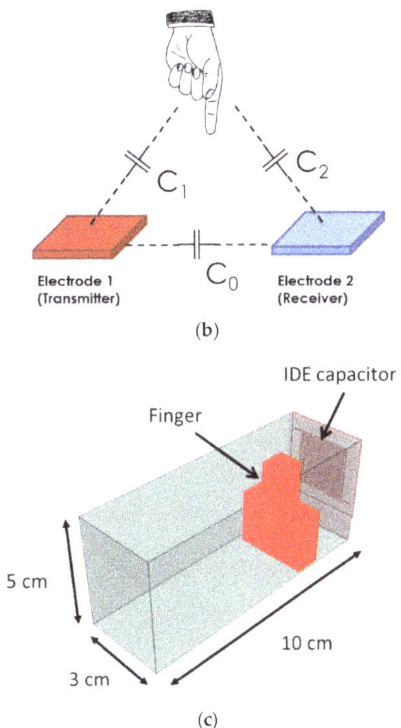

Figure 5. (a) Photo images showing the testing setup where a human finger is positioned at different positions above the IDE capacitor. (b) Illustration of the different capacitance paths as a human finger approaches the capacitor. C_0 is the intrinsic capacitance and $C_1 C_2$ are the parasitic capacitances. (c) FE model in the numerical simulation (finger and IDE capacitor indicated) with a size of 10 cm × 5 cm × 3 cm.

Figure 6a,b shows the capacitance versus distance of a human finger placed above the IDE capacitor up to 28 cm (both experimental and simulated indicated in legend), and time based capacitance measurements as the finger hovers above the IDE capacitor at a height of 1 3 cm, respectively. In this simulation, a simple numerical finger model was constructed (Appendix A). The main factors resulting in the discrepancy between the experimental and simulated values were due to the leaky fringe electrical field and the finger model difference. In addition, the simulated distance was limited at 10 cm due to calculation complexity beyond that. Nonetheless, the trend of the capacitance change agreed very well in both experimental and simulation, and the most sensitive region lay within 1 cm from the IDE capacitor. The disturbance in the field line resulted from the float electrical potential. When the distance between the finger model and sensor was within 1 cm, the leaky fringe electrical field around the finger became large, resulting in a significant capacitance change.

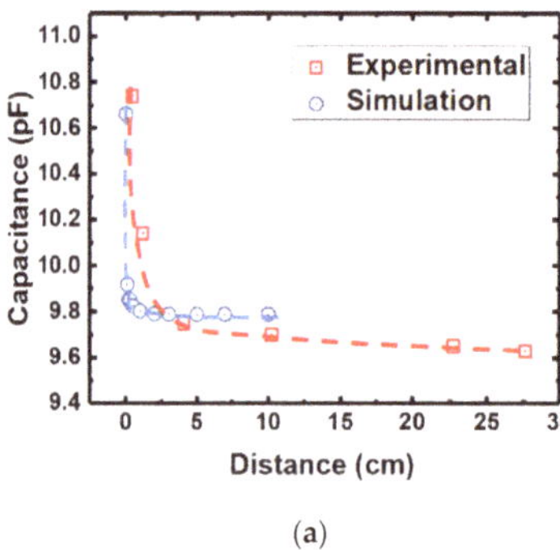
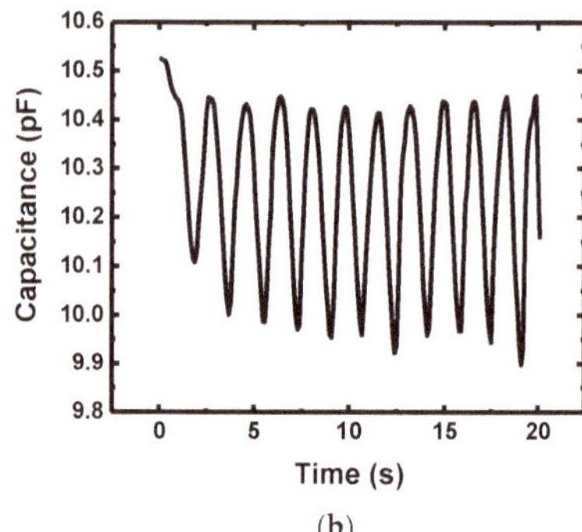

Figure 6. (**a**) Measured capacitance w.r.t. various distances a human finger was placed from the IDE capacitor, and (**b**) transient capacitance plot when a human finger hovered above the IDE capacitor in a cyclic manner at a height of ~1–3 cm above.

4. Demonstration of an IDE Capacitor Based Flexible Hybrid Respiratory System

Complementary Metal Oxide Semiconductor (CMOS) chips and printed circuit boards (PCBs) integrated with soft components allow them to be interfaced comfortably with the human body. The co-design of composite materials and electronic circuits/system in a monolithic form of a flexible hybrid electronic system provides the guideline for on-body wearables. In this section, we demonstrate a prototype hybrid flexible respiratory system by integrating the IDE capacitor to a rigid capacitance sensing circuit using external wires [29]. In subsequent measurements, the capacitance data measured by the chip is transmitted wirelessly via Wi-Fi protocol through a Raspberry Pi to a computer for readout. In the current setup, the IDE capacitor was connected to the external chip using tungsten wires. More reliable interconnects suitable for flexible system integration will be further investigated [30].

The human body respiratory detection was achieved by utilizing the proximity sensing effect of the IDE capacitor, where the sensor was attached to the human chest. The expansion and contraction of the chest caused change in the effective dielectric constant of the surrounding medium and disturbed the electrical field lines. This allowed for the respiratory motion to be picked up in a straightforward manner. Hence, the sensor was placed at a location where chest motion was obvious in this demonstration. Figure 7a shows the photo image of the integrated system attached to a human chest, while Figure 7b shows the logged capacitance values by the circuitry over 50 s. During the respiratory rate tracking, the sensor and electronics were fully covered up by clothing. Notwithstanding the slight motion of the clothing during breathing, we did not observe any distortion to the acquired waveform. Although noise was present in the collected capacitance data, the undulating trace indicated the respiratory motion was clearly visible over a range of ~0.2 pF. We demonstrate two different respiratory rates of 20/min and 60/min as shown in Figure 7b. In both cases, the sensor responded adequately, and the captured rates matched with manual counting. The implementation shown here is straightforward but effective, thus paving the way for such sensor to be implemented in a wearable medical device.

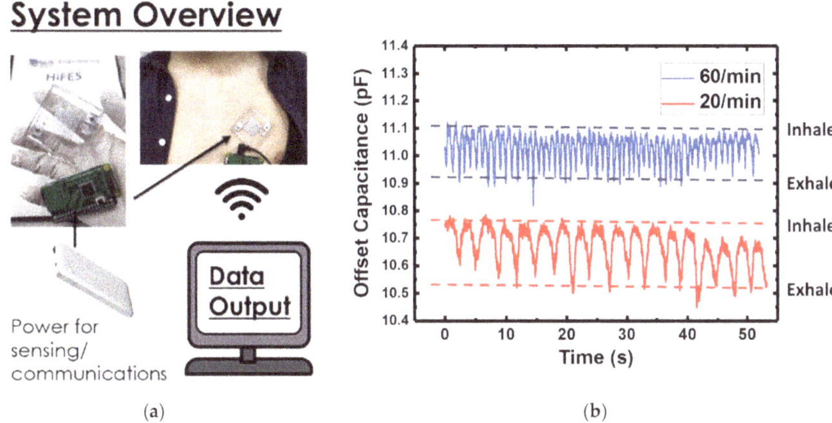

Figure 7. (**a**) System overview of the respiratory tracker with a photo image of the patch attached to a human chest, and (**b**) time dependence capacitance values measured by the capacitance chip as the human subject was breathing over 50 s. Two different respiratory rates were demonstrated as indicated in the legend.

5. Conclusions

In summary, we describe in detail the fabrication and electrical characterization of a soft PDMS-Galinstan based IDE capacitor. An FE simulation model was developed that allowed for dynamic boundary conditions to be applied, matching with experiments. With the Galinstan electrode completely encapsulated in the soft PDMS matrix, the IDE capacitor was mechanically robust to physical deformation as well as moisture. In addition to strain sensing in both the x and y axes, we show the proximity effect of such a class of soft sensors by modifying the surrounding dielectric medium. The disturbance of the electrical field lines by surrounding objects resulted in a change in effective dielectric constant, and consequently, a change in the capacitance. This effect allowed for spatial detection, and we showed experimentally that a sensing distance up to 28 cm can be achieved. The FE simulation elucidated the capacitance change mechanism and provide a guideline for more different applications. Finally, we demonstrate for the first time the use of an IDE capacitor based flexible hybrid electronics respiratory system utilizing the proximity effect. An accurate human breathing pattern was successfully tracked, paving the way for its use as a part of continuous health monitoring applications.

Author Contributions: Conceptualization, Y.L. (Yida Li), Y.L. (Yuxuan Luo), S.N., C.H.H., and A.V.-Y.T.; Methodology, Y.L. (Yida Li), Y.L. (Yuxuan Luo), and S.N.; Software, Y.L. (Yijie Liu) and Z.L.; Validation, Y.L. (Yida Li), Y.L. (Yuxuan Luo), and S.N.; Formal Analysis, Y.L. (Yida Li); Investigation, Y.L. (Yida Li), Y.L. (Yuxuan Luo), and S.N.; Resources, Z.L., C.H.H., and A.V.-Y.T.; Data Curation, Y.L. (Yida Li), Y.L. (Yuxuan Luo), and S.N.; Writing—Original Draft Preparation, Y.L. (Yida Li); Writing—Review and Editing, All; Supervision, C.H.H. and A.V.-Y.T.; Funding Acquisition, C.H.H. and A.V.-Y.T.

Funding: The work was funded by Singapore National Research Foundation's Returning Singapore Scientist Scheme (Grant Ref: NRF-RSS2015-003), Hybrid Integrated Flexible Electronic Systems (HiFES) Program (hifes.nus.edu.sg), E6Nanofab at the National University of Singapore (NUS), NUS AcRF R263000B55112, and LAkSA project R263-501-011-133.

Conflicts of Interest: The authors declare no conflict of interest.

Appendix A

The finite element (FE) model was built up and simulated using commercial software Abaqus for non-linear FE analysis. We constructed two FE models to simulate the effect on the capacitance of the sensor due to deformation and proximity. In the FE model, the PDMS substrate was modelled using a C3D8RH element. PDMS was modelled using a hyperelastic material model (Mooney–Rivlin model in Abaqus) and was incompressible. The material parameters used were equivalent to the initial Young's modulus = 1 MPa and Poisson's ration = 0.49, obtained experimentally. Since the channels were completely filled with a nearly incompressible fluid, the surface-based fluid cavity capability was used, and the pressure applied by the fluid on the surface of channel was determined using the cavity volume.

To obtain the capacitance of the deformed sensor, the steady-state linear electrical–mechanical analysis (piezoelectric elements in Abaqus) was performed on the deformed solid mesh, where the piezoelectric constants were set to zero and the dielectric constant parameter was 2.62. In order to obtain the capacitance, an electrical potential ΔU was applied on the interfaces between the elastomeric matrix and liquid metal at the left and right channels. The charge Q in the channel was calculated using numerical simulation and the capacitance was then obtained as:

$$C = \frac{Q}{\Delta U}$$

In the proximity sensing simulation, the human finger was modelled as a simple model with dimensions as shown in Figure A1. In this model, the human finger was regarded as a near perfect conductor compared with the PDMS and air. To determine the capacitance of the position sensor, the human was set to a floating electrical potential, where the net surface electrical charge was zero. The resulting capacitance as a result of the finger distance to the sensor was extracted using the same methodology as the deformation model described in the previous paragraph.

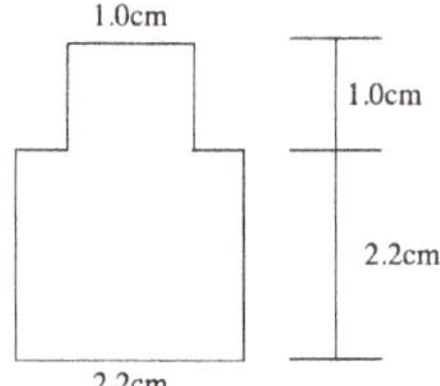

Figure A1. Shape and dimensions of the finger model. In this model, the thickness used was 1 cm.

References

1. So, J.-H.; Thelen, J.; Qusba, A.; Hayes, G.J.; Lazzi, G.; Dickey, M.D. Reversibly deformable and mechanically tunable fluidic antennas. *Adv. Funct. Mater.* **2009**, *19*, 3632–3637. [CrossRef]
2. Kim, J.; Banks, A.; Xie, Z.; Heo, S.Y.; Gutruf, P.; Lee, J.W.; Xu, S.; Jang, K.-I.; Liu, F.; Brown, G.; et al. Miniaturized flexible electronic systems with wireless power and near-field communication capabilities. *Adv. Funct. Mater.* **2015**, *25*, 4761–4767. [CrossRef]
3. Roberts, P.; Damian, D.D.; Shan, W.; Lu, T.; Majidi, C. Soft-matter capacitive sensor for measuring shear and pressure deformation. In Proceedings of the 2013 IEEE International Conference on Robotics and Automation, Karlsruhe, Germany, 6–10 May 2013; pp. 3529–3534.
4. Miyamoto, A.; Lee, S.; Cooray, N.F.; Lee, S.; Mori, M.; Matsuhisa, N.; Jin, H.; Yoda, L.; Yokota, T.; Itoh, A.; et al. Inflammation-free, gas-permeable, lightweight, stretchable on-skin electronics with nanomeshes. *Nat. Nanotechnol.* **2017**, *12*, 907–913. [CrossRef] [PubMed]
5. Xu, S.; Zhang, Y.; Jia, L.; Mathewson, K.E.; Jang, K.-I.; Kim, J.; Fu, H.; Huang, X.; Chava, P.; Wang, R.; et al. Soft microfluidic assemblies of sensors, circuits, and radios for the skin. *Science* **2014**, *344*, 70–74. [CrossRef]
6. Rogers, J.; Malliaras, G.; Someya, T. Biomedical devices go wild. *Science Advances* **2018**, *4*, eaav1889. [CrossRef] [PubMed]
7. Khan, Y.; Garg, M.; Gui, Q.; Schadt, M.; Gaikwad, A.; Han, D.; Yamamoto, N.A.D.; Hart, P.; Welte, R.; Wilson, W.; et al. Flexible hybrid electronics: Direct interfacing of soft and hard electronics for wearable health monitoring. *Adv. Funct. Mater.* **2016**, *26*, 8764–8775. [CrossRef]

8. Sheng, L.; Teo, S.; Liu, J. Liquid-metal-painted stretchable capacitor sensors for wearable healthcare electronics. *J. Med Boil. Eng.* **2016**, *36*, 265–272. [CrossRef]

9. Li, Y.; Luo, Y.; Nayak, S.; Liu, Z.; Chichvarina, O.; Zamburg, E.; Zhang, X.; Liu, Y.; Heng, C.H.; Thean, A.V.-Y. A stretchable-hybrid low-power monolithic ecg patch with microfluidic liquid-metal interconnects and stretchable carbon-black nanocomposite electrodes for wearable heart monitoring. *Adv. Electron. Mater.* **2019**, *5*, 1800463. [CrossRef]

10. Shay, T.; Velev, O.D.; Dickey, M.D. Soft electrodes combining hydrogel and liquid metal. *Soft Matter* **2018**, *14*, 3296–3303. [CrossRef] [PubMed]

11. Jin, H.; Matsuhisa, N.; Lee, S.; Abbas, M.; Yokota, T.; Someya, T. Enhancing the performance of stretchable conductors for e-textiles by controlled ink permeation. *Adv. Mater.* **2017**, *29*, 1605848. [CrossRef]

12. Cooper, C.B.; Arutselvan, K.; Liu, Y.; Armstrong, D.; Lin, Y.; Khan, M.R.; Genzer, J.; Dickey, M.D. Stretchable capacitive sensors of torsion, strain, and touch using double helix liquid metal fibers. *Adv. Funct. Mater.* **2017**, *27*, 1605630. [CrossRef]

13. Gao, W.; Emaminejad, S.; Nyein, H.Y.Y.; Challa, S.; Chen, K.; Peck, A.; Fahad, H.M.; Ota, H.; Shiraki, H.; Kiriya, D.; et al. Fully integrated wearable sensor arrays for multiplexed in situ perspiration analysis. *Nature* **2016**, *529*, 509–514. [CrossRef]

14. Liu, Y.; Pharr, M.; Salvatore, G.A. Lab-on-skin: A review of flexible and stretchable electronics for wearable health monitoring. *ACS Nano* **2017**, *11*, 9614–9635. [CrossRef] [PubMed]

15. Herbert, R.; Kim, J.-H.; Kim, Y.S.; Lee, H.M.; Yeo, W.-H. Soft material-enabled, flexible hybrid electronics for medicine, healthcare, and human-machine interfaces. *Materials* **2018**, *11*, 187. [CrossRef] [PubMed]

16. Schwartz, D.E.; Rivnay, J.; Whiting, G.L.; Mei, P.; Zhang, Y.; Krusor, B.; Kor, S.; Daniel, G.; Ready, S.E.; Veres, J.; et al. Flexible hybrid electronic circuits and systems. *IEEE J. Emerg. Sel. Top. Circuits Syst.* **2017**, *7*, 27–37. [CrossRef]

17. Sears, N.C.; Berrigan, J.D.; Buskohl, P.R.; Harne, R.L. Dynamic response of flexible hybrid electronic material systems. *Compos. Struct.* **2019**, *208*, 377–384. [CrossRef]

18. Tong, G.; Jia, Z.; Chang, J. Flexible hybrid electronics: Review and challenges. In Proceedings of the 2018 IEEE International Symposium on Circuits and Systems (ISCAS), Florence, Italy, 27–30 May 2018; pp. 1–5.

19. Abu-Khalaf, J.M.; Al-Ghussain, L.; Al-Halhouli, A.a. Fabrication of stretchable circuits on polydimethylsiloxane (pdms) pre-stretched substrates by inkjet printing silver nanoparticles. *Materials* **2018**, *11*, 2377. [CrossRef] [PubMed]

20. Rogers, J.A.; Someya, T.; Huang, Y. Materials and mechanics for stretchable electronics. *Science* **2010**, *327*, 1603–1607. [CrossRef] [PubMed]

21. Matsuhisa, N.; Inoue, D.; Zalar, P.; Jin, H.; Matsuba, Y.; Itoh, A.; Yokota, T.; Hashizume, D.; Someya, T. Printable elastic conductors by in situ formation of silver nanoparticles from silver flakes. *Nat. Mater.* **2017**, *16*, 834–840. [CrossRef] [PubMed]

22. Lazarus, N.; Meyer, C.D.; Bedair, S.S.; Nochetto, H.; Kierzewski, I.M. Multilayer liquid metal stretchable inductors. *Smart Mater. Struct.* **2014**, *23*, 085036. [CrossRef]

23. Fassler, A.; Majidi, C. Soft-matter capacitors and inductors for hyperelastic strain sensing and stretchable electronics. *Smart Mater. Struct.* **2013**, *22*, 055023. [CrossRef]

24. Staginus, J.; Aerts, I.M.; Chang, Z.-y.; Meijer, G.C.M.; de Smet, L.C.P.M.; Sudhölter, E.J.R. Capacitive response of pdms-coated ide platforms directly exposed to aqueous solutions containing volatile organic compounds. *Sensors Actuators B: Chem.* **2013**, *184*, 130–142. [CrossRef]

25. Ma, M.; Wang, Y.; Liu, F.; Zhang, F.; Liu, Z.; Li, Y. Passive wireless LC proximity sensor based on LTCC technology. *Sensors* **2019**, *19*, 1110. [CrossRef]

26. Smith, J.; White, T.; Dodge, C.; Paradiso, J.; Gershenfeld, N.; Allport, D. Electric field sensing for graphical interfaces. *IEEE Comput. Graph. Appl.* **1998**, *18*, 54–60. [CrossRef]

27. Dickey, M.D. Stretchable and soft electronics using liquid metals. *Adv. Mater.* **2017**, *29*, 1606425. [CrossRef]

28. Zhang, B.; Dong, Q.; Korman, C.E.; Li, Z.; Zaghloul, M.E. Flexible packaging of solid-state integrated circuit chips with elastomeric microfluidics. *Sci. Rep.* **2013**, *3*, 1098. [CrossRef]

29. Luo, Y.; Heng, C. An 8.2 µw 0.14 mm^2 16-channel cdma-like period modulation capacitance-to-digital converter with reduced data throughput. In Proceedings of the 2018 IEEE Symposium on VLSI Circuits, Honolulu, HI, USA, 18–22 June 2018; pp. 165–166.

30. Green Marques, D.; Alhais Lopes, P.; de Almeida, A.; Majidi, C.; Tavakoli, M. Reliable interfaces for egain multi-layer stretchable circuits and microelectronics. *Lab Chip* **2019**, *19*, 897–906. [CrossRef]

Article

Use of Superelastic Nitinol and Highly-Stretchable Latex to Develop a Tongue Prosthetic Assist Device and Facilitate Swallowing for Dysphagia Patients

Mahdis Shayan [1], Neil Gildener-Leapman [2], Moataz Elsisy [3], Jack T. Hastings [4], Shinjae Kwon [5], Woon-Hong Yeo [5,6], Jee-Hong Kim [7], Puneeth Shridhar [4], Gabrielle Salazar [4] and Youngjae Chun [3,4,8,*]

[1] Department of Cardiothoracic Surgery, School of Medicine, Stanford University, Falk Building, Room CV-035, 870 Quarry Rd, Palo Alto, CA 94304, USA; mahdis.shayan@stanford.edu

[2] Department of Surgery, Albany Medical College, 43 New Scotland Ave., Albany, NY 1228, USA; gildenn2@mail.amc.edu

[3] Department of Industrial Engineering, University of Pittsburgh, 522 Benedum Hall, 3700 O'Hara Street, Pittsburgh, PA 15261, USA; mme41@pitt.edu

[4] Department of Bioengineering, University of Pittsburgh, 522 Benedum Hall, 3700 O'Hara Street, Pittsburgh, PA 15261, USA; jackhastings@pitt.edu (J.T.H.); PUS8@pitt.edu (P.S.); gfsalazar2000@yahoo.com (G.S.)

[5] George W. Woodruff School of Mechanical Engineering, Institute for Electronics and Nanotechnology, College of Engineering, Georgia Institute of Technology, Atlanta, GA 30332, USA; skwon64@gatech.edu (S.K.); whyeo@gatech.edu (W.-H.Y.)

[6] Wallace H. Coulter Department of Biomedical Engineering, Parker H. Petit Institute for Bioengineering and Biosciences, Institute for Materials, Neural Engineering Center, Institute for Robotics and Intelligent Machines, Georgia Institute of Technology, Atlanta, GA 30332, USA

[7] Department of Otolaryngology, School of Medicine, University of Pittsburgh, 3550 Terrace Street, Pittsburgh, PA 15261, USA; je104@pitt.edu

[8] McGowan Institute for Regenerative Medicine, University of Pittsburgh, Pittsburgh, PA 15219, USA

* Correspondence: yjchun@pitt.edu; Tel.: +1-412-624-1193; Fax: +1-412-624-9831

Received: 14 October 2019; Accepted: 27 October 2019; Published: 30 October 2019

Abstract: We introduce a new tongue prosthetic assist device (TPAD), which shows the first prosthetic application for potential treatment of swallowing difficulty in dysphagia patients. The native tongue has a number of complex movements that are not feasible to mimic using a single mechanical prosthetic device. In order to overcome this challenge, our device has three key features, including (1) a superelastic nitinol structure that transfers the force produced by the jaws during chewing towards the palate, (2) angled composite tubes for guiding the nitinol strips smoothly during the motion, and (3) highly stretchable thin polymeric membrane as a covering sheet in order to secure the food and fluids on top of the TPAD for easy swallowing. A set of mechanical experiments has optimized the size and angle of the guiding tubes for the TPAD. The low-profile TPAD was successfully placed in a cadaver model and its mobility effectively provided a simplistic mimic of the native tongue elevation function by applying vertical chewing motions. This is the first demonstration of a new oral device powered by the jaw motions in order to create a bulge in the middle of the mouth mimicking native tongue behavior.

Keywords: dysphagia; swallowing; tongue; nitinol; superelastic; prosthesis

---

1. Introduction

Millions of Americans suffer from swallowing disorders primarily due to dysphagia every year [1,2]. Dysphagia occurs when the patient cannot properly transfer the food from the mouth to the esophagus or from the esophagus to the stomach. It leads to negative health outcomes and reduces quality of life due to gastrostomy dependence, aspiration pneumonia, malnutrition, as well as loss of physical and social pleasure experienced through eating.

Current treatment options for dysphagia include: postural strategies for head or body position, change in food bolus volume or viscosity, tonic muscle contraction including neuromuscular electrical stimulation (NMES), sensorial enhancement strategies, and gastrostomy tube alimentation and surgery [3]. A widely accepted rehabilitation approach is the Madison Oral Strengthening Therapeutic (MOST) system that provides pressure feedback for isometric tongue exercises [4]. Recently, people have studied methods to use human–computer interactions for feedback-loop rehabilitation [5,6]. However, when these therapies fail, there is no intervention available that could replace lost swallow functionality and address the decreased quality of life.

Therefore, we introduced a novel non-invasive prosthetic device that, in part, mimicked the elevation function of the native tongue. While some static devices such as palatal drop prostheses can increase intraoral pressures, all other prevailing oral appliances for defect compensation and swallowing aids are static devices [7]. However, the potential benefit of a dynamic prosthetic device is that it provides additional room when loading a food bolus when the device is in its relaxed state. Conversely, when the device is deployed, pressures would gradually increase to propel the food bolus towards the pharynx. Our unique device would provide immediate mechanical strength to propel the food bolus into the pharyngeal phase of swallowing. The design of the tongue prosthetic assist device (TPAD) was based on exploiting the excellent and unique mechanical properties of nitinol (i.e., superelasticity and high fatigue resistance) [8]. The nitinol biomaterials were used to compensate for the pressure that the restricted tongue muscles cannot provide.

In the present work, we have designed and manufactured a functional prototype, as shown schematically in Figure 1, through in vitro resistance evaluation, pressure measurement, and subsequently tested with human cadaver model. The design parameters used in developing TPAD were fully investigated and in vitro pressure measurement study results demonstrated the performance of the device.

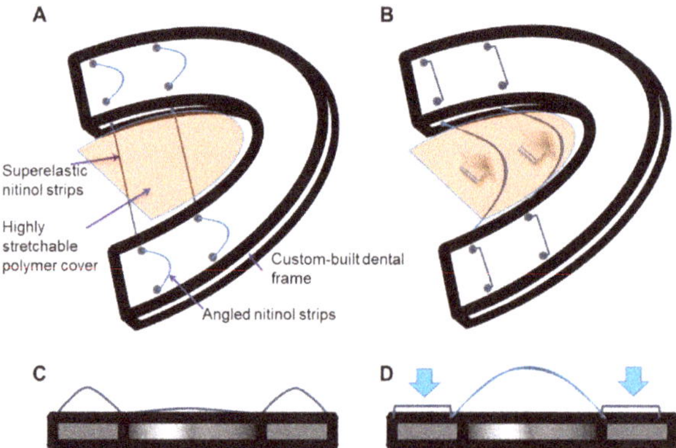

Figure 1. The schematic of the tongue prosthetic assist device (TPAD): dorsal–lateral view before applying the force (**A**) and after applying the force (**B**), front view before applying the force (**C**) and after applying the force (**D**).

2. Materials and Methods

2.1. Superelastic Nitinol and Acrylic Resin

Two materials were primarily used for the prototype construction, i.e., superelastic nitinol and denture acrylic resin. The superelastic nitinol biomaterials were used for the dynamic backbone in the tongue prosthetic assist device (TPAD). Mechanically drawn superelastic nitinol strips (width × thickness = 0.0240" × 0.0135", Confluent Medical, CA) provided a stable motion for creating a middle bulge when the mouth was closed, then, the nitinol strips returned to their initial position with their own elastic property. Another important material used in TPAD was acrylic resin that contains denture acrylic resin (Opti-Cryl, US Dental DEPOT, FL) and instant tray mix acrylic resin (Lang Dental, IL). This acrylic resin material was used to fabricate a custom-made dental model for tests.

2.2. Fabrication Processes of a TPAD Prototype

A structural frame for embedding modular segments was fabricated using both a disposable impression tray (Bosworth Tray Aways®, Bosworth, IL, USA) and dental impression kit (Flexitime, Heraeus Kulzer GmbH, Hanau, Germany). First, the impression of the cadaver's teeth was acquired by inserting the plastic tray containing putty material (i.e., dental impression kit) inside the cadaver's mouth and pressing it around the teeth. Over time, the putty material hardened, and a stiff negative impression was formed. The resulting impression mold was filled with Opti-Cryl acrylic resin in order to prepare the dental impression stone model after hardening the acrylic resin. The final acrylic tray was prepared by placing the mixture of acrylic resin monomer and hardener on top of the stone mold, then, the low-profile dental tray was subsequently separated after the completion of in situ polymerization.

Once the dental frame was successfully prepared, multiple holes were creased via a precision drilling process with the designated angles. Then, composite tubes were bent and subsequently fixed on the frame. Multiple nitinol strips were pre-shaped using stress-induced deformation processes, then inserted through the pre-embedded composite tubes. Finally, a highly stretchable latex covering membrane was attached on the nitinol backbone. Figure 2 shows the prototype of TPAD, indicating the device components and sizes.

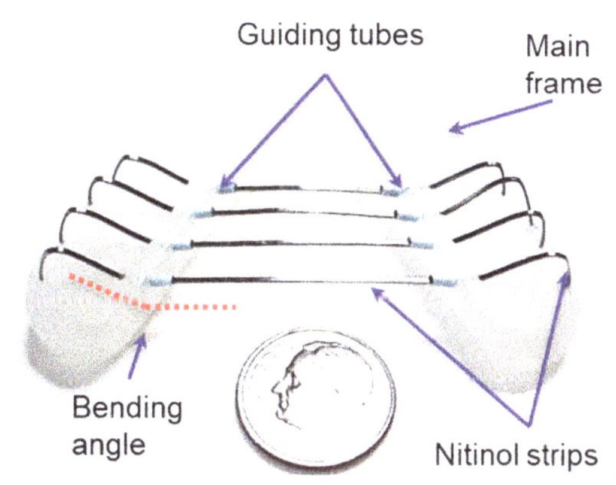

Figure 2. Information about TPAD, showing details of the components used in the prototype tongue indicating guiding tubes, bending angles, and device skeletal structure.

2.3. In Vitro Mechanical Property Measurements

2.3.1. Resistance Measurement between Nitinol Strip and Guiding Tubes

The resistance that occurred during the sliding motion of nitinol strips in guiding tubes were measured using a mechanical test system (FLC-5E, Starrett, Athol, MA, USA) that pulled the nitinol strip through the guiding tubes to characterize resistance. A single superelastic nitinol strip was first placed in the guide tube. The load cell attached to this mechanical test system measured the exerted resistance during the pulling process. Three different bending angles of tubes, such as 90°, 120°, and 150°, were used to quantify the bending angle associated resistance. Both the length and diameter of the guiding tubes were also varied and used for the resistance studies. Three tube lengths were 7, 9, and 11 mm, and two tube diameters were 5Fr (ID: 1.67 mm) and 6Fr (ID: 2.0 mm) sizes, as shown in Table 1. With the various parameters used in the prototype, resistance was measured during the sliding motion of the superelastic nitinol strip in the curved guiding tubes.

Table 1. Parameters and values in resistance testing of guiding tubes, and bending angles for device skeletal structure.

Parameter	Values Tested
Tube Bending Angle	90°, 120° and 150°
Tube Length	7, 9 and 11 mm
Tube Diameter	5 and 6Fr

2.3.2. Tongue Pressure Measurements

Two commercially available pressure measurement systems, the IOPI (Iowa Oral Performance Instrument, IOPI® Model 2.3, Redmond, WA, USA) and MOST (Madison Oral Strengthening Therapeutic, Swallow solutions Model 1.5, Madison, WI, USA) were used to measure the pressure produced by the TPAD. Both systems have been already used in exercise therapy to assist patients with oral dysphagia, therefore, the measured pressure levels were helpful to refine the prototypes. The MOST pressure measurement system has four pressure sensors (Figure 3A) and a MOST laptop interface (Figure 3B). The IOPI system contains a balloon sensor (Figure 3C) that is connected to the monitor (Figure 3D).

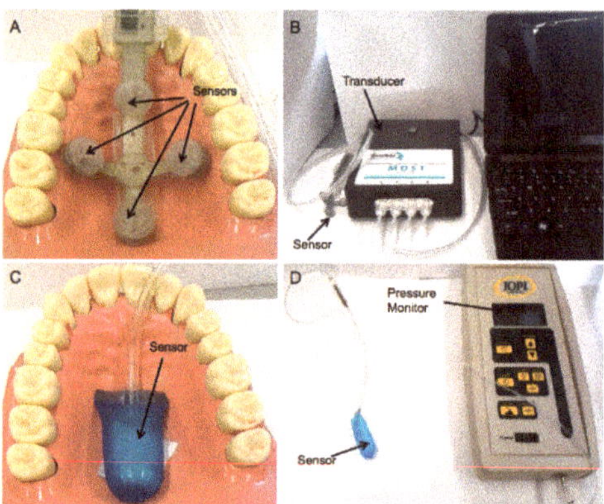

Figure 3. (**A**) MOST sensors placed in the palate of a mouth model, (**B**) MOST pressure measurement system, (**C**) IOPI sensor placed in the palate of a mouth model and, (**D**) IOPI pressure measurement system.

2.4. In Vitro Cadaver Test

Under the approval of the Committee for Oversight of Research and Clinical Training Involving Decedents (CORID ID # 433) at the University of Pittsburgh, two human cadaver heads were used to evaluate the performance of the TPAD. Because it is important to hold the cadaver heads firmly in order to perform both X-ray imaging and pressure measurement, the cadaver head was fixated within a metal frame. The custom-made dental tray was prepared from the dental impressions of the cadaveric models. The X-ray imaging technique was used in order to evaluate the movement of the TPAD in the cadaver model. Both the front and lateral images of the TPAD were recorded while the mouth was open and closed. The pressures produced by the TPAD were measured and recorded using both IOPI and MOST systems by placing the sensors on the palate of the mouth in the cadavers. Before and after the use of the cadaver head models, they were stored in ethyl alcohol (Fisher Scientific, Pittsburgh PA, USA).

2.5. Statistical Analysis

The values of measured pressures were expressed as the mean value ± standard deviation (SD). Statistical analysis was performed for each experiment using one-way analysis of variance (ANOVA) test and Tukey's honest significant difference (HSD) test by Minitab® 16.1.0 (©2010 Minitab Inc. State College, PA, USA) and a p-value < 0.01 was considered statistically significant.

3. Results

3.1. Resistance in the Guiding Tubes

Figure 4A,B represent the relationship between the length of the guiding tubes and resistance force during the nitinol wire movement. In the 5Fr catheter (Figure 4A), the lengthening of catheter tubes from 9 mm up to 11 mm significantly decreased the resistance force since lengthening the catheters provided a smoother surface for sliding the nitinol wire. The level of this decrease was the most in smaller bending angles (i.e., 90° and 120° bending angle) compared to 150° bending angle. Lengthening the catheter tubes from 9 mm up to 11 mm in 120° and 150° bending angles did not significantly reduce the resistance force while in the 90° bending angle, this length change significantly decreased the resistance force. As for the 6Fr catheters shown in Figure 6B, the 6Fr catheters had a larger diameter and the nitinol wire slid in it with less friction; therefore, the resistance force in the 6Fr catheter was lower compared with it in a similar condition in the 5Fr catheter. In addition, lengthening the 6Fr catheter tubes slightly decreased the resistance force. The resistance force on average was higher in the 5Fr catheter compared to the 6Fr catheter (p-value < 0.01), the smaller diameter of the 5Fr catheter compared to the 6Fr catheter limits the freedom of the nitinol wire. Therefore, the lowest resistance force was observed in the 6Fr guiding tube with 150° bending angle (i.e., 22.9 ± 3.48 mN).

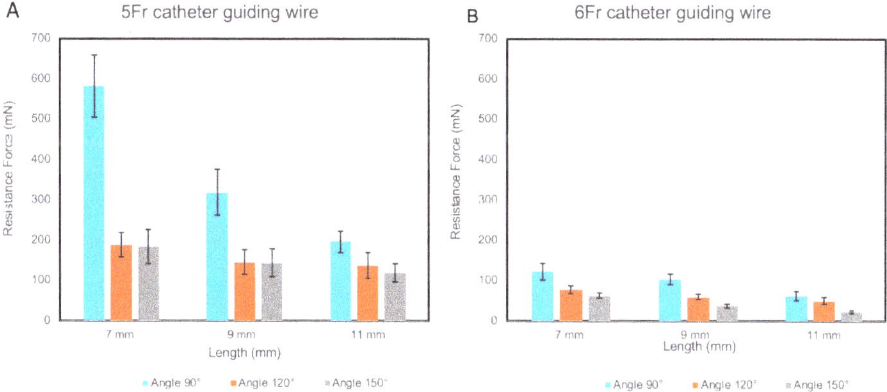

Figure 4. The relationship between the guiding tube length and the resistance during the nitinol wire movement in three different bending angles of 90°, 120° and 150°. (**A**) In a 5Fr catheter guiding tube and (**B**) in a 6Fr catheter guiding tube.

3.2. Prototype Fabrication

Figure 5 shows the prototype development process of TPAD. The acryl dental model was constructed as described in Section 2.2. Guiding tubes after bending with angles of 90°, 120°, and 150° were successfully fixed in the frame (Figure 5A), and subsequently superelastic nitinol strips were placed after pre-shaping as shown in Figure 5B. Initial bulge was created based upon the anatomy of the cadaver model to effectively measure the exerted pressure levels. Figure 5C,D represent the fabricated TPAD prototype after covering the stretchable membrane via a mechanical suturing method.

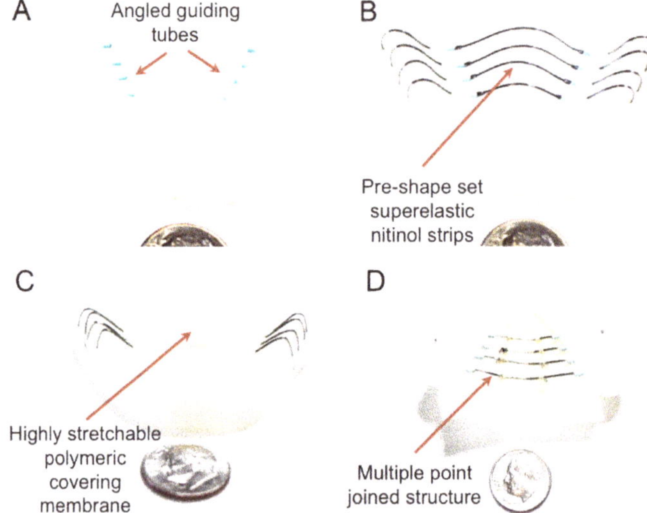

Figure 5. Fabrication steps of TPAD: (**A**) custom-built acrylic structural frame with catheter guiding tubes embedded into it, (**B**) nitinol strips placed in the mouth guard parallel to each other, (**C**) silicone layer covering on the parallel nitinol strips and, (**D**) the bottom view of the prototype.

3.3. In Vitro Assessment of the Mechanical Performance of TPAD

The performance of the TPAD was qualitatively assessed by applying the mechanical force to the dental contact areas as shown in Figure 6A–C. Figure 6A shows the middle area configuration of the TPAD before applying any force to the dental contact areas. Once the force was applied to the dental contact areas, the middle area began to bulge (Figure 6B). Then, the dental contact areas were fully depressed, making the curved nitinol strips flat. The height of the central area increased by up to four times when the device was actuated (Figure 6C). Plain film X-rays were also used to evaluate the motion of the device by visualizing the nitinol structure. Figure 6D shows the nitinol structure before the application of an external force. Once the dental contact areas were pushed down using aluminum bars, the central area began to bulge upwards (Figure 6E), which represents the elastic deformation of the nitinol structures during jaw closure. Figure 6F represents the fully actuated nitinol structure when the dental contact areas were fully pushed down. More elastic deformation occurred during this motion, but the nitinol structure recovered its original geometry when the external force was removed (i.e., fully elastic deformation).

3.4. TPAD Performance Evaluation using Cadaveric Head Model

Figure 7A,B show the representative front view X-ray images of the TPAD placed inside the cadaver mouth while the mouth was opened and closed, respectively. The device was successfully placed in the cadaver mouth due to its low-profile design as shown in Figure 7A. The dental contact areas were placed between mandibular and maxillary dentition. Three horizontally aligned wires on the top of the device were used as a maker of the top surface of the central area (the mechanical tongue). Once the mouth was closed, the dental contact areas were compressed and the three wires elevated centrally, the top surface made contact with the hard palate, increasing intraoral pressures, as the native tongue would typically do (Figure 7B). In the cadaver model, the force of the jaws was applied at a nearly 90 ° angle while in the in vitro performance test, the applied force was slightly inclined (~ 135 °) which resulted in a higher central bulge in the in vitro test.

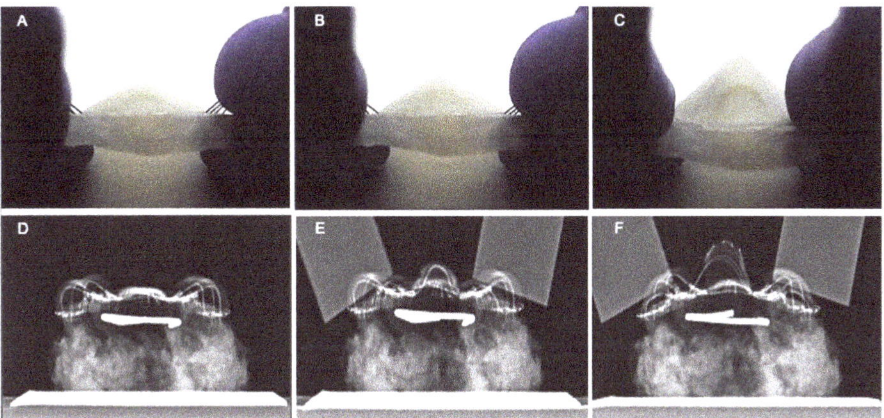

Figure 6. The representative TPAD in vitro performance: (**A**) before applying the force, (**B**) when it is half pressed, and (**C**) when it is totally pressed. X-ray images of the TPAD: (**D**) before applying the force, (**E**) after applying the force in the anterior position, and (**F**) in the posterior position.

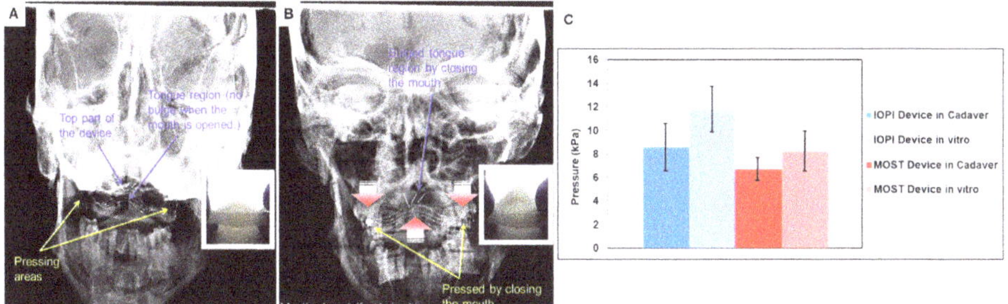

Figure 7. X-ray images of TPAD prototype placed in the cadaver mouth: front-view open mouth (**A**) and closed mouth (**B**). (**C**) Pressure measured by the Iowa Oral Performance Instrument (IOPI) and Madison Oral Strengthening Therapeutic (MOST) systems during the in vitro test.

The devices were also quantitatively evaluated by measuring the maximum exerted pressure levels during the biting motion. The pressure measured by MOST and IOPI did not always demonstrate similar pressure levels due to the different measurement mechanisms. The sensors of the MOST pressure measurement system are smaller and more sensitive compared to the one in the IOPI system and can record the maximum pressure with higher precision especially when the pressure is locally concentrated on a small area. The exerted tongue pressures by TPAD in the cadaver mouth were measured and compared with the in vitro laboratory pressure measurement results as shown in Figure 7. The pressures were decreased in the cadaver mouth by 25% in IOPI and 22.2% in MOST systems compared to the pressure produced during the in vitro laboratory tests, respectively.

4. Discussion

The tongue plays a significant role in swallowing, through preparing the bolus of food and pushing it posteriorly in order to transfer it from the oral to the pharyngeal phase of swallowing. Therefore, abnormal tongue performance can disturb swallowing whether at the oral preparatory or oral phase of swallowing [9–11]. The aim of the development of the TPAD was to augment the function of the impaired tongue by additional force on the food bolus to elevate it towards the palate and direct it posteriorly from the mouth to the pharynx. Our prototype uniquely demonstrated the concept of a simplistic prosthetic tongue through in vitro laboratory and cadaver testing. While these qualitative movements and exerted pressures were encouraging, our data did not prove clinical utility. The movements of the native tongue and pharynx are complex and coordinated, so the simplistic elevation movement of our device was only a partial mimic. To develop a device with potential clinical utility, several criteria were used in the design process.

First, the biting force of normal people was analyzed in order to design a "biting force" based mechanical prosthetic tongue device. People typically move their jaw in multiple directions while eating. Mastication is complicated and thus the exerted biting force is not constant; biting force is an indicator of the masticatory system and is dependent on the gender, age, craniofacial morphology, type of the food, and the measurement techniques [12]. The human temporomandibular joint (TMJ) and muscles of mastication are capable of moving the mandible in anterior to posterior, lateral, and vertical displacement. The lateral and anterior–posterior chewing movements are more difficult to transfer into wire deformation. For this reason, only the vertical biting force was used to design TPAD. The exerted biting force in vertical direction must overcome the resistance force of the tongue prototypes. Gay et al. reported the maximum biting force in ten different subjects are between 25 to 400 N, which are significantly greater force values than the maximum resistance force (i.e., ~2 N) that was measured in our prosthetic tongue devices [13]. Therefore, the force that is exerted through chewing process is sufficient to operate the TPAD by overcoming the resistance force of the tongue prototypes [14,15].

Secondly, the pressure levels produced by TPAD were evaluated. IOPI medical device results show that the average strength of tongue (i.e., pushing force by tongue) for different ages is in the range between 50 and 75 kPa [10,16]. The tongue strengths of our prototype shown in this study ranged from 6 to 12 kPa depending on the design of prototype, such as the angle and geometry of guiding tubes, as well as the type of measurement systems (i.e., IOPI and MOST). The use of thick nitinol strips in more locations increased the mechanical tongue strength showing the maximum mechanical tongue pressure exerted by the TPAD. While the pressure exerted by the prosthetic device is an important design factor to be considered, the device should have a low-profile design that has a sufficient spring-back force to recover the nitinol strips original geometry when the biting force is removed. In addition, the value of the exerted pressures was further reduced in the cadaver compared to the in vitro pressure measurement test of the same prototype (Figure 7). The first reason was that the cadaver model was not sufficiently soft and easy to manipulate to apply for the vertical direction biting force to the device mimicking human jaw motion. The second reason was that the central area of the prototype (i.e., bulging region) was difficult to align with the placed pressure sensors during the measurement. The cadaver model was especially useful to demonstrate device fit in the human mouth. We qualitatively assessed the proper fit using in vitro X-ray visualization technique with the cadaver model. The patient specific TPAD was placed on the bottom teeth and moved well with the biting motion of the cadaver. The device successfully transferred the vertical direction force to the central area that created an upward bulge mimicking typical tongue motion during swallowing. While the device worked properly with cadavers, additional ceramic materials were added to stabilize the device on the mandible in these specimens, since the teeth were irregular, and some were missing. The device could have equally been used as a type of dynamic palatal augmentation prosthesis, affixed to the maxillary dentition.

There are a number of clinical scenarios in which a TPAD-like device could be helpful to augment swallowing strength. These scenarios have all been cited in discussion of static palatal augmentation prostheses. In a patient with stroke-related dysphagia that maintains good airway protection, a TPAD could help augment the strength of swallow and treat delayed swallow. In anterior lateral sclerosis (ALS), the use of Palatal Augmentation Prosthesis (PAP) has been described [7]. In early bulbar onset ALS, the hypoglossal nerve may be affected first, with impaired oral control; subsequently laryngeal control and airway protection becomes impaired. There may be a window of opportunity for certain ALS patients to use a TPAD before the airway is at risk [17]. Perhaps a more ideal patient would be one with an isolated oral glossectomy defect. Those patients have been demonstrated to benefit from oral tongue static prosthetics, and many still maintain airway protection [18]. A more common patient might be an elderly sarcopenic patient with low muscular reserve after an acute illness. Within days into their hospitalization, they may develop tongue weakness and dysphagia [19]. Such patients may benefit from a TPAD type device during their rehabilitation to avoid prolonged hospitalization and potential gastrostomy.

A simplistic TPAD has been successfully designed, manufactured, and subsequently tested in vitro. The device mimicked the elevation of the native tongue that pushes the food bolus toward the palate and the back of mouth space. The unique superelastic property of nitinol wires used for the prototypes allows for a simple low-profile design. The device does not require any bulky component for transferring forces or recovering the original geometries. Even though our TPAD prototypes have not shown the equivalent strength of a native tongue, they were placed in the cadaver model with excellent fit. The results also demonstrate the functionality of the device showing the potential of the TPAD in clinical use. A limitation of this study was the small number of in vitro cadaver tests. The data presented in this manuscript can serve as a proof-of-concept for the use of new mechanical prosthetic tongue device fabricated using superelastic nitinol materials. More thorough in vitro studies using various designs are needed to demonstrate the optimal wire thickness and numbers, as well as the nitinol geometric design. Nevertheless, this study shows the first proof-of-concept stage toward a new tongue prosthetic device that utilizes superelastic nitinol. Future work will include the refinement of the prototype in order to increase the exerted pressure levels and device fatigue tests. Prospective prototypes will focus on anterior to posterior deployment of the pressure wave to encourage bolus transfer into the pharynx. In addition, soft food and viscous fluids will be used to evaluate swallowing capability with the device. Even with further refinements of the device design and functionality, one of the primary safety concerns that remains is to ensure controlled transfer of the food bolus from the oral cavity to the pharynx. We can envision that each prospective patient fitted for the TPAD would need instrumental evaluation by a speech pathologist. Patients with baseline aspiration and poor laryngeal protection would be contraindicated. Voluntary jaw contracture and coordination with a swallow would select against any patients with significantly altered mental status and low motivation for training on the TPAD.

5. Conclusions

In this work, we introduced a novel prosthetic device to increase an intraoral pressure during swallowing as the native tongue muscle generates a pressure to direct the food backward. The newly developed TPAD has been characterized in vitro using commercially available pressure measurement systems and human cadaver models. The experimental results have demonstrated the device feasibility for assisting in swallowing. Future studies will focus on a fatigue test, while examining the device performance in clinical conditions. In addition, we will study relevant swallowing behaviors with fluids and foods by adapting sequential motions of the tongue from anterior to posterior.

Author Contributions: Conceptualization, N.G.-L. and Y.C.; formal analysis, M.S., J.T.H., G.S., and Y.C.; funding acquisition, N.G.-L. and Y.C.; investigation, M.S., M.E., J.T.H., J.-H.K., G.S., and Y.C.; methodology, M.S., N.G.-L., M.E., and J.T.H.; supervision, N.G.-L., W.-H.Y., and Y.C.; validation, J.-H.K.; writing—original draft, M.S., N.G.-L., and Y.C.; writing—review and editing, M.E., S.K., W.-H.Y., P.S., and Y.C.

Funding: This study was supported by the Center for Medical Innovation (F_057-2013) and Central Research Development Fund at the University of Pittsburgh. W.-H.Y. acknowledges funding from the Nano-Material Technology Development Program through the National Research Foundation of Korea (NRF) funded by the Ministry of Science, ICT, and Future Planning (2016M3A7B4900044).

Acknowledgments: The authors gratefully appreciate the summer fellowship program from Swanson School of Engineering at the University of Pittsburgh to provide an opportunity for undergraduate students. APC charges for this article were fully paid by the University Library System, University of Pittsburgh.

Conflicts of Interest: The authors report no conflicts of interest.

References

1. Crary, M.A.; Groher, M.E. *Introduction to adult swallowing disorders*; Elsevier Science; Butterworth Heinemann: Oxford, UK, 2003.
2. Smith, B.S.; Adams, M. *Dysphagia: Risk Factors, Diagnosis and Treatment*; Nova Biomedical/Nova Science: New York, NY, USA, 2012.
3. Rofes, L.; Arreola, V.; Almirall, J.; Cabré, M.; Campins, L.; García-Peris, P.; Speyer, R.; Clavé, P. Diagnosis and management of oropharyngeal dysphagia and its nutritional and respiratory complications in the elderly. *Gastroenterol. Res. Pract.* **2010**, *2011*, 818979. [CrossRef]
4. Hewitt, A.; Hind, J.; Kays, S.; Nicosia, M.; Doyle, J.; Tompkins, W.; Gangnon, R.; Robbins, J. Standardized instrument for lingual pressure measurement. *Dysphagia* **2008**, *23*, 16–25. [CrossRef] [PubMed]
5. Bogaardt, H.; Grolman, W.; Fokkens, W. The use of biofeedback in the treatment of chronic dysphagia in stroke patients. *Folia Phoniatr. Logop.* **2009**, *61*, 200–205. [CrossRef] [PubMed]
6. Lee, Y.; Nicholls, B.; Lee, D.S.; Chen, Y.; Chun, Y.; Ang, C.S.; Yeo, W.-H. Soft electronics enabled ergonomic human-computer interaction for swallowing training. *Sci. Rep.* **2017**, *7*, 46697. [CrossRef] [PubMed]
7. Kikutani, T.; Tamura, F.; Nishiwaki, K. Case presentation: Dental treatment with pap for als patient. *Int. J. Orofac. Myol.* **2006**, *32*, 32–35.
8. Duerig, T.; Pelton, A.; Stöckel, D. An overview of nitinol medical applications. *Mater. Sci. Eng. A* **1999**, *273*, 149–160. [CrossRef]
9. Youmans, S.R.; Stierwalt, J.A. Measures of tongue function related to normal swallowing. *Dysphagia* **2006**, *21*, 102–111. [CrossRef] [PubMed]
10. Logemann, J.A. *Evaluation and Treatment of Swallowing Disorders*; College-Hill Press: San Diego, CA, USA, 1983.
11. Lazarus, C.L.; Logemann, J.A.; Pauloski, B.R.; Rademaker, A.W.; Larson, C.R.; Mittal, B.B.; Pierce, M. Swallowing and tongue function following treatment for oral and oropharyngeal cancer. *J. Speech Lang. Hear. Res.* **2000**, *43*, 1011–1023. [CrossRef] [PubMed]
12. Koc, D.; Dogan, A.; Bek, B. Bite force and influential factors on bite force measurements: A literature review. *Eur. J. Dent.* **2010**, *4*, 223–232. [CrossRef] [PubMed]
13. Gay, T.; Rendell, J.; Majoureau, A.; Maloney, F.T. Estimating human incisal bite forces from the electromyogram/bite-force function. *Arch. Oral Biol.* **1994**, *39*, 111–115. [CrossRef]
14. Braun, S.; Bantleon, H.-P.; Hnat, W.P.; Freudenthaler, J.W.; Marcotte, M.R.; Johnson, B.E. A study of bite force, part 1: Relationship to various physical characteristics. *Angle Orthod.* **1995**, *65*, 367–372. [PubMed]
15. Ferrario, V.F.; Sforza, C.; Zanotti, G.; Tartaglia, G.M. Maximal bite forces in healthy young adults as predicted by surface electromyography. *J. Dent.* **2004**, *32*, 451–457. [CrossRef] [PubMed]
16. Vanderwegen, J.; Guns, C.; Van Nuffelen, G.; Elen, R.; De Bodt, M. The influence of age, sex, bulb position, visual feedback, and the order of testing on maximum anterior and posterior tongue strength and endurance in healthy belgian adults. *Dysphagia* **2012**, *28*, 159–166. [CrossRef] [PubMed]
17. Braun, M.M.; Osecheck, M.; Joyce, N.C. Nutrition assessment and management in amyotrophic lateral sclerosis. *Phys. Med. Rehabil. Clin. N Am.* **2012**, *23*, 751–771. [CrossRef] [PubMed]
18. Balasubramaniam, M.K.; Chidambaranathan, A.S.; Shanmugam, G.; Tah, R. Rehabilitation of glossectomy cases with tongue prosthesis: A literature review. *J. Clin. Diagn. Res.* **2016**, *10*, ZE01–ZE04. [CrossRef] [PubMed]
19. Hathaway, B.; Baumann, B.; Byers, S.; Wasserman-Wincko, T.; Badhwar, V.; Johnson, J. Handgrip strength and dysphagia assessment following cardiac surgery. *Laryngoscope* **2015**, *125*, 2330–2332. [CrossRef] [PubMed]

MDPI

St. Alban-Anlage 66

4052 Basel

Switzerland

Tel. +41 61 683 77 34

Fax +41 61 302 89 18

www.mdpi.com

Materials Editorial Office

E-mail: materials@mdpi.com

www.mdpi.com/journal/materials